AF739189

# INTRODUCTION TO
# WASTE MANAGEMENT

Navodita Bhatnagar

TAP

Toronto Academic Press

# INTRODUCTION TO WASTE MANAGEMENT

**Navodita Bhatnagar**

**Toronto Academic Press**
4164 Lakeshore Road
Burlington ON L7L 1A4
Canada
**www.tap-books.com**

**Email:** orders@arclereducation.com

**© 2025 Toronto Academic Press**

**ISBN:** 978-1-77956-331-6

This book contains information obtained from highly regarded resources. Reprinted material sources are indicated and copyright remains with the original owners. Copyright for images and other graphics remains with the original owners as indicated. A Wide variety of references are listed. Reasonable efforts have been made to publish reliable data. Authors or Editors or Publishers are not responsible for the accuracy of the information in the published chapters or consequences of their use. The publisher assumes no responsibility for any damage or grievance to the persons or property arising out of the use of any materials, instructions, methods or thoughts in the book. The authors or editors and the publisher have attempted to trace the copyright holders of all material reproduced in this publication and apologize to copyright holders if permission has not been obtained. If any copyright holder has not been acknowledged, please write to us so we may rectify.

**Notice:** Registered trademark of products or corporate names are used only for explanation and identification without intent of infringement.

Toronto Academic Press publishes wide variety of books and eBooks. For more information about Toronto Academic Press and its products, visit our website at www.tap-books.com.

# ABOUT THE AUTHOR

**Navodita Bhatnagar** is a PhD holder in Environmental Biotechnology from Institute of Technology Carlow, Ireland. She finished her Master's in Technology in Biotechnology from Birla institute of technology and Sciences, Goa in 2015 and Bachelor's in technology from Guru Gobind Singh Indraprastha University, Delhi in 2013. She is an alumnus of Indo-German center for sustainability and a recipient of DAAD fellowship for the same which is essentially a collaborative interface to gather scientists in the field of environment sciences to come together and work for the common cause which is environment protection. Her field of expertise is waste management, conversion of biomass to energy, sustainable waste disposal and bioremediation.

# Table of Contents

# List of Figures

# List of Tables

# PREFACE

Most human activities generate waste. Despite this, waste production remains a major concern, as it has been since prehistoric times. Recently, both the rate and quantity of waste generation have increased. As the volume of waste grows, so does the variety of waste types. Unlike in the prehistoric times, when waste was simply a nuisance to be disposed of, proper management wasn't a significant issue due to small populations and abundant land. Back then, the environment could easily absorb the waste without degradation.

The significant increase in waste generation began in the sixteenth century with the Industrial Revolution, which prompted people to move from rural areas to cities. This migration led to a population explosion, causing a surge in the volume and variety of waste in cities. Materials like metals and glass began to appear in large quantities in municipal waste streams. The large urban populations led to indiscriminate littering and open dumps, which became breeding grounds for rats and other vermin, posing serious public health risks. These unhealthy waste management practices resulted in several epidemics with high death tolls.

In response, public officials in the nineteenth century started to dispose of waste in a controlled manner to safeguard public health. Most developed countries went through a period of environmental development and have now effectively addressed many of the health and pollution issues associated with waste generation. However, the increasing rate of urbanization and development in emerging countries is leading to a repeat of the same historical problems that developed countries have already faced.

### Content/Structure of the book

This book provides a comprehensive exploration of waste management through ten detailed chapters, each addressing a crucial aspect of the field.

Chapter 1 introduces the fundamental concepts of waste management. It covers the history of waste management practices, the significance of proper waste disposal, and the impact of waste on the environment and public health. Readers will gain an understanding of the basic principles and the evolution of waste management from ancient times to modern-day practices.

Chapter 2 covers the scientific underpinnings necessary for effective waste management. It explains the basics of geology and geotechnics, which are crucial for understanding landfill sites and waste containment. The chapter also covers toxicology, providing insights into how different types of waste can affect human health and the environment.

An overview of the legal framework governing waste management is presented in Chapter 3. It discusses major environmental laws and regulations, both at the national and international levels that influence waste management practices. The chapter emphasizes the importance of compliance and the role of regulatory bodies in enforcing these laws.

Chapter 4 focuses on municipal solid waste (MSW), which includes everyday items discarded by the public. It examines the sources, composition, and methods of collection and disposal of MSW. The chapter also explores modern techniques for managing municipal waste, such as recycling, composting, and incineration, and their environmental and economic impacts.

Hazardous waste, which poses significant risks to health and the environment, is the focus of Chapter 5. It defines what constitutes hazardous waste and discusses the regulations and safety measures required for its management. The chapter also covers the methods for the treatment, storage, and disposal of hazardous materials.

Chapter 6 addresses the challenges associated with managing medical waste, including waste generated by hospitals, clinics, and laboratories. It highlights the types of medical waste, such as sharps, pharmaceuticals, and infectious materials, and the protocols for their safe handling, treatment, and disposal to prevent contamination and spread of diseases.

The complexities of managing nuclear waste are explored in Chapter 7. It provides an overview of the types of nuclear waste, including low-level and high-level waste, and discusses the strategies for their long-term storage and disposal. The chapter also examines the environmental and safety concerns associated with nuclear waste management.

Electronic waste, or e-waste, is covered in Chapter 8, which addresses the growing problem of discarded electronic devices. It discusses the sources and components of e-waste, the environmental hazards posed by improper disposal, and the recycling processes that can recover valuable materials while mitigating environmental damage.

Chapter 9 focuses on strategies for reducing the volume and toxicity of waste generated. It explores various waste minimization techniques, such as source reduction, product redesign, and sustainable consumption practices. The chapter emphasizes the importance of reducing waste at its source to achieve more sustainable waste management practices.

The final chapter examines the management of waste generated from pharmaceuticals and personal care products (PPCPs). It discusses the environmental impact of PPCPs, including water contamination and effects on wildlife, and explores methods for the proper disposal and treatment of these substances to minimize their environmental footprint.

Together, these ten chapters provide a thorough look at the diverse aspects of waste management, making this book an essential resource for students, professionals, and anyone interested in understanding and addressing the challenges of Waste Management in the modern world.

## Audience

The audience for this book includes environmental science students, urban planners, policymakers, and professionals in sustainability and public health field. It serves as a foundational resource for anyone seeking a comprehensive understanding of waste management principles, practices, and challenges in both urban and rural settings. The book caters to undergraduate and graduate students studying Environmental Sciences, providing essential knowledge on waste reduction, recycling, composting, and waste-to-energy technologies. Urban planners and policymakers can utilize its insights to develop effective waste management strategies, addressing issues such as landfill management, hazardous waste disposal, and regulatory frameworks. Professionals in sustainability and public health sectors benefit from its emphasis on environmental impacts and health implications of improper waste management. Written in accessible language with case studies and practical examples, the book fosters critical thinking and problem-solving skills necessary for tackling contemporary waste management issues globally.

## Learning Features

- Start-of-chapter learning objectives
- Key terms in each chapter
- Summary Points at the end of each chapter
- Review Questions at the end of each chapter
- Case Studies and Activities in each chapter for solving real problems
- References at the end of each chapter

CHAPTER

# 1 Introduction to Waste Management

## LEARNING OBJECTIVES

**After studying this chapter, you will be able to:**

- Explain the history of waste management
- Introduce the concept of sustainability and present its role across the field of waste management.
- Describe the types of waste
- Understand the importance of effective waste management for a sustainable future
- Describe the waste processing and control
- Learn about environmental management regulation
- Introduce waste management economics and its components.

## KEY TERMS FROM THIS CHAPTER

| | |
|---|---|
| Biodegradable waste | Biomedical waste |
| Composting | Environmental impact |
| Hazardous waste | Incineration |
| Industrial solid waste | Open dumping |
| Pollution | Radioactive waste |
| Sustainability | Waste |

# INTRODUCTION

A necessary byproduct of most human activity is waste. The Asia-Pacific region's economic growth and rising living standards have increased the amount and complexity of waste produced, and industrial diversification and the expansion of healthcare facilities have added significant amounts of biomedical and industrial hazardous waste to the waste stream, which could have detrimental effects on the environment and public health.

Land occupation, resource depletion, amplified global warming from methane and other greenhouse gas emissions, contaminated waterways from landfilling, acidification, and harmful effects from air emissions during incineration are just a few of the direct and indirect environmental impact of waste disposal.

Waste's direct effects account for a considerable but relatively small portion of climate change, whereas resource depletion and other related effects are linked to indirect environmental impacts. This is primarily because many industries focus more on the output than the input, and the indirect effects of wastes are associated with the extraction and processing of various resources to produce different types of products. This demonstrates how, in comparison to its counterpart, the indirect effects of improper waste management can be more severe.

The term "waste management" refers to the various methods and techniques that recognize, regulate, and deal with the various kinds of waste from generation through disposal. When evaluated from a life-cycle perspective, which considers both direct effects like emissions and indirect effects like resource depletion, full implementation of waste management processes, including waste prevention and reuse, recycling can help avoid significant environmental impacts.

In terms of material resource efficiency, disposal options like landfills and incinerators are not the best practices for handling separately collected recyclables and mixed municipal solid waste. Therefore, it is crucial to assess the effects of these disposal methods to calculate the benefits that proper waste management practices will have on the environment, the economy, and society as a whole.

# 1.1. HISTORY OF WASTE MANAGEMENT

Humans have always produced waste. There may not have been much waste generated in places with low population densities. Even waste that was mostly biodegradable needed to be managed in areas with higher populations. There have been occasions when this has been reintroduced into groundwater, affecting areas such as Nor Loch. The residents of the village would get together and burn their trash in big dumps as part of a set monthly ritual practised by the Maya people of Central America (Barbalace, Roberta Crowell 2003).

The first known wastewater management system is located in present-day Syria (El Kowm). Located in the Fertile Crescent, the Mesopotamian "oasis" shows evidence of wastewater management beginning around 6500 BCE. The area is about 120km northeast of the ancient city of Palmyra. The site of El Kowm had vast urban planning centered around domestic wastewater drainage (Maysed. 2010).

There is a sophisticated gutter system within residences, as well as the connection of these gutter/drainage systems to larger systems within the city. The fluids were transported through these gutter systems from residences into ditches in the streets of the city. Additionally, there are other planned systems in the cities in surrounding areas, suggesting that there was diffusion of waste management techniques throughout the region. Many of these systems include settling chambers to mitigate the blockages that often occur in sewage systems. These chambers provide space for sediment deposits to build up without interfering with the flow of the sewer. In addition to these systems in Syria, there is also evidence of sophisticated waste management systems within other societies in the past.

Throughout their empire, the Romans employed highly developed aqueducts and waste disposal systems. The engineers of ancient Rome constructed an extensive network of sewers by using the Cloaca Maxima. There was severe pollution in the Tiber River where the Cloaca Maxima emptied. The Romans' drinking water, which they obtained from the Tiber, became contaminated due to this pollution. The aqueduct technology was eventually developed because of the need for fresh water. While the pipes inside the cities themselves were frequently made of ceramic, wood, and leather, the wider network of aqueducts typically used lead pipes. The quality of waste management techniques varied significantly amongst the socioeconomic groups. In Roman society, having plumbing and other water-based amenities, including access to sewer systems, was considered a mark of prestige. Only those who paid for it were able to access it. Archaeological sites and prehistoric writings also provide proof of the existence of the first waste management labour force in Europe. This force's responsibilities include gathering stored waste from homes without sewage connections and reselling it to farmers for a profit. It was also necessary for the households making use of these services to pay. The waste management systems of the Ancient Roman Empire were significantly more socially stratified and dependent on the socioeconomic standing of the populace than those of the Fertile Crescent civilizations. It is also possible to observe this waste management system stratification in Ancient Egypt (Eslamian, Saeid, 2014). During the Aztec Empire was another unusual method of managing waste and water.

### 1.1.1. Modern Era

After industrialization began and large population centers in England continued to grow urbanely, waste accumulated in the cities led to a sharp decline in sanitation and overall urban quality of life. The streets became choked with filth due to the lack of waste clearance regulations. Calls for the establishment of a municipal authority with waste removal powers were mooted as early as 1751 by Corbyn Morris in London, who proposed that .. "as the preservation of the health of the people is of great importance, it is proposed that the cleaning of this city, should be put under one uniform public management, and all the filth be ... conveyed by the Thames to proper distance in the country."

In the latter part of the 18th century, an organized solid waste management system was first implemented in London. A system for collecting waste and recovering resources was set up in the vicinity of the "dust yards." Coal ash, or "dust," was the primary component of municipal waste and had a marketable value for both soil improvement and brick production. Because of their high profitability, dust contractors were able to efficiently recover all the leftover waste after the "rag-and-bone men"—the unofficial sector of the street—had removed the easily sellable materials and items. Thus, this served as a prototype for structured solid waste management for the entire city. The dust-yard system was effectively operating until the mid-1850s, at which point the value of "dust" collapsed on the market. It had a significant role in helping England's transition to a solid waste management system that was institutionalized and managed by municipalities.

The first comprehensive legislation on the subject surfaced in the middle of the 1800s, propelled by progressively catastrophic cholera outbreaks and the growth of a public health discourse. The social reformer Edwin Chadwick's 1842 report The Sanitary Condition of the Laboring Population, which argued for the necessity of sufficient waste removal and management facilities to improve the health and wellbeing of the city's population, had a significant impact on this new focus. The miasmatic theory of disease transmission, which served as the foundation for Chadwick's proposals, was disproven after the turn of the 20th century.

The Nuisance Removal and Disease Prevention Act of 1846 marked the start of London's gradually developing system for regulated waste management. The first body to oversee sanitation in a rapidly growing city was the Metropolitan Board of Works, and the Public Health Act of 1875 mandated that all households dispose of their waste weekly in "moveable receptacles," which introduced the idea of a dustbin. The Public Works Department of the Ashanti Empire oversaw maintaining sanitation in Kumasi and its surrounding areas. They ordered citizens to maintain their compounds weed-free and clean, and they kept the streets clean every day.

The first incineration plants, also known as "destructors," were established as a result of the sharp rise in waste that needed to be disposed of. Manlove, Alliott and Co. constructed Nottingham's first incinerator in 1874. Ltd. to Alfred Fryer's design. But because of the copious amounts of ash, they generated and the way they drifted over the surrounding areas, these faced opposition (Gandy, Matthew (1994).

At the turn of the 20th century, similar municipal waste disposal systems appeared in other major cities in North America and Europe. The first US city to implement public-sector garbage management was New York City in 1895.

Originally, there were only open-bodied dump trucks used for garbage collection, drawn by a group of horses. The early 20th century saw the motorization of these vehicles, and the 1920s saw the introduction in Britain of the first close-body trucks that used a dumping lever mechanism to remove odors. These were quickly modified to include "hopper mechanisms," in which the waste was mechanically raised to the truck's roof after the scooper was filled at the floor level. In 1938, the Garwood Load Packer became the first truck to use a hydraulic compactor.

In the United States during the 1800s, horse manure frequently choked out cities. Even though the smell was acceptable for 19th-century sensibilities, walking through the streets without boots led to shoes that looked terrible. Dogs and pigs roamed free in many cities without trash collection services, eating the trash but releasing foul-smelling excrement. Disease was made easier by the fact that dead animals, especially horses, were left on the streets. (Nag &Vizayakumar, 2005; Pitchell, 2014; Gaudillat et al., 2018)

# 1.2. SUSTAINABILITY AND WASTE MANAGEMENT

Sustainability is the ability to better utilize resources in three important areas: the economy, environment, and society. It is characterized by meeting our needs to complete a task or activity effectively and productively that benefits the implementer and does not negatively affect the environment around us.

- Economic impact is represented in the financial benefits of adopting different plans in the operations involved in achieving a certain task.
- Environmental impact involves the output from the activities we perform, in the form of emissions, non-natural or non-native products released in the planet's ecosystem.
- Societal impact targets the effect of our activities on humans' quality of life. This involves their access to resources, employment and as well as the presence of a healthy ecosystem for them to live in.

For instance, the UK's third sector reaped an estimated 430 million pounds in benefits in 2012 as a result of recycling various types of residential and commercial waste. This procedure helped to create 11,000 full-time equivalent jobs and lessened the demand for highly water-intensive and greenhouse gas-producing industries like apparel. Furthermore, when calculating the waste produced by the apparel industry alone. The annual waste of clothing from each household is equivalent to 1000 bathtubs and 10,000 kilometers of carbon emissions from a modern 5-passenger car.

The concept of the 3Rs of waste management and other elements that will be added as part of the environmental pyramid govern this relationship, which some believe is the foundation upon which sustainability is based.

The three Rs of sustainability are reduce, reuse and recycle, which can be applied as follows:

- Waste Reduction is the most preferable to us as it provides the best sustainable outcome of the 3Rs of waste management.
- Waste Reuse whenever we find an inevitable source of waste, we look at opportunities to reuse, either for the same purpose or for another. This provides us with moderate sustainable payback.
- Waste Recycling is the least preferable as it provides the least sustainable returns.

Note: Putting money into these three areas would be wise. It's important not to confuse these methods' profitability with their viability; for instance, recycling certain materials may require up to ten times the energy required to produce them in the first place, but it can also be very profitable for a stand-alone company. Concurrently, saving the initial cost as well as the recycling cost would have been achieved by eliminating the need for such waste, if that were possible. (Nag &Vizayakumar, 2005; Pitchell, 2014; Gaudillat et al., 2018)

# 1.3. TYPES OF WASTE

Any material that is being disposed by recycling, burning, incineration, or disposal that is solid, liquid, or contained gaseous is considered a waste. It could be an old commercial product that needs to be disposed of because it can no longer be used for its intended purpose or a byproduct of a manufacturing process. There are numerous ways to classify waste; however, for the purposes of the applications discussed in this text, we will concentrate on three of the most well-known methods, which are illustrated in Table 1.1 below.

Table 1.1. Waste Categorization Approaches

| Characterization | Types of Waste |
|---|---|
| Based on physical properties, effects etc. | a) Solid wastes: Wastes in the form of solid i.e., local, commercial, and industrial waste.<br><br>b) Liquid wastes: Wastes in the form of liquid or watery. i.e., oils, chemicals, polluted water from ponds or rivers etc. |
| Based on the biological properties of wastes | a) Biodegradable wastes<br><br>b) Non-biodegradable wastes |
| Based on the effects of waste on human health and the environment | a) Hazardous wastes: Dangerous substances emitted from commercial, industrial and agricultural or economical use, which are unsafe to use for further purposes.<br><br>b) Non-hazardous wastes: Safe wastes emitted from commercial, industrial and agricultural or economical use, considered harmless to use for further purposes. |

Source: Ameer Mubaslat. Introduction to Waste Management. International Youth Ambassadors Foundation. ISBN: 978-9957-67-886-9.

Municipal solid waste (MSW) is the most well-known type of waste that will be discussed in this text because it makes up the majority of the waste dumped in the landfills and has needs strict management. Municipal solid wastes, or MSW, are made up of everyday items like batteries, paints, appliances, newspapers, food scraps, bottles, clothing, furniture, and packaging that we discard after using them. MSW is also known as trash.

MSW is produced by homes, businesses, schools, and hospitals. Paper, glass, metals, textiles, wood, plastics, and food wastes make up MSW. The two most popular techniques for managing MSW are landfilling and open dumping. Both approaches have drawbacks, including contaminated environments, the production of methane gas, which contributes to global warming, and labour disputes (Figure 1.1).

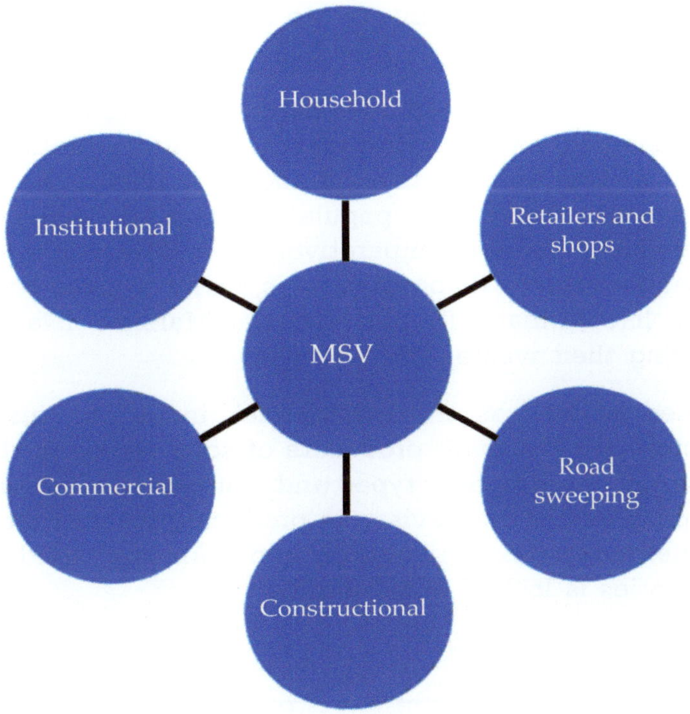


Figure 1.1. Sources of MSW.

Source: AmeerMubaslat. Introduction to Waste Management. International Youth Ambassadors Foundation. ISBN:978-9957-67-886-9.

Because of the chemical bonds between C–H–O in MSW, energy can be produced whenever these bonds are broken. This feature made it possible to extract methane from the biodegradable components of MSW and use it to further produce electricity.

Urban populations produced 2.01 billion tons of solid waste in 2016, with an average daily contribution of 0.74 kg per person. The annual generation of waste is predicted to rise by 70% between 2016 and 2050 due to increasing urbanization. By 2050, it is anticipated that waste production will have increased from roughly 2 billion tons in 2016 to 3 billion tons. A total of 1.9 billion tons of MSW is produced annually, with each individual contributing 218 kg of MSW to this enormous amount. This demonstrates the enormous potential of using MSW instead of conventional methods of landfilling and dumping.

### 1.3.1. Generation and Characteristics of Waste

A deep understanding of the types and amounts of waste being produced is essential for creation of reliable and affordable solid waste management plans. While waste characterization and quantification serve as the foundation for management and intervention in some of the region's more developed nations, systematic waste arising surveys receive less attention elsewhere, and little is known about the types, amounts, seasonal fluctuations, and future trends of waste generation. At the national level, some general trends and common elements are apparent, despite the lack of thorough or consistent data.

Generally speaking, the developed nations produce significantly more waste per person than the developing nations. However, there are situations where managing even modest amounts of waste presents a significant challenge. For instance, the South Pacific subregion's small islands' sparse populations and low levels of economic activity have made waste production there comparatively low. However, because of their small land areas and consequent lack of disposal options, many of these countries—particularly small atoll countries like Kiribati, Tuvalu, and the Marshall Islands—face significant challenges in managing their waste.

Residential homes as well as the commercial, industrial, institutional, building, and agricultural sectors are the main producers of solid waste in the region. Table 1.2 provides a breakdown of the different types and sources of solid waste. These sources are defined for the purposes of this review as producing waste that falls into four main categories: hazardous, industrial, agricultural, and municipal solid waste. Below, each of these waste categories is looked at in detail.

Table 1.2. Sources and Types of Solid Wastes

| Source | Typical Waste Generators | Types of Solid Wastes |
|---|---|---|
| Residential | Single and multifamily dwellings | Food wastes, paper, cardboard, plastics, textiles, leather, yard wastes, wood, glass, metals, ashes, special wastes (e.g., bulky items, consumer electronics, white goods, batteries, oil, tyres), and household hazardous wastes |
| Industrial | Light and heavy manufacturing, fabrication, construction sites, power and chemical plants | Housekeeping wastes, packaging, food wastes, construction and demolition materials, hazardous wastes, ashes, special wastes |
| Commercial | Stores, hotels, restaurants, markets, office buildings, etc. | Paper, cardboard, plastics, wood, food wastes, glass, metals, special wastes, hazardous wastes |

| | | |
|---|---|---|
| Institutional | Schools, hospitals, prisons, government centres | Same as commercial |
| Construction and demolition | New construction sites, road repair, renovation sites, demolition of buildings | Wood, steel, concrete, dirt, etc. |
| Municipal services | Street cleaning, landscaping, parks, beaches, other recreational areas, water and wastewater treatment plants | Street sweepings, landscape and tree trimmings, general wastes from parks, beaches, and other recreational area, sludge |
| Process | Heavy and light manufacturing, refineries, chemical plants, power plants, mineral extraction and processing | Industrial process wastes, scrap materials, off specification products, slag, tailings |
| All of the above should be included as "municipal solid waste." | | |
| Agriculture | Crops, orchards, vineyards, dairies, feedlots, farms | Spoiled food wastes, agricultural wastes, hazardous wastes (e.g., pesticides) |

Source: https://www.unescap.org/sites/default/files/CH08.PDF.

#### *1.3.1.1. Municipal Solid Waste*

Households, workplaces, hotels, retail establishments, schools, and other institutions produce municipal solid waste or MSW. Food waste, paper, plastic, rags, metal, and glass are the main constituents; however, small amounts of hazardous waste, such as used batteries, lightbulbs, and automobile parts, as well as abandoned medications and chemicals, are also frequently included in the collected waste.

MSW generation rates differ from city to city and season to season, but they are highly correlated with levels of economic activity and development. High-income countries (such as Australia, Japan, Hong Kong, China, Republic of Korea, and Singapore) produce between 1.1 and 5.0 kg/capita/ day; middle-income countries (such as Indonesia, Malaysia and Thailand) generate between 0.52 and 1.0 kg/capita/day, whilst low-income countries (such as Bangladesh, India, Viet Nam and Myanmar) have generation rates of between 0.45 and 0.89 kg/capita/ day. The MSW generation by the region's high-, middle-, and low-income nations is depicted in Figure 1.2.

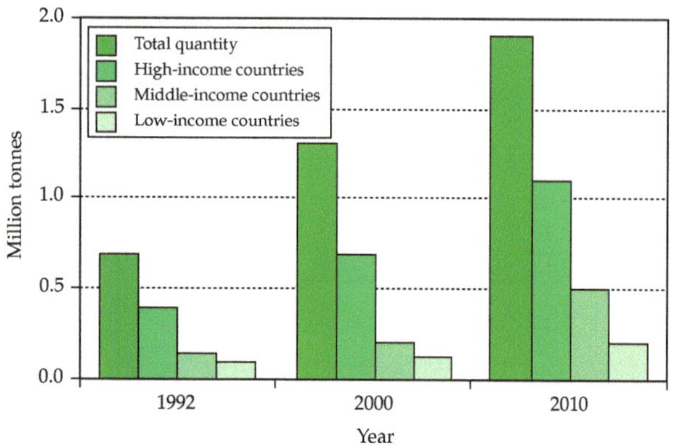


Figure 1.2. Municipal solid waste generation in different groups of countries in the region.

Source: https://www.unescap.org/sites/default/files/CH08.PDF.

The Asian and Pacific Region currently produces about 1.5 million tons of MSW per day, and by 2025, this amount is predicted to more than double. It's possible to argue that the current estimate of waste generation is overly conservative; the real levels are most likely more than twice this amount. The current share of each subregion in the waste produced by the region is shown in Figure 1.3.

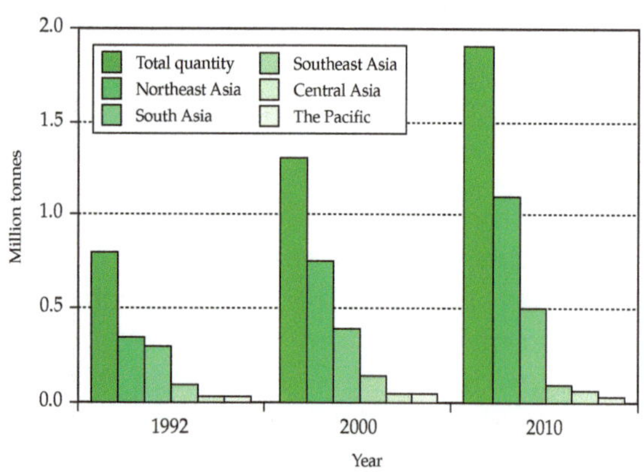


Figure 1.3. Estimated generation of municipal solid waste in different subregions.

Source: https://www.unescap.org/sites/default/files/CH08.PDF.

There are considerable regional variations in the composition of municipal solid waste (see Figure 1.4). In certain middle- and low-income countries, the waste generated has more than 70% organic content and more than 50% moisture content. Some municipal authorities include industrial waste and waste from construction and demolition as part of the municipal waste stream. There are also differences in the classification and reporting of different waste types.

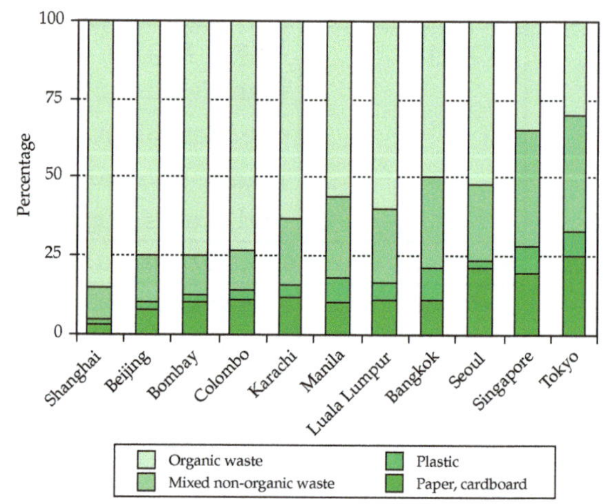


Figure 1.4. Approximate composition of municipal solid waste in selected cities of escape member countries.

Source: https://www.unescap.org/sites/default/files/CH08.PDF.

Certain differences between cities are related to fuel use and climate. Cities where coal is the primary energy source and where heating is required in the winter, like Beijing, Shanghai, Seoul, and Tokyo, have significantly more ash in their waste during those chilly months. Cities and towns (like Calcutta, Dhaka, and Hanoi) with unpaved or badly paved streets with a lot of dust and dirt from street sweeping have additional variances in the basic infrastructure. The amount of organic waste varies significantly between cities based on the quantity of trees and shrubs in public areas. Large, heavy waste materials, like abandoned cars, furniture, and packaging, are more common in higher-income nations like Singapore, Japan, the Republic of Korea, and Brunei Darussalam than they are in lower-income nations like Bangladesh, Cambodia, Myanmar, Nepal, Sri Lanka, and Vietnam. An example of the amounts and kinds of MSW produced in a few South Pacific subregional countries is given in Table 1.3.

Table 1.3. Quantities and Types of MSW Generated in Selected South Pacific Countries

| Selected Countries | Types of waste generated | % of total | Total Amount of MSW generated (tonnes per day) |
|---|---|---|---|
| Cook Islands | | | 4.75 |
| | Organic | 32% | |
| | Glass | 24% | |
| | Plastic | 12% | |
| | Metals | 10% | |
| Fiji (Suva, the capital) | | | 35.6 |
| | Metal | 10-16% | |
| | Glass | 5-10% | |
| | Plastic | 7-12% | |
| | Vegetative debris | 25-39% | |
| | Paper | 27-34% | |
| Vanuatu | | | 15 |
| | Vegetative debris | 35-40% | |
| | Wood | 25-30% | |
| | Paper | 10-12% | |
| | Plastic | 6-8% | |
| | Glass/Ceramic | 3-5% | |
| | Metals | 2-3% | |
| | Textile | 3-6% | |

Source: https://www.unescap.org/sites/default/files/CH08.PDF.

In squatter areas of many Asian and Pacific cities, where "wrap and throw" sanitation is practised or bucket latrines are emptied into waste containers, there is a significant amount of human faeces in the MSW. The latter is typical in many of the region's cities with minimal sewage systems, including Hanoi, Dhaka, and Calcutta.

#### *1.3.1.2. Industrial Solid Waste*

Like everywhere else, the Asia-Pacific Region collects a diverse array of materials with varying degrees of environmental toxicity when it comes to industrial solid waste. Paper, packaging materials, food processing waste, oils, solvents, resins, paints and sludges, glass, ceramics, stones, metals, plastics, rubber, leather, wood, cloth, straw, abrasives, and so on would usually fall into this category. Like municipal solid waste, the precise rates of generation are largely unknown due to the lack of an organized and regularly updated database on industrial solid waste.

The generation of industrial solid waste differs not only among developed nations but also among developing nations (refer to Figure 1.5). For instance, the People's Republic of China has a one to three-generation ratio between municipal and industrial solid waste. Nonetheless, this ratio is substantially lower in Pakistan, Bangladesh, and Sri Lanka. In developed, high-income nations like Japan and Australia, the ratio is one to eight. Nonetheless, the region generates 1 900 million tons of industrial solid waste annually, based on the average ratio for the area. With the current growth rates, this amount is predicted to increase significantly and double in less than 20 years. Such incremental growth will present extremely significant challenges because the current industrial solid waste collection, processing, and disposal systems in many countries are inadequate.

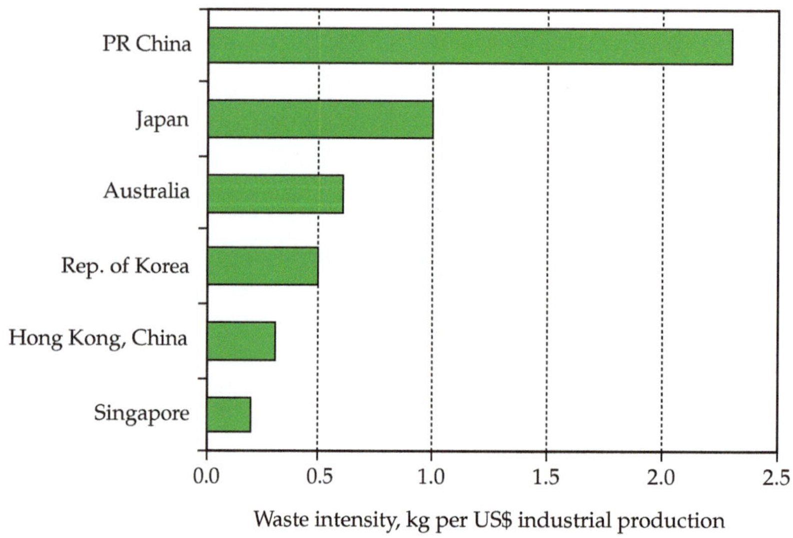


Figure 1.5. Waste intensity of industrial production in selected countries in the region.

Source: https://www.unescap.org/sites/default/files/CH08.PDF.

### *1.3.1.3. Agricultural Waste and Residues*

Increases in livestock manure, crop residues, and agro-industrial byproducts are inevitable outcomes of growing agricultural production. An estimate of the yearly production of agricultural residues and waste in a few chosen countries in the region is given in Table 1.4; the waste's effects on receiving inland and coastal waters are discussed regarding liquid and slurry waste. (Nag &Vizayakumar, 2005; Pitchell, 2014; Gaudillat et al., 2018)

Table 1.4. Approximate Estimate of Annual Production of Agricultural Waste and Residues in Selected Countries in the Region

| Country | Annual production, million tonnes | | |
|---|---|---|---|
| | Agricultural waste (manure/animal dung) | Crop residues | Total |
| Bangladesh | 15 | 30 | 45 |
| PR China | 255 | 587 | 842 |
| India | 240 | 320 | 560 |
| Indonesia | 32 | 90 | 122 |
| Malaysia | 12 | 30 | 42 |
| Myanmar | 28 | 4 | 32 |
| Nepal | 4 | 12 | 16 |
| Pakistan | 16 | 68 | 84 |
| Philippines | 20 | 12 | 32 |
| Rep. of Korea | 15 | 10 | 25 |
| Sri Lanka | 6 | 3 | 9 |
| Thailand | 25 | 47 | 72 |

Source: https://www.unescap.org/sites/default/files/CH08.PDF.

People's Republic of China, is the nation in the Asia-Pacific region that produces the most crop residues and agricultural waste followed by India. The production of rice, corn, and wheat alone results in about 587 million tons of residues annually in China (see Figure 1.6). The amounts of waste produced by Malaysia's production of rice, palm oil, rubber, coconut, and forest products are shown in Figure 1.7.

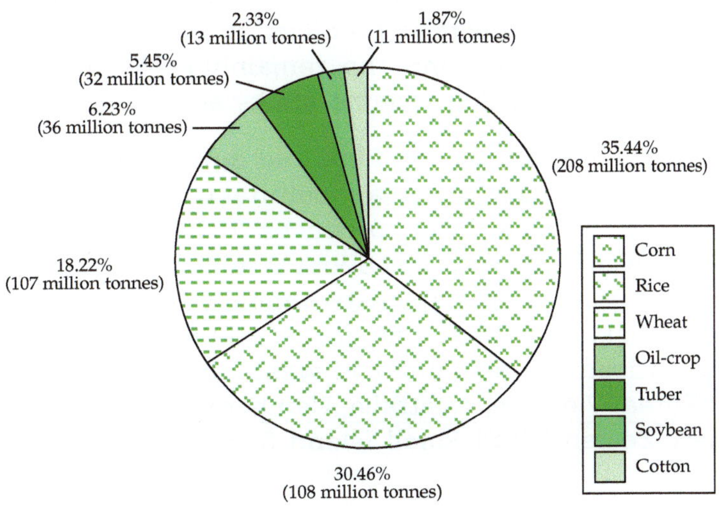


Figure 1.6. Proportionate annual production of agricultural waste in People's Republic of China.

Source: https://www.unescap.org/sites/default/files/CH08.PDF.

In Myanmar, the yearly production of animal waste is approximately 28 million tons, with over 80 of this coming from cattle husbandry. Crop waste and residues amount to about 4 million tons, of which more than half is rice husk.

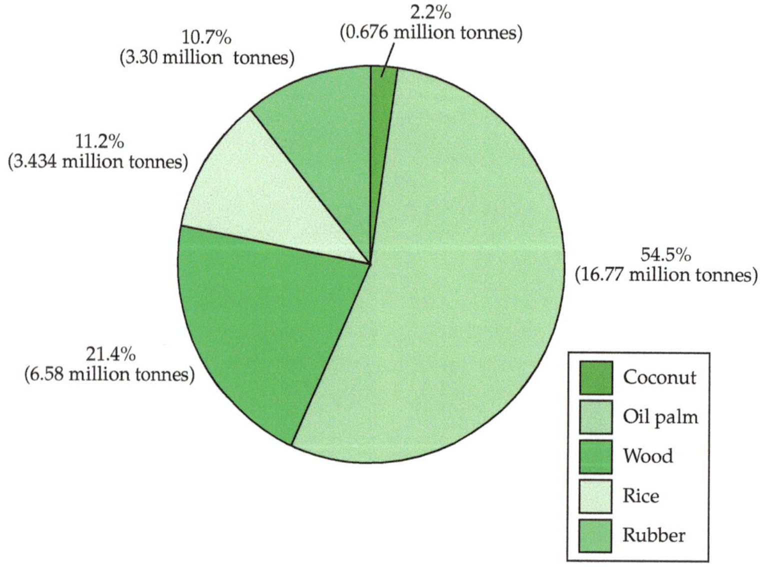


Figure 1.7. Proportionate annual production of agricultural waste in Malaysia.

Source: https://www.unescap.org/sites/default/files/CH08.PDF.

In Pakistan, approximately 56.22 million tons of various crop residues are produced, of which 12.46 million tons comes from cotton, 2.90 million tons from maize, 12.87 million tons from sugarcane, 8.16 million tons from rice, and 19.83 million tons from wheat. An additional 28 million tons of waste is produced in Pakistan, of which 58% is animal waste, 40% is sugarcane bagasse, and the remaining 2% are a mixture of tobacco, sunflower, castor seed, mustard, and sesame stalks as well as jute sticks.

Animal dung, paddy husk, straw, coir fibre and dust, bagasse, sawdust and off-cuts from the timber industry are all considered agricultural waste in Sri Lanka. The processing of coir, or the fibre from coconut husk, produces 700,000 tons of coir dust annually, whereas commercial rice milling produces about 2 million tons of paddy husk. Thailand produces over 25 million tons of animal waste annually in addition to approximately 4.6 million tons of paddy husk, 35 million tons of rice straw, and 7 million tons of bagasse. Massive amounts of agricultural residues and waste are also produced in other nations like the Republic of Korea, Vietnam, Australia, Cambodia, Japan, Lao People's Democratic Republic, Nepal, New Zealand, and Small Island States in the South Pacific.

### *1.3.1.4. Hazardous Waste*

The Asian and Pacific Region uses huge amount of toxic chemicals thus creating a lot of hazardous waste due to the rapid development of the hospital and healthcare facilities, industry, and agriculture. Approximately 110,000 different toxic chemical types are currently sold commercially. One thousand more new chemicals are introduced to the market annually for industrial and other purposes.

The dependability of data on the amounts and kinds of hazardous waste produced at the national level limits the availability of solid data on the generation of hazardous waste for the Asia-Pacific Region. This is caused by several factors, such as a lack of skilled workers to perform the required assessment, industries' unwillingness to provide process information, including waste arising data, and a lack of understanding of how hazardous the generated waste is. When data is available, there are major challenges in trying to make international comparisons because hazardous waste is defined and classified differently in each of the region's countries.

A wide range of industrial, agricultural, and manufacturing processes, nuclear establishments, hospitals, and healthcare facilities produce the majority of hazardous waste as byproducts. Chemical, petrochemical, petroleum, metals, wood treatment, pulp and paper, leather, textiles, and energy production facilities (coal-fired and nuclear power plants and petroleum production plants) are the main high-volume producers of industrial hazardous waste. Hazardous waste is produced by small and medium-sized businesses such as auto and equipment repair shops, textile factories, electroplating and metal finishing shops, dry cleaners, hospitals, and pest control companies. The main categories of hazardous waste produced in the Asia-Pacific region are pesticide, organophosphate, herbicide, urea and fungicide-bearing waste, waste solvents, and waste containing chlorine. Particularly, the region uses a lot of solvents, which leads to the production waste solvents.

The types, amounts, and origins of hazardous waste differ greatly between nations and are impacted by the volume and variety of industrial activity. Table 1.5 offers a cautious assessment of the hazardous waste generation trends in the past, present, and future for several chosen nations. It is imperative to emphasize that these approximations are based on data that could be deemed insufficient and unconfirmed. A World Bank study estimated the toxic releases of hazardous waste in the Asia-Pacific region in the lack of trustworthy regional data, and it forecasted sharp annual increases in the production of hazardous waste in China, India, Indonesia, the Philippines, and Thailand. The study's even more important finding was that there will be a rise in the intensity of hazardous waste generation per unit of output.

Table 1.5. Conservative Estimate of Annual Production of Hazardous Waste in Selected Countries and Territories in the Asian and Pacific Region

| Country/Territory | Estimated annual production, tonnes x $10^3$ | | |
|---|---|---|---|
| | 1993 | 2000 | 2010 |
| Australia | 109 | 275 | 514 |
| Bangladesh | 738 | 1 075 | 1 560 |
| PR China | 50 000 | 130 000 | 250 000 |
| Hong Kong, China | 35 | 88 | 165 |
| India | 39 000 | 82 000 | 156 000 |
| Indonesia | 5 000 | 12 000 | 23 000 |
| Japan | 82 | 220 | 415 |
| Malaysia | 377 | 400 | 1 750 |
| Mongolia | 15 | 26 | 45 |
| Nepal | 130 | 260 | 450 |
| New Zealand | 22 | 62 | 120 |
| Pakistan | 786 | 1 735 | 3 100 |
| Philippines | 115 | 285 | 530 |
| Papua New Guinea | 25 | 45 | 80 |
| Rep. of Korea | 269 | 670 | 1 265 |
| Singapore | 28 | 72 | 135 |
| Sri Lanka | 114 | 250 | 460 |
| Thailand | 882 | 2 215 | 4 120 |
| Viet Nam | 460 | 910 | 1 560 |

Source: https://www.unescap.org/sites/default/files/CH08.PDF.

For the amounts of petroleum waste produced in nations that extract or process crude oil, such as Brunei Darussalam, China, India, Indonesia, Malaysia, Japan, and the Republic of Korea, better and more trustworthy data is available. Malaysia produces 0.71 metric tons of petroleum waste annually, while the combined production of waste oil in the following countries is estimated to be 10.55 million liters: Fiji, Solomon Islands, Papua New Guinea, Federated States of Micronesia, Samoa, Vanuatu, Tonga, Marshall Islands, Nauru, Cooks Islands, Kiribati, and Tuvalu. (Nag &Vizayakumar, 2005; Pitchell, 2014; Gaudillat et al., 2018)

## 1.3.2. Environmental Impacts of Waste

Over the past ten to fifteen years, the Asian and Pacific Region has seen significant economic growth and urbanization, which has increased the amount of MSW generated in many cities, including Bangkok, Beijing, Mumbai, Calcutta, Colombo, Dhaka, Hanoi, Jakarta, Kuala Lumpur, Manila, and Shanghai. In addition to causing long-term environmental and public health issues, careless and open dumping on the edges of numerous cities in the region has degraded important land resources. The events that took place in July 2000 at the garbage dump in Quezon City, which is located outside of Manila and collapsed, killing hundreds of people, are evidence of the dire consequences that can result from careless dumping. The dump was "seven stories high."

The region's surface and groundwater supplies have been contaminated by careless dumping, and open burning of waste has a major negative impact on urban air pollution. A large amount of the region's contribution to the greenhouse effect is caused by the uncontrolled release of methane, which is created as a byproduct of the breakdown of organic wastes.

The amount of waste treatment and management facilities has not increased in proportion to the rise in potentially dangerous industrial, biomedical, and nuclear waste. While the careless disposal of spent chemicals, paints, oils, batteries, and carcinogens like asbestos can have serious negative effects on both human health and the environment, the uncontrolled dumping of biomedical waste has the potential to spread pathogens, or organisms that cause disease. Along with biocides and toxic effluents from sawmills and timber processing areas, several pollution incidents have also been reported from industrial waste, abattoirs, or food processing plants. Table 1.6 lists the overall effects of various waste categories on the air, water, and soil in a few chosen countries across different subregions.

Table 1.6. Impacts of Various Categories of Wastes on Water, Soil and Air in Selected Countries of Different Subregions

| | Agricultural wastes and residues | | | Municipal wastes | | | Industrial wastes | | | Hazardous wastes | | |
|---|---|---|---|---|---|---|---|---|---|---|---|---|
| | Water Pollution | Land Pollution | Air Pollution | Water Pollution | Land Pollution | Air Pollution | Water Pollution | Land Pollution | Air Pollution | Water Pollution | Land Pollution | Air Pollution |
| Australia | ⊕ | ⊕ | ❍ | ⊕ | ⊕ | ❍ | ⊕ | ⊕ | ❍ | ⊕ | ⊕ | ❍ |
| Bangladesh | • | ⊕ | ❍ | • | ⊕ | ⊕ | • | ⊕ | ❍ | • | ⊕ | ❍ |
| Brunei Darussalam | ❍ | ❍ | ❍ | ⊕ | ❍ | ❍ | ❍ | ❍ | ❍ | ❍ | ❍ | ❍ |
| Cambodia | • | ⊕ | ❍ | • | ⊕ | ❍ | • | ⊕ | ❍ | • | ⊕ | ❍ |
| PR China | • | ⊕ | ⊕ | • | ⊕ | ❍ | • | ⊕ | ❍ | • | ⊕ | ❍ |
| Cooks Islands | ⊕ | ⊕ | ❍ | ⊕ | ⊕ | ❍ | ⊕ | ⊕ | ❍ | ⊕ | ⊕ | ❍ |
| Fiji | ⊕ | ⊕ | ❍ | • | ⊕ | ❍ | • | ⊕ | ❍ | • | ⊕ | ❍ |
| Hong Kong, China | ⊕ | ⊕ | ❍ | ⊕ | ⊕ | ❍ | ⊕ | ⊕ | ❍ | ⊕ | ❍ | ❍ |
| India | • | ⊕ | ❍ | • | ⊕ | ❍ | • | ⊕ | ⊕ | • | ⊕ | ❍ |
| Indonesia | • | ⊕ | ❍ | • | • | ❍ | • | • | ❍ | • | ⊕ | ❍ |
| Japan | ⊕ | ❍ | ❍ | ⊕ | ❍ | ❍ | ⊕ | ⊕ | ❍ | ⊕ | ⊕ | ❍ |
| Kazakhstan | ⊕ | ❍ | ❍ | • | ❍ | ❍ | ⊕ | ❍ | ❍ | ⊕ | ❍ | ❍ |
| Kiribati | ⊕ | ❍ | ❍ | ⊕ | ❍ | ❍ | ⊕ | ❍ | ❍ | ⊕ | ❍ | ❍ |
| Lao People's Democratic Republic | • | ⊕ | ❍ | • | • | ⊕ | ⊕ | ⊕ | ❍ | ⊕ | ⊕ | ❍ |
| Malaysia | • | ⊕ | ❍ | • | ⊕ | ❍ | • | • | ❍ | • | • | ❍ |
| Maldives | ⊕ | ❍ | ❍ | ⊕ | ❍ | ❍ | ⊕ | ❍ | ❍ | ⊕ | ❍ | ❍ |
| Marshall Islands | ⊕ | ❍ | ❍ | ⊕ | ⊕ | ❍ | ⊕ | ⊕ | ❍ | ⊕ | ❍ | ❍ |
| Micronesia Federated States of | ⊕ | ❍ | ❍ | ⊕ | ⊕ | ❍ | ⊕ | ❍ | ❍ | ⊕ | ⊕ | ❍ |
| Mongolia | ⊕ | ⊕ | ❍ | • | ⊕ | ❍ | ⊕ | ❍ | ❍ | ⊕ | ❍ | ❍ |
| Myanmar | ⊕ | ⊕ | ❍ | • | • | ❍ | ⊕ | ⊕ | ❍ | ⊕ | ⊕ | ❍ |
| Nepal | ⊕ | ⊕ | ❍ | ⊕ | ⊕ | ❍ | ⊕ | ⊕ | ❍ | ⊕ | ⊕ | ❍ |
| New Zealand | ⊕ | ❍ | ❍ | ⊕ | ⊕ | ❍ | ⊕ | ⊕ | ❍ | ⊕ | ⊕ | ❍ |
| Pakistan | ⊕ | ⊕ | ❍ | • | ⊕ | ❍ | • | ⊕ | ⊕ | • | ⊕ | ⊕ |
| Papua New Guinea | ⊕ | ⊕ | ❍ | • | ⊕ | ❍ | ⊕ | ⊕ | ❍ | ⊕ | ⊕ | ❍ |
| Philippines | • | ⊕ | ❍ | • | • | ❍ | • | • | ❍ | • | • | ❍ |
| Rep. of Korea | ⊕ | ⊕ | ❍ | ⊕ | ⊕ | ❍ | ⊕ | ⊕ | ❍ | ⊕ | ⊕ | ❍ |
| Samoa | ⊕ | ⊕ | ❍ | ⊕ | ⊕ | ❍ | ⊕ | ⊕ | ❍ | ⊕ | ⊕ | ❍ |
| Singapore | ❍ | ❍ | ❍ | ❍ | ❍ | ❍ | ❍ | ❍ | ❍ | ❍ | ❍ | ❍ |
| Solomon Islands | ⊕ | ⊕ | ❍ | ⊕ | ⊕ | ❍ | ⊕ | ⊕ | ❍ | ⊕ | ⊕ | ❍ |
| Sri Lanka | ⊕ | ⊕ | ❍ | • | ⊕ | ❍ | • | ⊕ | ❍ | • | ⊕ | ❍ |
| Tonga | ⊕ | ⊕ | ❍ | ⊕ | ⊕ | ❍ | ⊕ | ❍ | ❍ | ⊕ | ❍ | ❍ |
| Tuvalu | ⊕ | ❍ | ❍ | ⊕ | ⊕ | ❍ | ⊕ | ❍ | ❍ | ⊕ | ❍ | ❍ |
| Thailand | • | ⊕ | ❍ | • | • | ❍ | • | • | ⊕ | • | • | ⊕ |
| Vanuatu | ❍ | ❍ | ❍ | ⊕ | ⊕ | ❍ | ⊕ | ⊕ | ❍ | ⊕ | ⊕ | ❍ |
| Viet Nam | • | • | ❍ | • | • | ❍ | • | • | ⊕ | • | • | ❍ |

Source: https://www.unescap.org/sites/default/files/CH08.PDF.

# 1.4. THE IMPORTANCE OF EFFECTIVE WASTE MANAGEMENT FOR A SUSTAINABLE FUTURE

Our planet is currently experiencing a severe environmental crisis, with waste management, pollution, and climate change among the main issues. Our world is creating more waste than it can manage due to industrialization and an ever-growing population. To guarantee that our planet has a sustainable future, effective waste management is essential. It is not only about keeping our surroundings orderly; it's also about lessening the harm that waste causes to the environment and to people's health. This section will discuss the value of efficient waste management and how it can contribute to a cleaner, more environmentally friendly future. Now let's get started and learn how even small steps toward appropriate waste management can have a big impact.

## 1.4.1. The Impact of Improper Waste Management

The environment and public health are greatly impacted by improper waste management. Improper disposal of waste can cause serious pollution of the air, water, and soil, resulting in significant damage to the environment. Additionally, it may harm marine life and wildlife, which may have a domino effect on the food chain. One of the most pressing issues of our day is climate change, which is exacerbated by the release of greenhouse gases from the breakdown of waste.

People's health can also be significantly impacted by improper waste management. Skin irritation, respiratory disorders, and other health problems can result from exposure to hazardous waste. Medical waste should be disposed of properly to prevent disease outbreaks and serious harm to the public's health. By putting into practice efficient waste management techniques, all these negative effects can be reduced. (Nag &Vizayakumar, 2005; Pitchell, 2014; Gaudillat et al., 2018)

## 1.4.2. The Benefits of Effective Waste Management

There are several advantages to efficient waste management for the economy, the environment, and human health. Effective waste management can stop diseases from spreading, cut down on pollution, and protect the environment. It can also strengthen the local economy and open job opportunities. Reusing waste can save energy, reduce greenhouse gas emissions, and lessen the need for raw materials. We can lessen the need for landfill sites, which pose risks to the environment and public health, by reducing the production of waste.

The sustainability of the environment can also be enhanced by efficient waste management. It can assist in the transition to a circular economy, in which waste is viewed as a resource and recycled or reused. We can lessen our impact on the environment and contribute to the development of a cleaner, greener future by implementing sustainable waste management techniques.

### 1.4.3. Waste Management Options-Reduce, Reuse, and Recycle

The three R's—Reduce, Reuse, and Recycle—are the most efficient ways to manage waste. The best way to manage waste is to reduce waste production. It entails cutting back on usage, staying away from single-use items, and choosing sustainable options to lessen the quantity of waste produced. Reusing waste entails putting waste materials to new uses. It may entail giving things a makeover, mending, or new use. Reusing waste materials to create new products is known as recycling waste. Recycling can preserve natural resources, cut greenhouse gas emissions, and save energy.

The three Rs provide a thorough approach to waste management that can lower waste generation, preserve resources, and advance sustainability. We can lessen our impact on the environment and help create a greener future by adhering to the three R's.

### 1.4.4. The Role of Government in Waste Management

An important part of waste management is played by the government. Policies and rules that promote environmentally friendly waste management techniques can be created and implemented by the government. They may also offer resources and money to help with waste management programs. To find creative solutions for waste management, governments can also spend money on research and development.

Through the development of awareness campaigns and educational initiatives, the government can further encourage recycling and waste reduction. By offering incentives and assistance, they can persuade companies and sectors to embrace environmentally friendly waste management techniques. Governments can also collaborate with business and civic groups to create efficient waste management programs.

### 1.4.5. Implementing Effective Waste Management in Your Community

Community involvement is necessary for efficient waste management. It is possible to decrease waste generation, preserve resources, and advance sustainability by implementing sustainable waste management techniques. Here are some ways you can implement effective waste management in your community:

- Use reusable bags, bottles, and containers.
- Avoid single-use products and packaging.
- Sort your waste and recycle as much as possible.
- Compost organic waste.

- Donate or sell items you no longer need.
- Support businesses that practice sustainable waste management.

By following these simple steps, you can make a significant impact on waste management in your community.

### 1.4.6. Innovative Waste Management Solutions

Innovative waste management solutions are emerging to tackle our waste management challenges. Here are some of the innovative waste management solutions that are gaining traction:

- Waste to Energy: Converting waste into energy is a sustainable waste management solution. It involves converting waste into electricity, heat, or fuel, reducing the need for fossil fuels.
- Biodegradable Plastics: Biodegradable plastics can reduce the amount of plastic waste that ends up in landfills or the environment. They are designed to break down into natural components, reducing their environmental impact.
- Upcycling: Upcycling involves transforming waste materials into new products, which can reduce waste production and conserve resources.
- Circular Economy: The circular economy is a sustainable economic model that aims to eliminate waste and promote the reuse of resources. It involves treating waste as a resource and designing products and systems that are regenerative and sustainable.
- Innovative waste management solutions offer a promising future for waste management. By adopting these solutions, we can move towards a greener and more sustainable future.

### 1.4.7. The Future of Waste Management and Sustainability

Sustainability and waste management have a bright future. An increasing number of people are implementing sustainable waste management practices because of increased awareness of the effects of waste on the environment and human health. To lower waste production, preserve resources, and advance sustainability, businesses and governments are also investing in sustainable waste management solutions.

Waste management will change in the future to accommodate a circular economy, in which waste is recycled or treated as a resource. It will entail implementing cutting-edge waste management strategies like the circular economy, waste to energy, biodegradable plastics, and upcycling. To achieve a cleaner and greener future, waste management in the future will require the involvement of businesses, governments, and the entire community.

## 1.4.8. Challenges and Opportunities in Waste Management

The problems associated with waste management include the growing population, industrialization, and consumerism, all of which increase the amount of waste generated. Effective waste management in some areas is also hampered by a lack of resources and infrastructure. Implementing sustainable waste management solutions may also be hampered by their high cost.

Nonetheless, waste management offers opportunities as well. There is a growing need for sustainable waste management solutions as people become more conscious of how waste affects the environment and public health. To find cutting-edge waste management solutions, businesses and governments are investing in research and development. The circular economy presents a chance to develop a waste-free, resource-reuse-focused sustainable economic model. (Nag &Vizayakumar, 2005; Pitchell, 2014; Gaudillat et al., 2018)

# 1.5. WASTE PROCESSING AND CONTROL

Wastes is gathered from various sources and disposed of during the waste management process. Waste is collected, transported, treated, analyzed, and disposed of during this process.

## 1.5.1. Current Waste Management Practices

Table 1.7 shows that low-, middle-, and high-income countries in the Asian and Pacific region currently use very different solid waste management practices. The subsequent subsections provide an overview of these practices' efficacy and scope of application.

Table 1.7. Comparison of Typical Solid Waste Management Practices

| Activity | Low Income | Middle Income | High Income |
|---|---|---|---|
| Source reduction | No organized programs, but reuse and low per capita waste generation rates are common. | Some discussion of source reduction, but rarely incorporated in any organized program. | Organized education programs are beginning to emphasize source reduction and reuse of materials. |
| Collection | Sporadic and inefficient. Service is limited to high visibility areas, the wealthy, and businesses willing to pay. | Improved service and increased collection from residential areas. Larger vehicle fleet and more mechanization. | The collection rate greater than 90. Compactor trucks and highly machined vehicles are common. |
| Recycling | Most recycling is through the informal sector and waste picking. Mainly localized markets and imports of materials for recycling. | The informal sector is still involved, some high technology sorting and processing facilities. Materials are often imported for recycling. | Recyclable material collection services and high-technology sorting and processing facilities. Increasing attention towards long-term markets. |

| | | | |
|---|---|---|---|
| Composting | Rarely undertaken formally even though the waste stream has a high percentage of organic material. | Large composting plants are generally unsuccessful, some small-scale composting projects are more sustainable. | Becoming more popular in both backyard and large-scale facilities. Waste stream has a smaller portion of compostables than low and middle-income countries. |
| Incineration | Not common or successful because of high capital and operation costs, high moisture content in the waste, and high percentage of inerts. | Some incinerators are used but experiencing financial and operational difficulties; not as common as high-income countries. | Prevalent in areas with high land costs. Most incinerators have some form of environmental controls and some type of energy recovery system. |
| Landfilling | Low-technology sites, usually open dumping of waste. | Some controlled and sanitary landfills with some environmental controls. Open dumping is still common. | Sanitary landfills with a combination of liners, leak detection, leachate collection system, and gas collection and treatment systems. |
| Costs | Collection costs represent 80 to 90 of the municipal solid waste management budgets. Waste fees are regulated by some local governments, but the fee collection system is very inefficient. | Collection costs represent 50 to 80 of the municipal solid waste management budgets. Waste fees are regulated by some local and national governments, more innovation in fee collection. | Collection costs can represent less than 10 per cent of the budget. Large budget allocations to intermediate waste treatment facilities. Upfront community participation reduces costs and increases options available to waste planners (e.g., recycling and composting). |

Source: https://www.unescap.org/sites/default/files/CH08.PDF.

### *1.5.1.1. Municipal Solid Waste*

A conceptual framework known as the waste management hierarchy was created to help direct and prioritize waste management choices made by individuals and organizations alike. Waste prevention is given priority, then reuse, recycling, recovery, and disposal.

- Collection and Transfer

Municipal solid waste (MSW) is collected in many of the region's cities using a range of containers, from used grocery bags and plastic drums or bins to old kerosene cans and rattan baskets. Certain cities have established (officially or unofficially) neighborhood dumping sites on the sides of the road where waste is gathered in bags and loose.

For the cities in the region, waste collection—and, when necessary, waste transfer—often represents the largest expense associated with solid waste management. Numerous collection methods are employed, such as door-to-door collection and indirect collection, which involves putting containers, skips, or communal bins in front of markets, in neighborhoods, and other suitable places. Collection and transfer services are capital-intensive and highly mechanized in high-income industrialized nations like Singapore, Japan, New Zealand, Australia, and the Republic of Korea. These nations use standardized collection vehicles, compactors, and containers to provide collection services to most urban and even rural areas, with collection rates ranging from 90%. Regulations control the separation of recyclables at the source and the subsequent collection of these materials. This process is made easier by the availability of colored bins or bags or by the creation of local recycling centres. While many of these cities have hired private waste collection companies and given private sector businesses and industrial establishments responsibility for collecting and disposing of their own solid waste, a sizable portion of these cities still maintain direct municipal control over some aspects of the collection process.

Waste collection and transfer in the region's middle-class and lower-class nations are typically labor-intensive tasks performed by employees who work directly for the municipal government. Low levels of mechanization are employed for waste collection; handcarts and tractor-trailers are used to gather the waste from neighborhood dumping sites and communal bins. The lack of compactors in the collection vehicles and containers forces the transportation of loose waste, which limits the collection system's capacity and makes the systems relatively inefficient.

While collection rates of well over 50% are attained in Bangkok, Mumbai, Delhi, Jakarta, Kuala Lumpur, Manila, and Shanghai, they are noticeably less than 50% in some cities of the lower and middle-income countries, such as Dhaka, Calcutta, and Hanoi. In contrast, over 90% of collections occur in Hong Kong, China, Seoul, Singapore, Sydney, and Tokyo. Naturally, there are differences in the services provided by collection agencies in wealthy and impoverished areas, and in many cities, these services are not available to the poor, particularly those living in unofficial settlements.

Decentralized pre-collection has shown to be successful in raising collection rates in certain cities. For instance, waste from each Kampong is transported to a transfer station or temporary storage location so that the city service can collect it. Kampongs, or villages, are formally in charge of primary collection in Indonesian cities. Similar systems are used in Delhi and Chennai, which has led to comparatively good collection systems. In other places, the MSW collection and transportation system is vulnerable due to the absence of effective transfer facilities. Transfer stations are used in cities like Tokyo, Singapore, and Sydney to collect waste from a city subdivision, compact it to increase transportation efficiency, and then move it to larger haulage vehicles to be delivered to the city's disposal sites. Furthermore, recycling and reuse materials are separated at transfer stations to act as material recovery centers. Few cities in developing nations have well-designed transfer stations with enough space, tools, and vehicles to handle and process the waste that is collected.

The privatization of collection services is growing. Currently, private waste collection companies handle more than 20% of the collection services in the entire region. Particularly in Australia, Hong Kong, China, Malaysia, the Republic of Korea, Singapore, and

Thailand, this practice is becoming more popular. Cost-cutting is the primary driver of privatization in Singapore and other countries; over the past ten years, the cost of collecting and disposing of waste has increased threefold to over US$700 million. The Ministry of the Environment (ENV) established a private limited company in 1994 with a $250 million authorized capital fund. The Ministry was able to focus on its regulatory role of protecting public health and environmental standards through legislation and licensing controls when SEMAC assumed control of the ENV's collection unit in April 1996.

In the cities and towns of the poorer developing nations, the effectiveness of solid waste management is severely limited by financial constraints and a lack of technical expertise. Low waste collection rates are caused by several factors that affect collection efficiency, including a lack of storage bins, collection vehicle shortages, nonexistent or insufficient transfer stations, traffic congestion, and a lack of public compliance. Trenches of trash are frequently left uncollected in some cities, and illegal deposits are made in drains, canals, and open spaces. The issue is further compounded by the lack of coordination and overlap of responsibilities between various levels of local government and government agencies.

- Material Recovery, Reuse and Recycling

Recent years have seen a notable improvement in the rate of recovery of recyclable materials from MSW in many of the region's countries, including Singapore, Japan, and the Republic of Korea. Overall resource recovery in the region increased from less than 10% of all MSW in 1988 to 30% in 1998; a large portion of this growth can be attributed to higher rates of recovery for metals, plastics, glass, paper, and paperboard.

Paper and paperboard make up the largest category (nearly 60% of the total tonnage of materials recovered), which frequently obscures the significance of recovery rates for other materials. For instance, recovered aluminum only makes up around 3% of the total tonnage of recovered materials, but it is far more valuable economically than paper products combined.

Japan is one of the Asian and Pacific region's top recyclers of materials from the MSW stream. In Japan, nearly half of the waste paper is recycled or retrieved, with the rate of retrieval rising from roughly 48% in 1990 to roughly 56% in 1997. Comparably, between 1990 and 1997, the percentage of aluminum and steel cans that were recycled increased from 40% to 60% and from 45% to 70%, respectively, while the percentage of glass that was recycled increased from 48% to 57% in the same time frame. Significant amounts of paper and cardboard, plastics, glass, and metals are recycled in other nations in the region, including Australia, Bangladesh, the People's Republic of China, Malaysia, New Zealand, the Philippines, the Republic of Korea, Singapore, and Thailand. Table 1.8 provides an example of the categories of materials recycled in Singapore in 1997. Recycling is often done at the source, especially in the more developed economies (i.e., government, and non-governmental organizations) and the commercial sector actively supports it at the home, company, and industry levels. In other places, like Vietnam, unofficial recycling networks have thrived despite the government's lack of official support or promotion. (Nag &Vizayakumar, 2005; Pitchell, 2014; Gaudillat et al., 2018)

In developing nations, recycling waste mostly depends on scavengers or waste pickers gathering materials informally. Up to 2% of people in Asian and Pacific cities are thought to make ends meet by salvaging materials from trash to sell them for recycling or reuse, or their use. Large communities of waste scavengers can be found in some cities: in the Philippines, about 15,000 squatters earn a living by picking through the Smoky Mountain municipal trash dump; in Bangalore, there are reportedly between 20,000 and 30,000 scavengers; and in Jakarta, there are roughly 15,000 to 20,000 waste pickers. Scavenger co-ops are becoming more popular in some of the region's nations, such as Indonesia, India, and the Philippines, and some of these communities are highly organized.

Table 1.8. Various Categories of Materials Recycled from MSW in Singapore in 1997

| Waste type | Estimated quantity in tonnes in 1997 | | | Recycling rate (per cent) |
|---|---|---|---|---|
| | Total waste disposed | Total waste recycled | Total waste output | |
| Food waste | 1 085 000 | 24 700 | 1 109 700 | 2.2 |
| Paper/cardboard | 576 000 | 324 000 | 900 000 | 36.0 |
| Plastics | 162 000 | 35 300 | 197 300 | 17.9 |
| Construction debris | 126 000 | 188 000 | 314 000 | 59.9 |
| Wood/timber | 249 000 | 34 800 | 283 800 | 12.3 |
| Horticultural waste | 75 400 | 67 600 | 143 000 | 47.3 |
| Earth spoils | 75 400 | – | 75 400 | – |
| Ferrous metals | 75 400 | 893 000 | 968 400 | 92.2 |
| Non-ferrous metal | 14 000 | 76 000 | 90 000 | 84.4 |
| Used slag | 120 000 | 135 000 | 255 000 | 52.9 |
| Sludge (Industry/ PUB) | 50 200 | – | 50 200 | – |
| Glass | 30 800 | 4 600 | 35 400 | 13.0 |
| Textile/leather | 25 200 | – | 25 200 | – |
| Scrap tyres | 5 600 | 5 700 | 11 300 | 50.4 |
| Others | 126 000 | 1 300 | 127 300 | 1.0 |
| Total | 2 976 000 | 1 790 000 | 4 586 000 | 39.0 |

Source: https://www.unescap.org/sites/default/files/CH08.PDF.

- Disposal Methods for MSW

Table 1.9 lists the different ways that municipal solid waste can be disposed of in the region's chosen countries and territories.

- Open Dumping

The most popular method of disposing of solid waste in the area is open dumping, which usually entails the uncontrolled disposal of waste without any control over leachate, dust, odor, landfill gas, or vermin. Open burning of waste at dump sites is a practice in certain cities. Waste is often dumped in coastal and inland wetlands and ravines, as is the case in Mumbai, Calcutta, Colombo, Dhaka, and Manila, or along the shoreline and into the sea, as in the case of Indonesia's Joyapura

Due to a lack of available land, waste is being dumped at extremely high levels; at the Quezon City dumpsite in the Philippines, the waste thickness was over 12 meters and occasionally exceeded 20 meters.

Table 1.9. Disposal Methods for Municipal Solid Waste in Selected Countries of the Region

| Country/ Territory | Disposal methods | | | | |
|---|---|---|---|---|---|
| | Composting (per cent) | Open dumping (per cent) | Land filling (per cent) | Incineration (per cent) | Others* (per cent) |
| Australia | 10 | – | 80 | 5 | 5 |
| Bangladesh | – | 95 | – | – | 5 |
| PR China | 10 | 50 | 30 | 2 | 8 |
| Cook Islands | – | 60 | 30 | – | 10 |
| Fiji | – | 90 | – | – | 10 |
| Hong Kong, China | – | 20 | 60 | 5 | 15 |
| India | 10 | 60 | 15 | 5 | 10 |
| Indonesia | 15 | 60 | 10 | 2 | 13 |
| Japan | 10 | – | 15 | 75 | – |
| Kazakhstan | – | 85 | – | – | 15 |
| Rep. of Korea | 5 | 20 | 60 | 5 | 10 |
| Maldives | – | 90 | – | – | 10 |
| Malaysia | 10 | 50 | 30 | 5 | 5 |
| Mongolia | 5 | 85 | – | – | 10 |
| Myanmar | 5 | 80 | 10 | – | 5 |
| Nepal | 5 | 70 | 10 | – | 15 |
| New Zealand | 5 | – | 85 | – | 10 |
| Pakistan | 5 | 80 | 5 | – | 10 |
| Philippines | 10 | 75 | 10 | – | 5 |
| Papua New Guinea | – | 80 | – | 5 | 15 |
| Samoa | – | 80 | – | – | 20 |
| Singapore | – | – | 30 | 70 | – |
| Sri Lanka | 5 | 85 | – | – | 10 |
| Thailand | 10 | 65 | 5 | 5 | 15 |
| Viet Nam | 10 | 70 | – | – | 20 |

Source: https://www.unescap.org/sites/default/files/CH08.PDF.

An additional risk associated with uncontrolled dumpsites is the accumulation of landfill gas, primarily methane, which has resulted in fire outbreaks and negative health effects for workers and nearby residents.

In the South Pacific subregion's Small Island Developing States, a major issue with solid waste disposal is now the lack of available land. Old cars and refrigerators are thrown into the lagoons of French Polynesia, and municipal waste is bundled into wire gabions and used to build sea walls in the Marshall Islands, demonstrating how often dumping at sea has been used as a solution. In the latter instance, loose waste materials and leachate were able to enter the ocean straight through the gabions.

- Landfilling

In the Asia-Pacific region, cities in both high- and low-income nations have come to accept the most desirable disposal option: dumping solid waste at a fully or semi-engineered sanitary landfill. As for the disposal of solid waste, other cities in Australia, the People's Republic of China, Japan, the Republic of Korea, Malaysia, and Thailand have adopted controlled tipping or sanitary landfilling. Bandung, Singapore, Hong Kong, China, Seoul, Chennai, and Tokyo do have well-designed and reasonably operated sanitary landfills. Disused tin mines are used by Kuala Lumpur for MSW landfills located all over the city.

Several landfills in the area have developed electricity-generating facilities, putting the gas generated by their waste to beneficial use. In Ho Chi Minh City, work is currently underway on a landfill/biogas power generation facility; others are slated for Chennai and possibly Colombo.

One of the biggest obstacles to landfill siting in the region's densely populated cities and towns is the availability of land. For instance, severe land constraints in China, Singapore, and Hong Kong have prompted the development of intricate engineering infrastructure solutions to guarantee high standards of operational and maintenance control. These solutions have also made it possible to develop landfills that are suitable for use in coastal, offshore, and mountainous regions. The two landfills currently in use in Singapore are getting close to capacity, and an offshore landfill at Pulau Semakau, which will cost S$840 (US$500) million, is almost finished. The landfill, which spans 7 km and covers 350 hectares of sea, will meet Singapore's waste disposal needs through 2030 and beyond. The waste is placed in cells, which will ultimately rise to a height of fifteen meters above sea level.

- Composting

While organic waste is widely composted on a small scale in the area, efforts to expand composting to a larger scale to reduce the amount of municipal solid waste that needs to be disposed of or to generate income through the sale of compost have not been very successful. Many composting facilities in the area are not operating at maximum efficiency nor are they producing compost with a viable market. The lack of material segregation results in compost contamination with plastic, glass, and toxic residues, and the high operating and maintenance costs make compost more expensive than commercial fertilizers. In these conditions, not much of the compost generated is appropriate for use in agriculture.

One common example is the forced-air composting facility located in Hanoi. The plant currently processes only 20% of its planned capacity, and because the product is too contaminated with plastic and glass, the municipal authorities have not been able to convince nearby farmers to accept it for free. Research and pilot projects are being used to actively promote small-scale neighborhood composting. These programs have been in place in Indonesia for more than ten years, and in Cipinang Besar and Watam (East Jakarta), a few small private businesses have been founded to provide compost to golf courses and estate gardens. Ho Chi Minh City has two small composting plants, while Bandung has a box-style windrow composting plant established next to an existing dumpsite. Small-scale vermicomposting of organic waste is done in open boxes or containers and is popular in the People's Republic of China, India, Indonesia, and the Philippines. In this process, worms and microorganisms are used to turn organic materials into nutrient-rich compost.

On a slightly bigger scale, Australia, Bangladesh, the People's Republic of China, India, the Philippines, and Thailand are testing the composting of organic MSW with agricultural waste and sludge from municipal sewage treatment plants. The availability of land, the high costs of operation, maintenance, and transportation, as well as the incomplete segregation of waste materials, continue to be significant obstacles to the widespread use of co-composting.

- Incineration

The incineration of MSW is still an expensive and technically unsuitable method of disposing of waste for a large portion of the Asian and Pacific region. High capital, operating, and maintenance costs coupled with stricter air pollution control regulations have hindered the growth of waste incineration facilities. Furthermore, a large proportion of the MSW produced in the region's low- and middle-income nations has high organic and moisture contents and a comparatively low combustible fraction.

For example, the waste must be dried on-site for five days before it is suitable for combustion, so the Indonesian city of Surabaya imported an incinerator that is currently running at two-thirds of its design capacity. It is estimated that the cost of waste incineration in this case is approximately ten times higher than the cost of open dumping/landfilling in other Indonesian cities, even without the cost of air pollution control mechanisms.

Only a few nations, including Australia, the People's Republic of China, Hong Kong, China, Indonesia, Japan, Singapore, and the Republic of Korea, have modern, full-scale incinerators in operation. These nations have high levels of combustible MSW, which in certain cases has increased due to moisture-reducing compaction at transfer stations.

With assistance from abroad, Beijing, Shanghai, Guangzhou, Shenyang, Beijing, and Beijing have all started building MSW incinerators. Because the incinerators in Hong Kong, China could not meet air pollution standards, they were closed; however, plans are being made to build new plants.

#### *1.5.1.2. Industrial Solid Waste*

The techniques used to dispose of industrial solid waste include open dumping, landfilling (including sanitary and semi-engineered landfilling), and incineration. These techniques are essentially the same as those used to dispose of MSW.

Non-hazardous industrial solid waste is accepted at open dumps or landfills alongside municipal solid waste in many countries, such as Bangladesh, the People's Republic of China, India, Indonesia, Malaysia, the Philippines, and Thailand (although where facilities are available, potentially hazardous industrial solid waste is disposed of either in secure landfills or is incinerated).

Industrial waste is often buried in dump pits inside or close to the location of the industrial facility from which it originated, or it is dumped on private land in developing nations with inadequate waste management infrastructure.

### 1.5.1.3. Agricultural Waste and Residues

Table 1.10 lists the main techniques for disposing of agricultural waste in a few countries in the region.

The most common method of disposal in many traditional, sedentary agricultural systems is the land application of raw or composted agricultural waste, which allows farmers to recycle important nutrients and organics back into the soil. Comparably, communities that raise fish in Bangladesh, the People's Republic of China, India, Indonesia, Malaysia, the Philippines, Thailand, and Vietnam frequently combine fish farming with other agricultural pursuits like raising livestock, growing fruits and vegetables, and cultivating paddy fields.

Table 1.10. Disposal Methods of Agricultural Waste and Residues in Selected Countries in the Region/Area

| Country | Disposal methods of agricultural waste and residues | | | | | | |
|---|---|---|---|---|---|---|---|
| | Land application | Fish farming | Composting | Biogas production | Utilization as | | |
| | | | | | Fuel | Animal feed | Building materials |
| Australia | ○ | ❐ | ● | ○ | ❐ | ○ | ❐ |
| Bangladesh | ○ | ❐ | ❐ | ❐ | ● | ○ | ● |
| Cambodia | ○ | ○ | ○ | ○ | ● | ○ | ● |
| PR China | ● | ● | ● | ● | ● | ● | ● |
| India | ● | ○ | ● | ● | ● | ● | ● |
| Indonesia | ● | ● | ● | ○ | ● | ● | ○ |
| Japan | ○ | ○ | ● | ○ | ○ | ○ | ○ |
| Lao People's Democratic Republic | ○ | ○ | ○ | ❐ | ● | ● | ● |
| Malaysia | ○ | ○ | ● | ○ | ● | ● | ○ |
| Myanmar | ○ | ○ | ○ | ○ | ○ | ○ | ● |
| Nepal | ❐ | ❐ | ❐ | ❐ | ○ | ○ | ❐ |
| New Zealand | ○ | ❐ | ○ | ○ | ○ | ○ | ❐ |
| Pakistan | ❐ | ❐ | ○ | ○ | ○ | ○ | ❐ |
| Philippines | ○ | ○ | ● | ● | ● | ● | ● |
| Rep. of Korea | ○ | ○ | ● | ○ | ○ | ○ | ○ |
| Sri Lanka | ❐ | ❐ | ○ | ○ | ○ | ○ | ○ |
| Thailand | ● | ● | ● | ● | ● | ● | ● |
| Viet Nam | ● | ● | ● | ● | ● | ● | ● |

Source: https://www.unescap.org/sites/default/files/CH08.PDF.

Anaerobic digestion is a common method used in many agriculturally based economies to produce biogas from agricultural wastes. The anaerobic digesters' slurry is used as liquid fertilizer, a supplement to cattle and pig feed, and a medium for soaking seeds before germination. Biogas, which contains about 60% methane, is mainly used directly for cooking, heating, and lighting.

Hazardous waste incinerators have been developed in Australia, Japan, Hong Kong, China, Malaysia, the Republic of Korea, Singapore, and Thailand; purpose-built sanitary landfills have been developed to receive hazardous waste in Australia, Japan, Malaysia, New Zealand, and the Republic of Korea. Several nations in the South Pacific subregion, including Bangladesh, the People's Republic of China, India, Mongolia, the Philippines,

Pakistan, Sri Lanka, and numerous island states, typically dispose of hazardous waste alongside MSW in open dumps or attempt to store especially toxic wastes in sealed containers.

There has been progress in several nations, such as Singapore, Australia, Japan, Hong Kong, China, Republic of Korea, and Japan, on techniques for detoxifying hazardous waste and then immobilizing it by forming it into bricks and other useful materials.

Around 250,000 to 300,000 tons of commercially viable hazardous industrial waste are produced annually by the petrochemical, chemical, and non-ferrous industries along Thailand's Eastern Seaboard, where a significant hazardous waste management program is currently in place. At the center of the petrochemical and chemical industries of Thailand, the Map Ta Phut Industrial Estate is home to a hazardous waste treatment plant run by the Industrial Estate Authority of Thailand.

The Bukit Nanas Integrated Waste Treatment Facility in Malaysia is the first all-inclusive treatment facility in the nation, featuring a secure landfill, stabilization, high-temperature incineration, and physical and chemical treatment. To handle hazardous waste from the JABOTABEK (Jakarta, Bogor, Tangeran, and Behasi) industrial area, a centralized facility for treating hazardous waste has also been developed in West Java, Indonesia. The facility increased the amount of hazardous waste it treated from 9.7 tons to 29 tons between 1994 and 1997. However, during the political and economic crisis of 1998–1999, industrial production fell, which also caused a decline in the amount of waste treated, which dropped to 16.6 tons in 1998 before rising to 18.8 tons in 1999.

Japan has sophisticated systems in place to handle and get rid of the 500 million tons of hazardous waste that its industries produce annually. To lower the net amount of waste needing treatment and disposal, recycling and material recovery are encouraged, and landfills specifically designed to handle hazardous waste have been created. Nonetheless, burning is the most often used method of disposal; there are about 3,840 hazardous industrial waste incinerators in the nation, and many of them have facilities for recovering energy to generate electricity or heat.

Most hazardous wastes produced by industries in Hong Kong, China, are received by the Chemical Waste Treatment Centre (CWTC), with some solid chemical wastes—such as asbestos, tannery off-cuts, and treatment residues—being co-disposed of at landfills.

#### *1.5.1.4. Biomedical Waste*

To meet the medical and healthcare needs of the expanding population, the number of hospitals and other healthcare facilities in the Asia-Pacific region has been rising. While medical and healthcare facilities and services have long been considered by city planners, until recently, they—as well as municipal waste management authorities—paid very little attention to the wastes produced by these facilities, which have the potential to be harmful to both human health and the environment.

However, in the recent years, there has been significant worry about the possibility that improper handling and management of clinical and biomedical waste could lead to the spread of pathogens and contaminate the environment. While most developed nations, like Australia, Japan, New Zealand, and Singapore, have implemented regulatory programs and guidelines to control waste from such institutions, these programs have not yet reached their full potential in developing nations, like Bangladesh, the People's Republic of China, India, Indonesia, Pakistan, and the Philippines.

National guidelines for the management of clinical and related wastes have been published by the National Health and Medical Research Council in Australia. State-level biomedical management guidelines have also been produced in a similar manner. Medical waste management guidelines have been prepared by a working group that was established by the Ministry of Health and Welfare in Japan. To rationalize and suggest practices for the management of healthcare waste in the nation, the Standards Association of New Zealand published the "New Zealand Standards on Healthcare -Waste Management." The Ministry of the Environment later produced the "Hospital Waste Management Manual," which included comprehensive guidelines for hospital waste handling and disposal as well as a standard format to help hospitals in preparing their written policies and procedures. In Singapore, guidelines for the management and safe disposal of hospital wastes were drafted in July 1988. Similarly, the Department of Health in the Philippines prepared guidelines in 1990 on effective and efficient methods of hazardous waste collection, storage, and disposal by hospitals, clinics, and research laboratories. Similarly, the Ministry of Health in Malaysia prepared preliminary guidelines for the management of hospital waste in 1988.

Hospital waste is typically collected and disposed of with other domestic waste, and hospital and waste management authorities are not well-informed about the risks associated with infectious biomedical waste, despite China's National Environmental Protection Agency recently developing the Solid Waste Pollution Prevention and Control Law and the Regulations on Management of Hazardous Wastes.

Hospitals across the region have implemented on-site incinerators to dispose of clinical waste; however, these facilities are frequently ill-maintained and poorly designed, and workers are not always aware of the risks. (Nag &Vizayakumar, 2005; Pitchell, 2014; Gaudillat et al., 2018)

#### *1.5.1.5. Radioactive Waste*

There is a paucity of information on radioactive waste disposal procedures, and few comprehensive national surveys have been carried out. In 46 operational nuclear power plants in Japan, low-level radioactive waste is collected and temporarily stored in 2,000-liter drums kept on-site in storehouses. Eight drums are packaged together in special enclosed containers and transported by land and sea to the Rokkasho-mura Burial Centre in Aomori Prefecture for long-term storage. Four nuclear research centres in Indonesia produce low-level radioactive waste, which is conditioned into cement matrices and then shipped to the RWMC (Radioactive Waste Management Centre) in Serpong for permanent burial. Permanent land burial is a method used by the Republic of Korea, India, Pakistan, and other countries in the region to dispose of radioactive waste.

#### *1.5.1.6. Transboundary Movement of Hazardous Waste*

The Asia-Pacific Region is facing significant pressure to stop being a preferred location for the disposal of hazardous waste, especially since industrialized nations' industries are under domestic pressure to dispose of their waste in an expensive and controlled manner. A total of 30.5 million tons of hazardous waste were sent by industrialized nations to countries in the Asia-Pacific region between 1994 and 1997. The People's Republic of China saw its first known instance of imports of this kind in September 1994. By the first quarter of 1995, Chinese customs had recorded 22 different incidents involving about 3,000 tons of foreign hazardous waste. Chinese customs discovered nearly one case of mislabeled hazardous waste per week between 1995 and 1996; most of these cases came from the United States, the Republic of Korea, and Japan in particular. 640 tons of Californian waste, including used syringes, decaying animal corpses, and toxic sludge, were discovered dumped in a Beijing suburb in June 1998.

Quantity of hazardous waste from developed countries to be dumped in India also increased during the same period. Thousands of tons of toxic waste is illegally transported to India for disposal or recycling despite a court order in New Delhi that forbids the importation of hazardous materials. Over 1,450 tons of hazardous waste, including scrap lead batteries, zinc, and copper ash, were exported by Australia to India in 1995. In contrast, between October 1996 and January 1997, about 569 tons of lead battery waste were imported into India via Mumbai's main seaport.

The subregional Waigani Convention on the transboundary movement of hazardous and radioactive wastes has raised awareness, which has prevented industry attempts to use the Pacific islands as hazardous waste dumps (in conjunction with power co-generation) from succeeding. Nevertheless, other nations in the region, such as Thailand, Bangladesh, Pakistan, Indonesia, the Philippines, and the Philippines, have developed into massive dumps for hazardous waste from industrialized nations exporting goods both inside and outside of the region.

### 1.5.2. Critical Problems and Shortcomings

When it comes to waste management, many Asian and Pacific nations face serious challenges. Numerous common deficiencies have been noted, such as inadequate funding, poor long-term planning, careless handling and disposal of hazardous and biomedical wastes, inadequate recycling and reuse, ineffective legislation and institutions, a shortage of skilled workers, and inadequate monitoring and enforcement. In addition, there is a lack of political support and government priority for taking action.

The consensus in many nations, especially concerning industrial waste, is that the industry and municipalities cannot be forced to introduce costly and sophisticated waste treatment and disposal technologies in order to impede economic growth. In many of the region's countries, the policy's shortsightedness and its consequences for long-term issues and waste treatment costs are still unknown or poorly understood.

The political systems of the nations, the government agencies, and the insufficient application of environmental laws are all major causes of concern. Many government agencies that handle waste have a complex mix of administrative, commercial, industrial, and operational duties and responsibilities. These organizations are extremely inefficient due to several factors, including their highly bureaucratic organizational structure, a lack of accountability and transparency in decision-making, low pay, corruption, nepotism and/or the hiring of unqualified staff, challenging and complex procurement procedures, and the powerful influence of political authorities on waste management-related technical decisions. This is frequently linked to a high degree of reliance on the allotted budget and a lack of adequate management systems. Government officers are often forced into corruption by the extremely low salaries they receive.

In many countries, there are still few private-sector waste management incentives and initiatives, and the relevant authorities are rarely willing to see the privatization of waste management services. Funding that could be utilized to set up a waste management system run by private contractors is generally lacking. Furthermore, the public's participation and awareness in waste management are extremely low. This is a result of a lack of knowledge of the advantages of proper waste collection, treatment, and disposal has for both the economy and public health.

A large portion of the current waste collection, treatment, and disposal infrastructure and facilities have not kept up with recent economic growth. The primary cause of the increasing inefficiency of waste management practices in the area is not a lack of knowledge, but rather a lack of funding and administrative support. The current standards for waste collection, treatment, and disposal are still very low, especially in smaller cities and rural areas (where consumption patterns have also changed). Inappropriate technology adoption leads to a host of issues in the area. Numerous instances of plant failure exist.

Particularly regarding the collection, handling, and disposal of municipal solid waste, there are numerous issues in Asian and Pacific cities. Because waste collection services depend on an inadequate number of vehicles—many of which are outdated, poorly maintained, and unreliable—they are frequently irregular. When travelling to the disposal site, open cars lose some of their load. The scarcity of replacement parts for collection vehicles is another serious issue. Collection workers sort materials or other items from the waste awaiting collection to make extra money (sometimes they even have contracts with junk dealers), i.e., e. they spend a lot of time on this activity and disregard their primary responsibility. Waste transportation typically wastes a lot of time; collection trucks can occasionally take several hours to get from the city to the landfill or dump site due to congested streets and heavy traffic. Again, open dumping at the side of the road, in waterways, rivers, and beside railroads is quite common. Collection is erratic; biomedical and industrial hazardous waste are frequently dumped in the same locations, which are often uncontrolled open dumping sites.

The best results from waste management techniques come from a comprehensive and integrated approach to the collection and disposal of all waste that is produced. However, despite growing loads that strain the already constrained resources of waste collection and disposal agencies, waste management is currently given relatively low priority in many of the region's countries. (Nag &Vizayakumar, 2005; Pitchell, 2014; Gaudillat et al., 2018)

# 1.6. ENVIRONMENTAL MANAGEMENT REGULATION

The term "environmental externalities" describes the economic idea of unadjusted environmental effects of production and consumption that have an impact on enterprise costs and consumer utility outside of the market framework. Negative externalities lead to private production costs that are typically less than the "social" cost. The "polluter/user-pays" principle aims to encourage individuals and businesses to incorporate externalities into their financial plans and budgets.

Goals and rules are typically established by national laws or legislative frameworks (like those in the European Union), but they can also be decided by a company's strategic planners and decision-makers. Governments and international regulators play a role in this process. Their contributions aid in the adoption of environmental waste management guidelines and the accounting for externalities. We will specifically look into the role of the government and the work of the International Standardization Organization in environmental management.

## 1.6.1. Government Institutions and Waste Management

There are three levels of government in every nation that are involved in waste management: local government, government agencies, and national government.

### *1.6.1.1. Local Governments*

Waste management services, such as waste removal, storage, and disposal, are mandated by local governments. Municipalities, in collaboration with industry and other stakeholders, are primarily responsible for expanding recycling at the municipal level. Municipalities oversee diverting organic waste from landfills and composting, as well as providing extra bins for separation at the source. Regarding regulations, local governments have the authority to establish guidelines for solid waste management, disposal, compacting, and waste separation.

### *1.6.1.2. Government Agencies*

Government agencies are the main regulatory body for waste-related activities, except for those that require approval from the national authority. These agencies are typically represented by an environmental protection agency. It must encourage and guarantee the application of national standards and norms. They have the authority to establish industry waste management plans, designate contaminated land, list waste management operations, designate a priority waste, register waste transporters, and set provincial norms and standards.

#### *1.6.1.3. National Governments*

National governments are ultimately in charge of making sure that the various local government organizations are conducting their operations in the most suitable and efficient manner and that national waste plans are carried out. In terms of mandatory provisions, among many things, they are responsible for:

- Establishing the National Waste Management Strategy.
- Setting national norms and standards.
- Establishing and maintaining a National Contaminated Land Register.
- Establishing and maintaining a National Waste Information System.
- Preparing and implementing a National Integrated Waste Management Plan.

### 1.6.2. International Standardization

The ISO is the most well-known organization involved in developing waste management policies (International Organization for Standardization). Furthermore, it oversees the accreditation of numerous organizations engaged in waste and environmental management. Its goal is to create standards that are universally acknowledged and respected to promote and facilitate both domestic and international trade and commerce. With the help and participation of its 164 national standards body members, ISO can accomplish this goal.

As a result, there is typically global support for these standards. ISO 14000 standards will be our focus for this course. The ISO Technical Committee, ISO/TC 207, is responsible for developing the ISO 14000 standards. The ISO 14001 standards include guidelines and requirements for environmental systems. Additional standards in the family concentrate on strategies like life cycle analysis, audits, communications, labeling, and labeling; they also address environmental issues like climate change.

The main purpose of ISO 14001 is to define the requirements for the Environmental Management Systems and provide guidance for its implementation. Key elements of this standard include:

- Environmental policy
- Planning
- Implementation and Operation
- Checking and Corrective Action Management review

The primary benefit of putting ISO 14001 into practice is that it offers a methodical approach to managing a facility's environmental impact. Additionally, it offers the framework necessary for an organization to achieve very high environmental performance standards, demonstrating the organization's full commitment to reducing the effects of waste, pollution, and energy use.

The other key benefits of implementing ISO 14001 include:

- Cost Savings, due to increased environmental efficiency.

- Cutback of environmental responsibilities.
- Cutback of environmental risk.
- Increased organization ability to compete in the global marketplace.

Numerous other organizations, such as the Green Building Council, EDGE, and other international regulatory bodies, have programs that can assist you in evaluating and developing your waste management plans and operations.

# 1.7. WASTE MANAGEMENT ECONOMICS

You will learn about waste management economics and its elements in this section. It effectively illustrates the various financial gains that can be attained through waste management.

## 1.7.1. Circular Economy

To gain a deeper comprehension of waste and environmental economics, let us first introduce the notion of circular economy. The concept of the "circular economy" is essentially similar to nature, which views everything as wholesome food for something else. Carbon dioxide and nutrients are used by plants to grow and release oxygen. Animals produce $CO_2$, consume oxygen, and need nutrients. With this closed-loop system, nothing is wasted.

The linear economy, which has been in place for a long time and entails consuming inputs for processes and discarding the results without realizing their potential, is typically contrasted with the circular economy (Figure 1.8).

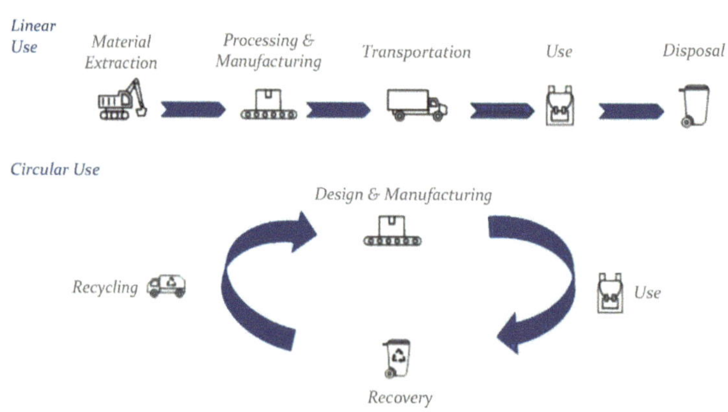


Figure 1.8. Linear and circular material cycles.

Source: AmeerMubaslat. Introduction to Waste Management. International Youth Ambassadors Foundation. ISBN:978-9957-67-886-9.

Considering waste management and sustainability in relation to the circular economy, it treats economy as a component of the society, which is a component of the environment, all three interacting in a closed ecosystem. This relation is illustrated in Figure 1.9.

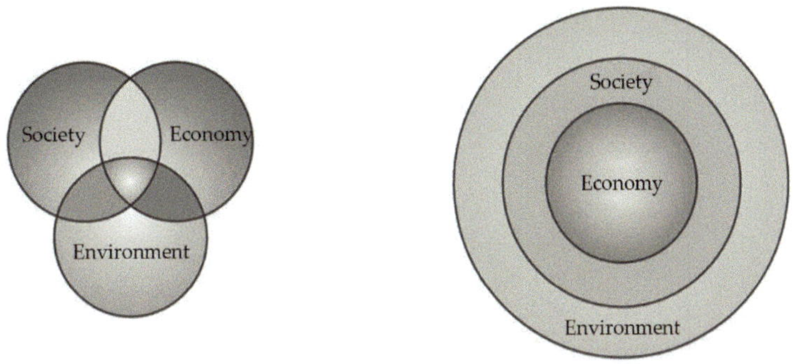


The conventional economic view of the interaction bewen economy, society and environment

The green economics paradigm: economy operates within social relationships and the whole of society is embedded within the natural world

Figure 1.9. Green economy paradigm.

Source: Ameer Mubaslat. Introduction to Waste Management. International Youth Ambassadors Foundation. ISBN:978-9957-67-886-9.

Technical and biological materials are the two types of material that are handled in the circular economy when it comes to waste management.

- This is the cycle of technical materials: raw materials are mined, products are made, and then the products are used until they are no longer useful. For instance, since copper mining is expected to end by 2040 and processing recycled copper requires 10–20% less energy than processing copper ore, it is preferable to mine copper that is no longer needed (post-consumer copper) rather than extracting new material from the ground.
- Biological materials that are farmed or gathered, then transported and processed before being consumed. Adding biogas or biochemical extraction and composting processes to the biological resources completes this cycle.

In this regard, Circular economy and therefore waste management economics utilize two characteristics which are substitution and dematerialization.

- Substitution: using different resources to achieve the same goal.
- Dematerialization: using less of a resource to serve the same economic function in a society.

Lastly, as the circular economy is a crucial component of sustainability, along with sustainable energy, waste management, and agriculture, we will employ it based on its ability to advance sustainable benefits across our operations.

## 1.7.2. Supply and Demand

An explanation of how buyers and sellers of resources interact is provided by the law of supply and demand theory. The relationship between a good or product's price and people's willingness to buy or sell it is defined by the theory. People tend to be more willing to supply than to demand when prices rise, and the opposite is true when prices fall (Figure 1.10).

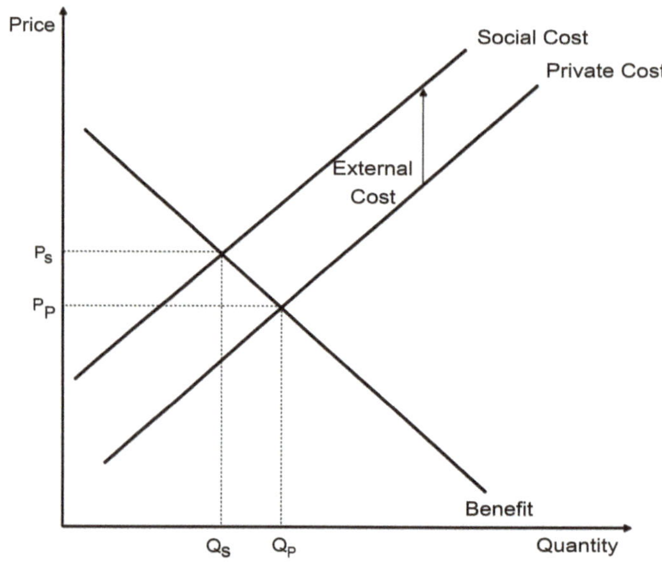


Figure 1.10. Supply and demand curves adapted to externalities.

Source: AmeerMubaslat. Introduction to Waste Management. International Youth Ambassadors Foundation. ISBN:978-9957-67-886-9.

The difference between both curves represents the economic and environmental advantage gained from adopting sustainable approaches. The environmental approach to supply and demand implies that the price set for manufacturing a certain product is way too cheap because it does not consider the externalities which push the supply curve in that direction.

In certain nations, externalities offer a means of reaping the rewards of waste management strategies that emphasize reduction, reuse, and recycling. These strategies open new markets for diverse materials and decrease various taxes associated with extended producer responsibility.

Utilizing the byproducts of various wastes, as we demonstrated with the examples of technical and biological waste, is another form of economic gain. Since examples are free, allow me to provide one more.

Sorting technical wastes and selling raw materials and sorted wastes like plastic, electronics, and batteries can be profitable for municipalities or waste management service providers. Additionally, they stand to gain from the composting of biological waste, which produces electricity through methane production that they can sell to the electric utility company.

Facilities stand to gain the most from altering their operations to use fewer resources and switching to more reusable items when it comes to purchases, additionally, if they produce enough garbage to be traded or recycled. The concepts of environmental economies and circular economy are becoming increasingly popular in waste management economics because they offer a great framework for understanding sustainable waste management goals and the long-term benefits it offers for both industrial and residential operations.

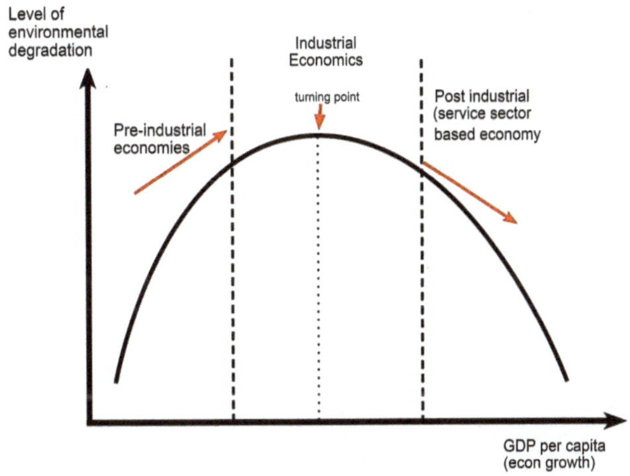


Figure 1.11. Kuznets curve.

Source: AmeerMubaslat. Introduction to Waste Management. International Youth Ambassadors Foundation. ISBN:978-9957-67-886-9.

The Kuznets curve, which illustrates the relationship between per capita income and environmental impact, is shown in Figure 1.11. This curve condenses economic data that indicates the nation is growing. Their detrimental influence on the environment increases until they become aware of these effects and can impose restrictions and provide incentives to lessen the effects. Following that turning point, they keep expanding their economy while lowering their environmental impact.

(Nag &Vizayakumar, 2005; Pitchell, 2014; Gaudillat et al., 2018)

# CASE STUDY

## SUSTAINABLE WASTE MANAGEMENT IN INDORE

Indore, a fast-growing city in India, has emerged as a model for sustainable waste management practices. Over the past few years, Indore has consistently ranked as the cleanest city in India, thanks to the efficient waste management system put in place by the municipal corporation. This case study explores the background, challenges faced, solutions implemented, and key learnings from Indore's successful waste management system.

Source: https://earth5r.org/sustainable-waste-management-in-indore-a-case-study/.

### *Background*

Indore, with a population of over 3.2 million people, generates around 1,100 metric tons of waste daily. Before 2016, the city struggled with waste management, leading to unhygienic conditions, increased pollution, and negative impacts on public health.

However, the launch of the Swachh Bharat (Clean India) campaign in 2014 led the Indore Municipal Corporation (IMC) to undertake a comprehensive transformation of its waste management system. This involved an overhaul of existing infrastructure, policies, and community engagement initiatives to create a more efficient and environmentally friendly waste management system.

### *Challenges Faced*

- **Lack of waste segregation at the source:** Indore faced issues with mixed waste, which hindered the recycling and disposal process. Unsegregated waste resulted in inefficient waste collection and processing, causing further strain on the waste management system.

- **Inefficient waste collection and transportation system:** With limited resources and vehicles, the city's waste collection and transportation system could not keep up with the growing population and waste generation.

Open dumping and burning of waste: The absence of adequate waste processing facilities led to the practice of open dumping and burning of waste, which contributed to air and land pollution.

- **Inadequate public awareness and participation:** Citizens were not fully aware of the importance of waste segregation, recycling, and proper disposal, resulting in low participation rates and disregard for waste management rules.
- **Limited infrastructure for waste processing and disposal:** The city's waste processing and disposal infrastructure was unable to cope with the increasing waste generation, leading to unmanaged landfills and environmental degradation.

### *Solutions Implemented*

- **Segregation at the source:** The IMC implemented a mandatory waste segregation policy, requiring households to separate waste into wet (biodegradable) and dry (recyclable) categories. This allowed for more efficient waste collection and processing, as well as increased recycling rates.
- **Door-to-door waste collection:** A fleet of over 600 GPS-enabled vehicles was deployed to collect segregated waste daily from all households and commercial establishments. This ensured timely and efficient waste collection, preventing littering and illegal dumping.
- **Waste processing and disposal:** The city established a state-of-the-art waste processing facility capable of handling 1,000 metric tons of waste daily, including a 15 MW waste-to-energy plant and a 200 TPD (tons per day) composting plant. These facilities enabled the city to process and dispose of waste more effectively, reducing the environmental impact of waste disposal.
- **Public awareness and participation:** The IMC launched numerous awareness campaigns, involving local celebrities, schools, and religious institutions, to educate the public on the importance of waste segregation and cleanliness. This resulted in increased community involvement and support for the waste management program.
- **Strict monitoring and enforcement:** Regular inspections, fines, and incentives were introduced to ensure compliance with waste management rules. This helped maintain the cleanliness of the city and encouraged citizens to adhere to waste segregation and disposal guidelines.

### Results Achieved

- **Waste segregation:** Over 90% of households in Indore now segregate their waste, significantly improving the efficiency of waste collection and processing, and reducing the burden on landfills.
- **Waste processing:** The city's waste processing facility successfully manages 1,000 metric tons of waste daily, with a 95% waste recovery rate. This has led to a substantial reduction in landfill usage and has minimized the environmental impact of waste disposal.
- **Cleanliness:** Indore has consistently ranked as the cleanest city in India in the annual Swachh Survekshan survey since 2017. This highlights the success of the city's waste management system and the active participation of its residents in maintaining cleanliness.
- **Health and environment:** Cases of vector-borne diseases have dropped by 60% since the implementation of the waste management system, and air quality has improved due to reduced open burning of waste. This has led to a healthier environment and improved overall quality of life for Indore's residents.

### Key Learnings

Political will and administrative commitment are crucial for the successful implementation of waste management systems. Indore's transformation was made possible by strong leadership and a dedicated municipal corporation committed to addressing the city's waste management challenges.

Public awareness and participation play a significant role in ensuring the success of waste management initiatives. By actively involving the community and raising awareness about the importance of waste segregation and proper disposal, Indore was able to achieve a high level of public participation and support.

Strict monitoring and enforcement mechanisms help ensure compliance with waste management rules and regulations. Indore's approach to enforcing waste segregation and disposal guidelines, combined with regular inspections and penalties, proved to be effective in maintaining the city's cleanliness.

Investing in modern waste processing infrastructure can significantly improve the efficiency of waste management systems and reduce environmental impact. Indore's investment in a state-of-the-art waste processing facility allowed the city to process and dispose of waste more effectively, leading to a substantial reduction in landfill usage and associated environmental issues.

Indore's transformation into a clean, sustainable city serves as an inspiring example for another urban centres in India and around the world.

By adopting a comprehensive, integrated approach to waste management, Indore has successfully addressed its waste management challenges and set a benchmark for sustainable urban living. The city's experience provides valuable insights and lessons for other municipalities looking to improve their waste management systems and promote environmental sustainability.

## CLASS ACTIVITY

Bring some waste materials from home. This can be kitchen waste (e.g., vegetable peels, used tea leaves, some food leftovers) and discarded articles of use, such as empty jars of plastic or/and glass, used paper, torn cloth etc.

- Bury those items in the school garden, or in a bucket or flowerpot, and cover them with sufficient soil.
- Leave the items buried for about three weeks after which take those items out for review.
- You will find that certain materials have changed their form, i.e., they have decayed, decomposed or are in the process of decomposition. On the other hand, some other materials are almost in the same form.
- Those materials that have decayed or decomposed are biodegradable materials, whereas those in which there was no change are non-biodegradable materials.
- Think about the reasons of this variation and write.

## SUMMARY

- Waste is an unavoidable by-product of most human activity. Economic development and rising living standards in the Asian and Pacific Region have led to increases in the quantity and complexity of generated waste, whilst industrial diversification and the provision of expanded healthcare facilities have added substantial quantities of industrial hazardous waste and biomedical waste into the waste stream with potentially severe environmental and human health consequences.
- Waste disposal leads to direct and indirect environmental impacts, such as land occupation, resource depletion, amplification of global warming due to methane and other greenhouse gas emissions, waters intoxication due landfilling, as well as acidification and toxic effects from emissions to air in the case of incineration.
- Waste management is defined as the different approaches and procedures designed and implemented to identify, control and handle the different types of waste from generation and until disposal. Full implementation of waste management processes, including waste prevention and reuse, and recycling wherever possible, has and can further help avoid considerable environmental impacts when assessed from a life-cycle perspective – considering direct effects such as emissions and indirect effects such as resource depletion.
- Sustainability is defined by meeting our needs to complete a certain task or activity in an efficient and productive manner that benefits the implementer and does not negatively affect the environment around us, in an approach that allows us to better utilize resources in three key areas, they are: economy, environment, and society.
- Waste is any solid, liquid, or gaseous material that is being discarded by disposal, recycling, burning or incineration. It can be a byproduct of a manufacturing process or an obsolete commercial product that can no longer be used for its intended purpose and requires disposal.
- Municipal solid waste (MSW) is generated from households, offices, hotels, shops, schools and other institutions. The major components are food waste, paper, plastic, rags, metal and glass, although demolition and construction debris are often included in collected waste, as are small quantities of hazardous waste, such as electric light bulbs, batteries, automotive parts and discarded medicines and chemicals.
- The principal types of hazardous waste generated in the Asian and Pacific Region, include waste solvents, chlorine-bearing waste and pesticide-organophosphate-herbicide-urea-fungicide-bearing waste. Solvents are extensively used in the region and, consequently, large quantities of waste solvents are produced.
- In the waste management process, the waste is collected from different sources and are disposed of. This process includes collection, transportation, treatment, analysis and disposal of waste.
- The waste management hierarchy is a conceptual framework designed to guide and rank waste management decisions at both the individual and organizational

levels. It gives top priority to waste prevention, followed by re-use, recycling, recovery and finally disposal.

## REVIEW QUESTIONS

1. What is waste management?
2. Where is the first wastewater management system located?
3. What are the 3 R of sustainability?
4. List the sources of MSW.
5. What is the role of government in waste management?

## REFERENCES


1. Abe, F. (2010). Management, Recycling and Reuse of Waste Composites. Boca Raton, FL: CRC Press.
2. Barbalace, R. C. (2003). The history of waste. EnvironmentalChemistry.com. Retrieved from: http://environmentalchemistry.com/yogi/environmental/wastehistory.html (accessed on 31 July 2024).
3. California Integrated Waste Management Board. (2010). Resource Guide to Integrated Waste Management (1st ed.). Sacramento, CA: California Integrated Waste Management Board.
4. Callan, S., & Thomas, J. M. (2010). Environmental Economics & Management: Theory, Policy, and Applications. Mason, OH: South-Western Cengage Learning.
5. Cheremisinoff, N. P. (2003). Handbook of Solid Waste Management and Waste Minimization Technologies. Amsterdam: Butterworth-Heinemann.
6. Eslamian, S. (2014). Handbook of Engineering Hydrology: Environmental Hydrology and Water Management. Boca Raton, FL: CRC Press.
7. Gandy, M. (1994). Recycling and the Politics of Urban Waste. London: Earthscan.
8. Gaudillat, P., Antonopoulos, I. S., Canfora, P., & Dri, M. (2018). Best Environmental Management Practice for The Waste Management Sector Learning from Frontrunners. Luxembourg: Publications Office. https://doi.org/10.2760/50247.
9. Kumar, S. (2014). Integrated Waste Management. Rijeka, Croatia: INTECH.
10. Ludwig, C., Hellweg, S., Stucki, S., & Hellweg, S. (2013). Municipal Solid Waste Management: Strategies and Technologies for Sustainable Solutions. Berlin: Springer. https://doi.org/10.1007/978-3-642-55636-4.
11. Mays, L. (2010). Ancient Water Technologies. https://doi.org/10.1007/978-90-481-8632-7.
12. McDougall, F. R. (2008). Integrated Solid Waste Management: A Life Cycle Inventory. New York, NY: Wiley-Blackwell.

13. Morsy, M. M., & Saleh, M. S. (1996). Industrial and non-industrial solid waste utilization in road building. Transportation and Communications Research.
14. Nag, A., & Vizayakumar, K. (2005). Environmental Education and Solid Waste Management. New Delhi: New Age International.
15. Pichtel, J. (2014). Waste Management Practices: Municipal, Hazardous, and Industrial. Boca Raton, FL: CRC Press.
16. Waste Economics Team, DEFRA. (2011). The Economics of Waste and Waste Policy (2nd ed.). London: Department for Environment, Food and Rural Affairs.

CHAPTER

# 2 Essentials of Geology, Geotechnics, and Toxicology

## LEARNING OBJECTIVES

- Define the geologic cycles
- Describe the index properties of soils
- Explain the hydrogeology
- Identify the essentials of toxicology

## KEY TERMS FROM THIS CHAPTER

Atlantic multidecadal oscillation (AMO)
Climate cycle
Earth's surface
Geologic cycle
Geotechnics
Igneous rock
Climate
Earth system
Geochemical cycle
Geology
Hydrologic cycle
Mantle dynamics

# INTRODUCTION

Geology, geotechnics, and toxicology play a crucial role in effective waste management, ensuring environmental protection and public health. Geology provides insights into the composition and structure of the Earth, which is essential for selecting suitable locations for waste disposal sites. Understanding the geological features of an area helps in assessing the natural barriers that may contain waste and prevent contamination of groundwater and soil. Geotechnics involves the application of engineering principles to analyze soil and rock mechanics, crucial for the design and construction of secure waste containment systems, such as landfills. This includes evaluating the stability of waste disposal sites, designing liners to prevent leachate leakage, and ensuring the structural integrity of waste storage facilities. Toxicology is essential for assessing the potential health risks associated with exposure to hazardous substances found in waste. It involves studying the toxic effects of chemicals and waste byproducts on humans and ecosystems, determining safe levels of exposure, and developing strategies for monitoring and mitigating these risks. Together, these disciplines enable the development of comprehensive waste management strategies that minimize environmental impact, ensure the safety of disposal sites, and protect public health from hazardous waste-related toxins. Effective waste management, therefore, relies on a multidisciplinary approach integrating geological, geotechnical, and toxicological expertise to address the challenges of waste disposal and contamination.

# 2.1. GEOLOGIC CYCLES

The intricate relationships between the tectonic, hydrologic, rock, and biological cycling of elements known as the biogeochemical cycle are collectively referred to as the geologic cycle. These different sub-cycles interact with one another and may result in weather, landslides, earthquakes, volcanic activity, flooding, groundwater flow, and other natural hazards and processes that are significant to environmental geology. Every other geologic sub-cycle has an impact on the rock cycle. For instance, the pressure and heat required for some or all the minerals in a rock to recrystallize and change the rock's composition are provided by tectonic processes.

## 2.1.1. Rock Cycle

A fundamental idea in geology, the rock cycle describes changes in the three primary types of rocks—sedimentary, metamorphic, and igneous—over geologic time. When a type of rock is pushed beyond its equilibrium physiological conditions, it undergoes changes. For instance, an igneous rock like basalt that is subducted beneath a continent may melt or disintegrate when exposed to the atmosphere. Owing to the forces of plate tectonics, the water cycle, and the rock cycle, rocks are constantly changing as they encounter new environments. The relationships between the three different types of rocks and the ways in which their processes vary over time are explained by the rock cycle. Because of this cyclical nature, rock alters both the biogeochemical cycle on planets where life exists and the geologic cycle.

### *2.1.1.1. Transition to Igneous Rock*

Magma can form when rocks are pushed far below the surface of the Earth. Magma cools and turns into an igneous rock when the conditions are no longer suitable for its continued liquid state. Known as intrusive or plutonic rock, it cools extremely slowly inside the Earth and produces rocks with a coarse-grained texture, like granite. Extrusive or volcanic rocks are the result of a rapidly cooling magma (also known as lava) that has been exposed to the atmosphere on the surface of the Earth due to volcanic activity. These fine-grained rocks can cool down so quickly that no crystals can form, producing natural glass-like obsidian; basalt is the most common type of fine-grained rock. Igneous rocks can form when any of the three primary types of rocks—sedimentary, metamorphic, and igneous rocks—melt into magma and cool.

- Secondary Changes

Although more than one of these alterations typically occurs in the same rock, epigenetic change (secondary processes occurring at low temperatures and low pressures) can be grouped under several headings, each of which is typical of a group of rocks or rock-forming minerals. Silicification, or the substitution of crystalline or crypto-crystalline silica for minerals, is most frequently observed in felsic rocks, like rhyolite, though it can also occur in serpentine, etc. The process known as "kaolinization" best demonstrated by granites and syenites, is the breakdown of feldspars, the most prevalent minerals in igneous rocks, into kaolin along with quartz and other clay minerals. Serpentinization, which is common in peridotites but occurs in most mafic rocks, is the transformation of olivine to serpentine (with magnetite). A variety of diabases, diorites, and greenstones exhibit chloritization, which is the conversion of augite (biotite or hornblende) to chlorite. In uralitization, secondary hornblende takes the place of augite. This group of rocks also undergoes epidotization, which is the process by which biotite, hornblende, augite, or plagioclase feldspar develops into epidote.

#### *2.1.1.2. Transition to Metamorphic Rock*

Metamorphism is the process wherein a rock undergoes physical or chemical changes in response to extreme heat and pressure. The effects on massive rock masses over a broad area are referred to as regional metamorphism, and they are usually connected to mountain- building events that occur within orogenic belts. These rocks frequently display foliation, or discrete bands with varying mineralogy and hues. When an igneous intrusion contacts a body of rock, it causes the surrounding country rock to become heated which is another major type of metamorphism. Through contact metamorphism, the intense heat of the magma and/or the addition of magma fluids that change the surrounding rock's chemical composition (metasomatism) cause the rock to change and re-crystallize. The processes of metamorphism can alter any pre-existing type of rock.

#### *2.1.1.3. Transition to Sedimentary Rock*

The weathering and erosion processes can cause rocks which are exposed to the atmosphere to become unstable in different ways. The original rock is broken down into smaller pieces by weathering and erosion, which also removes dissolved material. This broken material builds up and gets covered over by new material. A sedimentary rock is composed of merged sand grains, each of which is still a member of the class of rocks from which it originated. There are three ways that sedimentary rocks can form: precipitate sedimentary rock, which is formed by evaporation of a mineral-bearing solution, biogenic sedimentary rock, which is formed by the accumulation and lithification of material generated by living organisms, and clastic sedimentary rock, which is formed by lithification of these buried smaller fragments. Clastic rocks can be created from organic material, such as plant remains, or from pieces of larger rocks of any kind that have broken apart due to processes like erosion. Minerals from chemicals dissolved in all other rock types of deposit to form biogenic and precipitate rocks.

### 2.1.2. Water (Hydrologic) Cycle

Water is continuously circulated throughout the Earth-atmosphere system as part of the hydrologic cycle. The movement of water from the ground to the atmosphere and back again is the fundamental component of the water cycle. Surface water evaporates due to solar radiation. In the sky, the water vapor subsequently creates clouds. Water vapor condenses and descends to the earth as rain, snow, sleet, or hail, depending on the temperature and meteorological conditions. On earth's surface, some precipitation travels from high to low places and into bodies of surface water. We call this surface runoff. Other precipitation trickles down to the earth, where it condenses into groundwater. Water undergoes phase changes during this process, but the overall quantity of water particles doesn't change. To put it another way, even if 100 grams of water could be collected and brought to a boil, 100 grams of steam would remain. Similarly, 100 grams of steam collected and condensed would still equal 100 grams of water (Figure 2.1).

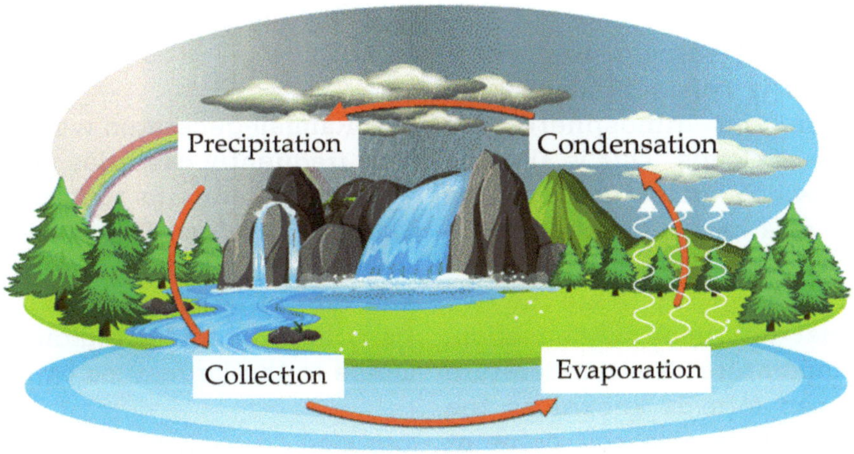


Figure 2.1. Water cycle diagram.

Source: https://byjus.com/biology/water-cycle/.

Water changes its state through a variety of processes from evaporation, melting and freezing, to sublimation, condensation, and deposition. All these changes require the application of energy.

### 2.1.2.1. Stages of the Water Cycle

Aside from the principal steps listed in the water cycle diagram above, there are numerous processes involved in the movement of water. The various phases of the water cycle are listed below. (Fetter, 2001)

- Evaporation

A substance changes states during evaporation, going from a liquid to a gas. In meteorology, water is the element that is most worrisome. Evaporation requires energy to occur. Any source of energy, including the sun, the atmosphere, the Earth, and living things can provide it. Everybody has firsthand experience with evaporation. The body sweats and secretes water onto the skin when it heats up from exercise or the temperature of the surrounding air. The idea is to make the liquid evaporate so that the body uses its heat to remove heat and cool itself. Stepping out of a shower or swimming pool has the same effect. Evaporation of water from skin removes heat from the body thus giving the sensation of coolness.

- Transpiration

The process by which water leaves plants through their stomata is called transpiration. On the underside of leaves, stomata are tiny holes that connect to the vascular tissues of plants. For most plants, transpiration is a passive process that is mostly regulated by soil moisture content and atmospheric humidity. Just 1% of the water that transpires through a plant is utilized by the plant to grow. The atmosphere absorbs the remaining 99%.

- Condensation

The process that turns water vapor into a liquid state is called condensation. Condensation in the atmosphere can manifest as dew or clouds. This is also the mechanism by which water accumulates on the exterior of a can or bottle of non-insulated cold beverage. The air temperature and the dew point temperature are the two temperatures that determine condensation, not a single temperature. The temperature at which dew can form is known as the "dew point," which is also the point at which air gets saturated and loses its ability to hold water vapor. Water vapor condenses with any further cooling. When the dew point and air temperature are the same, foggy conditions frequently arise. The opposite of evaporation is condensation. When condensation happens, the extra energy is released as heat energy because water vapor has a higher energy level than liquid water. Hurricane formation is aided by this heat release.

- Precipitation

The vapors will condense into water droplets above 0 degrees Celsius. It cannot, however, condense in the absence of dust or other contaminants. As a result, water vapor sticks to the surface of the particle. It emerges from the clouds and lands on the earth below when enough droplets combine. We refer to this process as precipitation (or rainfall). When there is an abnormally low air pressure or extreme cold, the water droplets freeze and fall as hail or snow.

- Infiltration

The process of infiltration allows rainwater to seep into the earth. The material that the water has seeped into determines the degree of absorption. For example, rocks will hold onto less water than soil. Rivers or streams can be followed by groundwater. However, occasionly, it may simply sink farther, creating aquifers.

- Runoff

Rainfall-generated water follows gravity and frequently cascades down the sides of hills and mountains to form rivers if it does not form aquifers. This process is referred to as runoff. When the rate of snowfall exceeds the rate of evaporation or sublimation, ice caps form in colder regions. The poles are home to the largest ice caps on the planet.

#### 2.1.2.2. Implications of Water Cycle

- The water cycle has a tremendous impact on the climate. For instance, the greenhouse effect will cause a rise in temperature. Without the evaporative cooling effect of the water cycle, the temperature on Earth would rise drastically.
- The water cycle is also an integral part of other biogeochemical cycles.
- The water cycle affects all life processes on earth.
- It is also known that the water cycle purifies the air. For example, water vapor needs to adhere to dust particles for precipitation to occur. Raindrops in contaminated cities carry pollutants and water-soluble gases with them when they descend from the clouds in addition to dust. It is also known that biological agents like smoke, particles of industrial soot, and bacteria can be carried by raindrops.

### 2.1.3. Plate Tectonics

The theory of plate tectonics, which examines the dynamics of Earth's outer shell, or lithosphere, transformed the study of Earth sciences by offering a consistent framework for comprehending the processes that create mountains, volcanoes, and earthquakes, as well as the evolution of the planet's surface and the reconstruction of its former continents and oceans. In the 1960s, the theory of plate tectonics was developed. The theory states that Earth is composed of two layers: the plastic (partially molten, moldable) asthenosphere layer, which is layered on top of a rigid outer layer called the lithosphere, which is normally around 100 km (60 miles) thick. The lithosphere is divided into several small plates, six or seven medium-sized regional plates, and seven very large continental and oceanic plates. These plates interact along their borders, where they converge, diverge, or slip past one another. They move relative to one another, usually at rates of 5 to 10 cm (2 to 4 inches) per year. Although they can occur in the interior of the plates, these interactions are thought to be the primary cause of most earthquakes and volcanic activity on Earth. Mountains rise where plates converge, or push together, and continents split and oceans form, or pull apart, because of plate motions. The Earth's geography has changed significantly over millions of years because of the continents' passive drift with the plates in which they are embedded.

The foundation of plate tectonics theory is a comprehensive synthesis of geophysical and geologic data. Its adoption represents a true scientific revolution, comparable in its implications to the discoveries of the genetic code in biology and quantum mechanics in physics, and it is now nearly universally accepted. The theory of plate tectonics has given scientists an overall framework to explain the past geography of continents and oceans, the processes governing the creation and destruction of landforms, and the evolution of Earth's crust, atmosphere, biosphere, hydrosphere, and climates. It also incorporates the much older idea of continental drift and the concept of seafloor spreading. It became clear in the late 20th and early 21st centuries that plate-tectonic processes had a major impact on the chemical and physical conditions that support the evolution of life on Earth, as well as the composition of the planet's oceans and atmosphere. They are also a major contributor to long-term climate change.

#### *2.1.3.1. Principles of Plate Tectonics*

Plate tectonic theory is essentially elegantly simple. The rigid, 50–100 km (30–60 mi) thick layer that makes up the Earth's surface is made up of a series of large and small plates. The Greek word lithos, which means "rock," refers to the collection of these plates as the lithosphere. The lithosphere is a layer of partially molten rock that slides over and rests on an underlying layer of plastic rock. This layer is called the asthenosphere, derived from the Greek word asthenos, which means "weak." The lithosphere-asthenosphere boundary is a zone of detachment, which permits plate movement. The lithospheric plates interact along their boundaries, diverging, converging or slipping past one another as they travel across the surface of the Earth under the influence of forces that are still poorly understood. Numerous major processes that shape the terrestrial surface, such as volcanism, earthquakes, and orogeny (the formation of mountain ranges), occur at plate boundaries, whereas the interiors of the plates are thought to remain essentially unaltered.

Convection in the Earth's mantle, the pull of large, ancient crust fragments into the mantle, or a combination of the two, can be the driving forces behind plate tectonics. To delve deeper into the mechanisms behind plate-driving.

### *2.1.3.2. Types of Plate Boundaries*

The movement of the plates relative to one another distinguishes three different types of plate boundaries. A variety of surface phenomena are linked to them.

The different types of plate boundaries are:

- Ranges beneath the less dense continent, or ocean-to-ocean subduction wherein older, cooler, denser oceanic crust slips beneath the less dense oceanic crust. Deep marine trenches are typically associated with subduction zones, and the basins that develop along the active boundary are often called "foreland basins."
- Earthquakes trace the path of the downward-moving plate as it descends into asthenosphere, a trench forms, and as the subducted plate is heated it releases volatiles, mostly water from hydrous minerals, into the surrounding mantle. The addition of water lowers the melting point of the mantle material above the subducting slab, causing it to melt. The magma that results typically leads to volcanism at zones of ocean-to-ocean subduction a deep trench forms in an arc shape. The upper mantle of the subducted plate then heats, and magma rises to form curving chains of volcanic islands e.g., the Aleutian Islands, the Mariana Islands, and the Japanese island arc.
- At zones of ocean-to-continent subduction mountain ranges form, e.g., the Andes, and the Cascade Range.
- At continental collision zones, there are two masses of continental lithosphere converging. Since they are of similar density, neither is subducted. The plate edges are compressed, folded, and uplifted forming mountain ranges, e.g., the Himalayas and the Alps. Closure of ocean basins can occur at continent-to-continent boundaries.
- Transform boundaries (conservative boundaries or strike-slip boundaries) occur where plates are neither created nor destroyed. Instead, two plates slide, or perhaps more accurately grind past each other, along transform faults. The relative motion of the two plates is either sinistral (left side toward the observer) or dextral (right side toward the observer). Transform faults occur across a spreading centre. Strong earthquakes can occur along a fault. The San Andreas Fault in California is an example of a transform boundary exhibiting dextral motion.
- Other plate boundary zones occur where the effects of the interactions are unclear, and the boundaries, usually occurring along a broad belt, are not well defined and may show various types of movements in different episodes. (Grove, et al., 2012).

### *2.1.3.3. Driving Forces of Plate Motion*

The relative strength of the asthenosphere and the relative density of the oceanic lithosphere allows the tectonic plates to move. The initial source of the energy needed to propel plate tectonics through convection or extensive upwelling and doming is the dissipation of heat from the mantle. Consequently, the excess density of the oceanic lithosphere sinking in subduction zones is a potent source that generates plate motion. This oceanic lithosphere is initially less dense than the underlying asthenosphere when the new crust forms at mid-ocean ridges, but it becomes denser with age as it conductively cools and thickens. At subduction zones, the old lithosphere's higher density compared to the underlying asthenosphere enables it to sink into the deep mantle, where it supplies much of the force responsible for plate movement. The tectonic plates can easily move in the direction of a subduction zone due to the asthenosphere's weakness.

- Mantle Dynamics

Large-scale convection currents in the upper mantle that can travel through the asthenosphere were the main theory for explaining the forces behind tectonic plate motions for a significant portion of the first quarter of the 20th century. When Arthur Holmes and a few others introduced this theory in the 1930s, it was instantly acknowledged as the answer to the theory's acceptance, which had first been addressed in Alfred Wegener's early 20th-century publications. Though it was accepted, the leading theory up until the significant discoveries of the early 1960s still predicted a static Earth with moving continents, so the theory was hotly contested in the scientific community. Seismic tomography, which provides two and three-dimensional images of the Earth's interior, reveals that the mantle's lateral density distribution varies. These variations in density can be thermal (due to heat energy-induced thermal expansion and contraction), mineral (due to variations in mineral structures), or material (due to variations in rock chemistry). Mantle convection resulting from buoyancy forces is the expression of this fluctuating lateral density.

In geodynamics, the direct and indirect relationships between mantle convection and plate motion are still being investigated. For tectonic plates to move, this energy needs to be transferred in some way to the lithosphere. Primary (through the large-scale convection cells) and secondary are the two basic categories of mechanisms that are thought to exist related to the dynamics of the mantle that influence plate motion. The friction between the convection currents in the asthenosphere and the more rigid lithosphere above is seen as the secondary mechanism driving plate motion. This is because mantle material is infiltrating subduction zones at ocean trenches as a result of downward plate pulls. When basal tractions continue to act on the plate as it descends into the mantle in a geodynamic setting, slab pull may happen (though they may act more on the slab's upper and lower surfaces). Moreover, viscous mantle forces can drive plates through slab suction when broken-off slabs sink into the mantle.

- Plume Tectonics

A modified version of mantle convection currents is used in the plume tectonics theory that was adopted by many researchers in the 1990s. It claims that super plumes serve as the primary convection cells' drivers or replacements, rising from the deeper mantle. These theories originated from the works of Beloussov and van Bemmelen in the early 1930s. These authors initially rejected plate tectonics and framed the mechanism in terms of fixed frame movements that were vertical. Later, Van Bemmelen developed the idea in his Undation Models, utilizing Mantle Blisters to generate horizontal movements and deflect gravitational forces from the local crustal doming. Contemporary theories that envision stationary hot spots or mantle plumes that eventually get displaced by oceanic and continental lithosphere plates and leave their mark on the geological record are consistent with these theories, even though they refer to these phenomena as modulators rather than actual driving forces. The mechanism is still advocated to explain the break-up of supercontinents during specific geological epochs. It has followers amongst the scientists involved in the theory of Earth expansion.

- Surge Tectonics

According to a different theory, the mantle flows as a network of channels directly beneath the Earth's crust rather than in cells or massive plumes, which subsequently give the lithosphere basal friction. The 1980s and 1990s saw a rise in the popularity of this theory, known as surge tectonics. Based on three-dimensional computer modelling, recent research indicates that the strength of the lithosphere and mantle convection patterns interact to control plate geometry.

- Driving Forces Related to Gravity

Gravitational forces are invoked as incidental phenomena within a more general driving mechanism, like the above-discussed forms of mantle dynamics. Through slab pull along subduction zones, gravity is cited as the primary driving force in contemporary perspectives. One of the hypothesized driving forces for plate motion is gravitational sliding away from a spreading ridge; this suggests that plate motion is propelled by the higher elevation of plates at ocean ridges. The oceanic lithosphere gradually cools and thickens with age as it forms at spreading ridges from hot mantle material (and thus adds distance from the ridge). Because the cool oceanic lithosphere is much denser than the hot mantle material it is formed of, it gradually sinks into the mantle with increasing thickness to offset the increased load. The outcome is a small lateral incline that is farther away from the ridge axis.

This force, also known as the "ridge push," is thought of as a secondary force. This is misleading because tensional features predominate along ridges and there is no force pushing horizontally. Since spreading ridges are merely the most noticeable feature and the topography throughout the entire plate can vary greatly, it is more accurate to refer to this mechanism as gravitational sliding. A distinct topographical feature that can counteract or at least have an impact on the influence of topographical ocean ridges is produced by the flexural bulging of the lithosphere before it subducts beneath an adjacent plate, which is one of the other mechanisms producing this gravitational secondary force. It is also hypothesized that hot spots and mantle plumes press against the underside of tectonic plates.

- Slab pull: According to scientific opinion, the asthenosphere is either too rigid or too incompetent to directly induce motion along the lithosphere's base through friction. Therefore, the most common belief regarding the strongest force influencing the plates is slab pull. According to this theory, the weight of cold, dense plates sinking into the mantle at trenches is primarily responsible for plate motion. According to recent models, trench suction is also a significant factor. Though it is moving, there is an issue with the North American Plate not being subducted anywhere. The Antarctic, Eurasian, and African plates are all in agreement.
- Gravitational sliding away from mantle doming: Older theories state that the presence of massive asthenosphere/mantle domes, which cause the lithosphere plates

to gravitationally slide away from them, is one of the driving forces behind the plates. This essentially vertically oriented mechanism's secondary phenomenon is represented by this gravitational sliding. Its foundations are found in van Bemmelen's Undation Model. This can have an impact on different sizes, ranging from an entire ocean basin to a single island arc.

- Driving Forces Related to Earth Rotation

As a meteorologist, Alfred Wegener suggested, that tidal and centrifugal forces were the primary forces causing continental drift; however, since the idea was of continents piercing through oceanic crust, these forces were thought to be far too tiny to cause continental motion. However, the oceanic crust is suggested to be in motion with the continents in the context of plate tectonics (accepted since the seafloor spreading proposals of Heezen, Hess, Dietz, Morley, Vine, and Matthews during the early 1960s), which led to the reconsideration of the Earth rotation proposals.

In more recent literature, these driving forces are:

- Tidal drag due to the gravitational force the Moon (and the Sun) exerts on the crust of Earth
- Global deformation of the geoid due to small displacements of the rotational pole with respect to Earth's crust
- Other smaller deformation effects of the crust due to wobbles and spin movements of Earth's rotation on a smaller timescale
- Forces that are small and generally negligible are:
- The Coriolis force
- The centrifugal force, which is treated as a slight modification of gravity

For these mechanisms to be considered generally valid, systematic relationships between the orientation and kinematics of deformation and the Earth's geographical latitudinal and longitudinal grid should be present worldwide. These systematic relations studies conducted in the first half of the twentieth century and the second half of the nineteenth century highlight the exact opposite: that the deformation grid was fixed relative to the Earth's equator and axis, that the plates had not moved in time, and that gravitational driving forces were primarily acting vertically, causing only local horizontal movements. As a result, many of the relationships identified during this pre-plate tectonics era were cited by later research to bolster their theories.

#### *2.1.3.4. Possible Tidal Effect on Plate Tectonics*

Of the many forces discussed above, tidal force is still highly debated and defended as a possible principal driving force of plate tectonics. The other forces are only used in global geodynamic models not using plate tectonics concepts or proposed as minor modulations within the overall plate tectonics model. In 1973, George W. Moore of the USGS and R. C. Bostrom presented evidence for a general westward drift of Earth's lithosphere with respect to the mantle, based on the steepness of the subduction zones (shallow dipping towards the east, steeply dipping towards the west). They concluded that tidal forces (the tidal lag or friction) caused by Earth's rotation and the forces acting upon it by the Moon are a driving force for plate tectonics. According to Alfred Wegener's theory, the Moon's gravity slightly pushes

Earth's surface layer back westward as Earth spins east beneath it. Since 1990, Doglioni et al., have primarily promoted this theory (Doglioni 1990). This was the case in a more recent study from 2006, where scientists examined and supported these concepts. According to Lovett (2006), this finding might also help explain why there are no plate tectonics on Venus or Mars because neither planet has a moon, and Mars' moons are too small to have a major impact on the planet's tidal forces. On the other hand, it is evident that numerous plates are moving north and east, and the Pacific Ocean basins' predominant westward motion is due only to the spreading center's eastward bias (which is not a predicted manifestation of such lunar forces). The authors of the same paper acknowledge that there is a small westward component to all the plate motions in relation to the lower mantle. However, they showed that the increased dominance of the steadily growing and accelerating Pacific plate is responsible for the westward drift, which has only been observed for the past 30 Ma. There is still disagreement over this, but a recent study by Hofmeister et al., the theory that Earth's rotation and the Moon's interaction serve as the primary forces driving the plates was brought back in 2022.

#### *2.1.3.5. Relative Significance of Each Driving Force Mechanism*

A plate's motion vector is a function of all the forces acting on it. The question that needs to be addressed is how much each process adds to the overall motion of each tectonic plate. The impact of the different processes that actively drive each individual plate results in the diversity of geodynamic settings and the characteristics of each plate. Considering the relative speeds of each plate as well as the data pertaining to the importance of each process to the overall driving force on the plate are two approaches to addressing this issue. The fact that lithospheric plates associated with subducting (descending) plates move at a significantly faster rate than other plate types is one of the most important correlations found to date. For example, the Pacific plate moves at a much faster rate than the plates of the Atlantic basin, which are essentially 'welded' to neighboring continents rather than subducting, surrounded by zones of subduction known as the Ring of Fire. Thus, it is believed that, except for the plates that are not being subducted, the forces associated with the down going plate—slab pull and slab suction—are what propel plates in their respective directions. However, a recent study has contradicted this view, finding that the actual motions of the Pacific Plate and other plates associated with the East Pacific Rise correlate more with a mantle convection upwelling whose horizontal spreading along the bases of the various plates drives them along via viscosity-related traction forces than with either slab pull or slab push. The mechanisms behind plate motion are still being actively studied in the fields of geophysics and tectonophysics. (Fetter, 2001)

### 2.1.4. Geochemical Cycle

The geochemical cycle is the evolutionary route that individual elements or groups of elements take on the surface and in the crustal and subcrustal zones of the Earth. Within the framework of geochemical cycles, geochemical differentiation (i.e., heat-assisted elemental recombination processes and the natural separation and concentration of elements by Earth processes. For the lithosphere (i.e., the crust and upper mantle), the crystallization of magma at depth or on the surface initiates the geochemical cycle. The igneous rock is then broken down by weathering and surface alteration, which is followed by the material being transported and deposited as sediment. This sediment eventually undergoes metamorphosis and lithification, leading to melting and the production of new magma. This perfect cycle could be broken at any time. With each cycle's progression, each element might be impacted differently. For instance, as igneous rock weathers, minerals rich in silicon, such as quartz and feldspar, are mostly carried as sediment, but minerals containing iron, magnesium, and calcium also break down and are carried in solution. Because of the precipitation of calcium or magnesium carbonates, some of the resulting sedimentary rocks are dominated by quartz and feldspar, while others are dominated by calcium and magnesium. Elements like sodium stay in solution until they precipitate in extremely cold or extremely hot conditions. When sedimentary rocks start to partially melt, elements separate based on their melting characteristics, volatiles are released into the atmosphere, and chemically separated bodies start to physically move. Long-term, or secular, changes take place in the geochemical cycle, even though it appears to be in a steady state over the short term. Thus, during geologic time, for instance, continents and oceans have changed.

#### *2.1.4.1. Differentiating Biogeochemical Cycles*

The terms geochemical cycle and biogeochemical cycle are sometimes used synonymously because they both pertain to the Earth's reservoirs. On the other hand, a geochemical cycle describes the chemical interactions that occur in crustal and subcrustal reservoirs, such as the deep earth and lithosphere, whereas a biogeochemical cycle describes the chemical interactions in surface reservoirs, such as the atmosphere, hydrosphere, lithosphere, and biosphere.

#### *2.1.4.2. Earth System*

The Earth is essentially closed off to matter, but it is open to radiation from space and the sun. It adheres to the law of conservation of mass, which states that matter cannot be created or destroyed. As a result, even after being changed and migrated, matter is still the same as it was when the Earth was formed. This is true of all closed systems. The Earth system is made up of seven distinct reservoirs, which are divided into two categories: isolated reservoirs, which include deep Earth and space, and surface reservoirs, which include the atmosphere, hydrosphere, biosphere, pedosphere, and lithosphere. The interactions between the lithosphere, or the Earth's crust, and the deep earth, or the planet's mantle and core, are the subject of geochemical cycles.

### 2.1.5. Climate Cycle

The climate cycle refers to the long-term patterns and variations in climate, driven by a combination of natural and anthropogenic factors. These cycles encompass periods of warming and cooling, influenced by astronomical, atmospheric, and geological processes. One of the most well-known climate cycles is the Milankovitch cycle, which involves changes in the Earth's orbit, axial tilt, and precession, leading to glacial and interglacial periods over tens of thousands of years. Volcanic activity also plays a significant role, as eruptions can inject large quantities of ash and sulfur dioxide into the atmosphere, reflecting sunlight and temporarily cooling the planet.

Additionally, oceanic cycles such as the El Niño-Southern Oscillation (ENSO) and the Atlantic Multidecadal Oscillation (AMO) significantly impact global weather patterns and climate. ENSO, for instance, causes periodic warming (El Niño) and cooling (La Niña) of the central and eastern Pacific Ocean, affecting precipitation and temperature patterns worldwide.

Human activities, particularly since the Industrial Revolution, have increasingly influenced the climate cycle. The burning of fossil fuels, deforestation, and industrial processes have elevated atmospheric concentrations of greenhouse gases like carbon dioxide and methane, enhancing the greenhouse effect and contributing to global warming. This anthropogenic impact is superimposed on natural climate cycles, leading to unprecedented changes in climate patterns.

Understanding the climate cycle is crucial for predicting future climate changes and developing strategies to mitigate and adapt to their impacts. Climate models that incorporate both natural variability and human-induced factors are essential tools for scientists to project future climate scenarios. These models help policymakers in creating informed strategies for reducing greenhouse gas emissions, managing natural resources, and protecting vulnerable communities from the adverse effects of climate change. The study of past climate cycles, through paleoclimatology, also provides valuable insights into the Earth's climate system, helping to contextualize current changes within the broader spectrum of natural variability. Overall, the climate cycle is a complex interplay of natural and human factors that shapes the long-term patterns of our planet's climate.

# 2.2. INDEX PROPERTIES OF SOILS

The characteristics of soil known as index properties aid in the identification and classification of soil for general engineering applications. Usually, laboratory testing is used to determine these attributes. The soil's index property can be ascertained using either disturbed or remolded samples. The term "index properties" can also refer to the representation of engineering properties like shear strength, permeability, and compressibility. Engineering properties must be determined through a complex and time-consuming test. The geotechnical engineer may occasionally be interested in a preliminary evaluation of the engineering properties without carrying out extensive testing. Only after the soil's index properties are established is this feasible.

## 2.2.1. Soil Index Properties

The properties of soil index which are non-cohesive in nature are:

- Particle size distribution
- Particle shape
- Relative density
- Consistency
- Clay and clay mineral content
- Soil classification
  - Distribution of article size: This property describes the size and distribution of soil particle sizes. This is regarded as the most fundamental feature of the soil index. Well-graded soil is made up of a broad variety of particle sizes. Conversely, soil with a small range of particle sizes is poorly graded soil.
  - Particle shape: The degree to which soil particles are packed closely determines the concept of particle shape. Specifically, the density of coarse-grained soil reveals the stiffness and strength of the soil.
  - The formula of relative density: The ratio of the bulk density to the maximum soil density is known as the relative density of the soil. Relative density shows how much the soil could deform and how much its density could increase in certain situations. Sieves and sediment analysis are two

methods for grading soils.

- Consistency: The soil's potential resistance to deterioration, rupturing, and deformation is referred to as its consistency. The theory that the soil can exist in four states based on its moisture content is the basis for the entire concept of soil consistency. Firstly, the soil is a viscous liquid with little resistance; shear has its own strength. Its moisture content decreases as it ages and it begins to gain some shear strength, but it is still very moldable. The solid phase of plastic is another name for this stage. As the material dries, the ability to be molded significantly decreases and fissures start to appear. The soil eventually gets so dry that it breaks easily. The strength and consistency of the soil can also be determined by looking at the compression strength. Various terms, including soft, medium, stiff, very stiff, and hard, are used to categorize the soil. Consistency in non-cohesive soil is mainly determined by the distribution of soil particle size and shape. In contrast, cohesive soil's consistency is primarily determined by its water content.
- Clay and clay mineral content: Another crucial component of the soil index is the amount of clay and clay minerals. Particle size and mineral properties allow for the differentiation of clay minerals, which are fine silicate minerals. The two most common silicate minerals are magnesium octahedron and silicate tetrahedron, both of which are highly malleable in nature. As was previously mentioned, the silicate content is the only factor that determines how plastic the clayey soils are. Therefore, soils devoid of clay minerals are plastic in nature, and their moisture content varies significantly between liquid and semi-solid states. The three most prevalent types of clay minerals are montmorillonite, illite, and kaolinite.
- Water content: Water content is the next property on the soil index. For instance, fine-grained soil's characteristics change significantly depending on the amount of water present. The water content of soil influences its four states—liquid, plastic, solid, and semi-solid—in significant ways. After determining bulk density, dry density is computed using moisture content. Soil water contains dissolved organic and inorganic materials, making it not entirely pure water.
- Soil classification: In general, coarse-grained soil is classified using the soil classification letter G (G for gravel) or letter S (S for sand). The letter W is used to refer to well-graded soil, and the letter P is used to classify poorly graded soil. If the coarse-grained soils contain more than 12% silt or clay, the second letter, M, is used to indicate silt and C, clay. The same designation is applied to fine-grained soils also. The second letter H (high) or L (low), is used to indicate the plasticity of the soil. For example, in the gravel and sand mixture, GW denotes well-graded and clean soil, and on the other hand, SW denotes well-graded and clean sandy soil. In the case of fine-grained soils of low plasticity, the word ML denotes high content of silt, and for fine-graded soils of high plasticity, MH also represents silt. (Fetter, 2001)

# 2.3. HYDROGEOLOGY

The branch of geology known as hydrogeology (from the Greek hydro- meaning water and -geology, the study of the Earth) examines the distribution and flow of groundwater in the rocks and soil that make up the Earth's crust, mostly in aquifers. It's common to use the terms hydrogeology, geohydrology, and groundwater hydrology interchangeably. The study of hydrogeology focuses on the principles that control the flow of subsurface water, as well as the mechanical, chemical, and thermal interactions that occur when this water interacts with porous materials and the flow of particles, energy, and chemical components. The characteristics of rocks and sediments that regulate water flow, the relationship between surface and groundwater, and the effects of human activity on the quantity and quality of groundwater resources are all studied by hydrogeologists. Hydrogeology is an interdisciplinary field that addresses a wide range of environmental, geological, and engineering problems by utilizing knowledge from geology, physics, chemistry, mathematics, and engineering. A valuable natural resource is water. There wouldn't be any life on Earth without water. Our bodies are made up of two thirds water by weight. Water supplies are also necessary to sustain industrial activity and food production. The amount of precipitation is the primary factor that determines the density and distribution of vegetation (Fetter, 2001). Certain deserts can support agriculture, but only if water is brought in from outside the region or pumped out of the earth (Fetter, 2001). When dependable water sources were established, civilizations prospered; however, when these water supplies faltered, they fell apart. To keep the body's vital fluids balanced, an individual needs three liters (L) of drinkable water each day (Fetter, 2001).

Primitive people in arid lands existed with little more than this amount as their total daily consumption.

The over-exploitation of groundwater by uncontrolled pumping can cause some problems (Hiscock, 2005):

- detrimental effects on neighbouring boreholes and wells,
- land subsidence,
- saline water intrusion,
- and the drying out of surface waters and wetlands.

Uncontrolled uses of chemicals and the careless disposal of waste on land cause groundwater pollution (Hiscock, 2005).

Major sources of groundwater pollution:

- agrochemicals,
- industrial and municipal wastes,
- tailings and process wastewater from mines,
- oil field brine pits,
- leaking underground storage tanks,
- leaking pipelines,
- sewage sludge,
- and septic systems

## 2.3.1. Scope of Hydrogeology

The scientific study of groundwater's characteristics, distribution, and flow in the Earth's subsurface is known as hydrogeology. It is an interdisciplinary field that incorporates aspects of engineering, chemistry, physics, hydrology, and geology. The scope of hydrogeology includes the following:

- Study of groundwater occurrence and availability: Hydrogeologists investigate the presence, accessibility, and distribution of groundwater below the surface. They investigate the subsurface using a variety of methods, including well logging, drilling, and geophysical surveys.
- Groundwater flow and transport: Hydrogeologists investigate the subsurface movement and conveyance of groundwater. They simulate the movement of pollutants in groundwater and forecast the direction and speed of groundwater flow using numerical models.
- Aquifer characterization: Aquifers are geological formations that store and transfer groundwater. Hydrogeologists describe the characteristics of aquifers. They research the hydraulic characteristics of aquifers, including storage coefficient, transmissivity, and hydraulic conductivity.
- Groundwater quality: Hydrogeologists investigate the characteristics of groundwater, encompassing its chemical makeup and the existence of impurities. They sample and analyze groundwater using a variety of methods, including well logging, slug testing, and pumping tests.
- Groundwater management: An essential function of hydrogeologists is to manage groundwater resources. They create strategies for the sustainable use and preservation of groundwater resources by drawing on their expertise in hydrogeology. This covers groundwater contamination control, groundwater recharge management, and well field design.
- Interaction of groundwater with surface water: The relationship between groundwater and surface water—such as rivers, lakes, and wetlands—is the subject of study for hydrogeologists. Using their hydrogeology expertise, they create strategies for sustainable water resource management and comprehend the function of groundwater in sustaining surface water flow.

## 2.3.2. Hydrogeological Investigation

The process of researching the characteristics and behavior of water below the surface is known as hydrogeological investigation. It entails using a variety of instruments and methods to collect information about the hydrogeological system, including the local hydrology and geology, the amount and quality of groundwater, and the possibility of managing and developing water resources. Numerous applications require hydrogeological investigation, including the planning and design of groundwater supply systems, the identification of possible water sources for mining or industrial activities, the assessment of groundwater-related environmental impacts, and the assessment of potential climate change effects on groundwater resources. Numerous tasks, including geological mapping, hydrological data collection, aquifer testing, water quality analysis, and computer modeling of groundwater flow and transport, may be included in hydrogeological investigations. The outcomes of hydrogeological studies can be utilized to guide decisions regarding the sustainable management and utilization of groundwater resources.

There are several steps involved in a hydrogeological investigation, including:

- Defining the study area: The first step in a hydrogeological investigation is to define the area of study, including the location and boundaries of the study area.
- Collecting data: The next step is to gather information about the geology, hydrology, and hydrogeology of the study area. This may include collecting data on the geology of the area, the surface and subsurface hydrology, and the groundwater resources.
- Analyzing data: The collected data is then analyzed to understand the occurrence and movement of groundwater in the study area. This may involve analyzing geological and hydrological data, as well as data on the quality and quantity of groundwater resources.
- Developing a conceptual model: Based on the data collected and analyzed, a conceptual model of the groundwater system in the study area is developed. This model helps to understand how groundwater moves through the subsurface and how it is affected by various factors.
- Testing and refining the model: The conceptual model is then tested and refined through further data collection and analysis, to improve the understanding of the groundwater system.
- Reporting findings: The final step in a hydrogeological investigation is to report the findings of the study, including any recommendations for the management and use of groundwater resources in the study area.

### 2.3.3. Hydrogeology and Human Affairs

Hydrogeology is closely tied to human affairs in many ways. Here are a few examples:

- Water supply: Hydrology's most significant use is in the evaluation and management of groundwater resources to produce drinking water. Aquifers are studied and characterized by hydrogeologists, who also create models to forecast how aquifers will react to various pumping scenarios and calculate groundwater flow and recharge rates. Water managers use this data to decide how much water to allocate, where to put wells, and how much to pump.
- Contaminant transport:

Hydrogeologists are also essential in determining and controlling groundwater pollution. They investigate how pollutants migrate through groundwater, evaluate the likelihood that pollutants will find their way to sources of drinking water, and create plans for cleaning up contaminated areas. Environmental site assessments for industrial sites, landfills, and other contaminated sites frequently include hydrogeological investigations.

- Land use planning: Planning for land use should take hydrogeology into consideration, especially in regions where groundwater supplies are susceptible to pollution or misuse. Hydrogeological investigations can identify areas that are suitable for certain types of development (e.g., residential, commercial, and agricultural), as well as regions that need to be kept off-limits to development to preserve groundwater supplies.
- Climate change: Understanding the effects of climate change on groundwater resources also requires an understanding of hydrogeology. Groundwater flow patterns and recharge rates are likely to be impacted by changes in evapotranspiration and precipitation patterns. Hydrogeological studies can be used to determine which areas are more susceptible to drought or other effects, as well as to forecast how aquifers will react to these changes.

Overall, hydrogeology is an important field that contributes to our understanding of water resources and their interaction with human activities.

## 2.3.4. Hydrologic Cycle

The water on our planet Earth is found in three phases as solid, liquid and gas.

The hydrologic cycle, also known as the water cycle, is the process by which water moves through the earth's systems. The cycle includes the following steps:

- Evaporation: The process that turns water from a liquid into a gas, usually from the ground or from the surface of lakes, rivers, and oceans.
- Transpiration: The method by which plants absorb water and release it into the atmosphere.
- Condensation: The process by which water vapor in the atmosphere cools and changes back into liquid form, forming clouds.
- Precipitation: The process by which water falls from the atmosphere in the form of rain, snow, sleet, or hail.
- Infiltration: The process by which water seeps into the ground and is absorbed by soil and rock.
- Runoff: The process through which water escapes from the earth's surface and travels across it to reach streams, rivers, and oceans.

The hydrologic cycle is essential for controlling the amount and distribution of water in the ground and on the surface of the planet, which is necessary for life support and a variety of ecosystems (Figure 2.2).

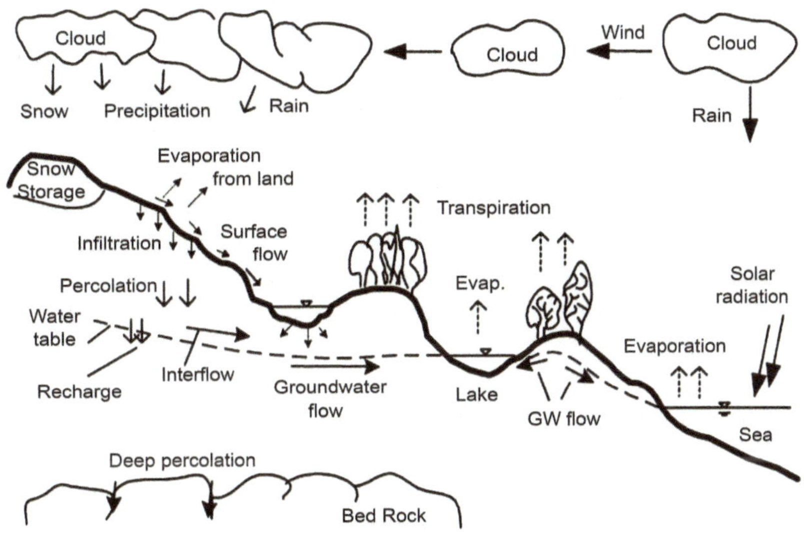


Figure 2.2. Hydrologic cycle.

Source: https://geologyscience.com/geology-branches/hydrogeology/.

## 2.3.5. Occurrence of Groundwater

The occurrence of groundwater is influenced by the following factors:

- Topography: The topography, or upward and downward tilts, of the land above them are followed by the water table, which is not horizontal. Since water seeps into streams, wetlands, and lakes lowers the water table near valleys, it is higher near hilltops and lower near valleys. The water table in an aquifer is primarily vertical and represents the topography of the surrounding region. This is because of the soil's, sediments,' and other porous materials' capillary action. Groundwater in the aquifer moves from areas of higher pressure to areas of lower pressure in both horizontal and vertical directions. The hydraulic gradient is the name given to the slope of the water table. The first factor affecting this is the rate at which water enters and exits the aquifer, and the second is the material's permeability. Sometimes the topography of a place has no bearing on the water table. A water table may occasionally meet the surface of the earth. Water table fluctuation is caused by the flattening of the water table and the gradual decline of the high water-table beneath the hills to the level of valleys in areas that experience monsoon climates, such as the Indian subcontinent, especially during dry seasons. Numerous factors, such as climate, land use, geology, etc., affect the water table in addition to topography.
- Climate: Recharge zones, which allow water to seep underground, are present everywhere in humid regions aside from adjacent floodplains and streams. Conversely, in arid areas, recharged areas are limited to the mountains, surrounding alluvial fans, and streams that have porous alluvium soil that allows water to seep through and replenish groundwater. In dry or arid regions,

groundwater is accessible at great depths, while in humid regions, it is found at shallow depths. The water table naturally rises in the wet season and falls in the dry. Climate change, for instance, can cause significant increases in the frequency and intensity of exceptional rainfall events in areas where groundwater recharge occurs. This can raise water tables to levels above previously recorded maxima, resulting in widespread "groundwater flooding" that damages crops and property. The water table in a region can also be impacted by land use. Most groundwater is created when rainwater runs off and seeps through the ground. Roads and other impermeable surfaces are common in urban areas. These stop water from leaking into the subsurface. Water becomes runoff and the water table lowers rather than reaching the saturation zone. Because more water seeps into the shallow aquifers, land that is irrigated using surface water sources has the biggest effects on groundwater, both quantitatively and qualitatively. The amount of water that filters below the zone of saturation is frequently determined by geology, which makes measuring the water table simple. More water can be retained in porous rocks than in dense ones. For instance, the water table of an area covered in light, porous pumice is easier to measure than the water table of an area covered in hard granite or marble, and the former can retain a fuller aquifer.

- Properties of Materials: Numerous characteristics determine how well a body of rock or sediment will serve as a groundwater resource. Porosity and hydraulic conductivity are the two key physical characteristics. Knowing an aquifer's transmissivity is crucial to understanding its capacity to produce groundwater.
  - Porosity of the Rock: Porosity can be ascertained by examining the configuration and form of soil particles. It's the quantity of void or air space that exists between soil particles. These areas experience infiltration, groundwater movement, and storage. The volume of pore space in a unit of material divided by its total volume is known as the porosity of soil. The porosity of the rock, or the percentage of gaps in each volume of rock, determines the total amount of water that can be contained in the rock. It is given as a percentage or decimal fraction. It is a measurement of the volume of groundwater held within the rocks.

It can be defined mathematically by the equation:

$$n = V_v \div V \times 100\%$$

Where, n = Porosity, expressed as a percentage; Vv = Volume of void space in a unit volume of geological materials, written as L3, cm3 or m3; V = Unit volume of earth material, including both voids and materials, read as L3, cm3 or m3; Porosity ranges from zero to around 60%.

The kind of rock that holds the water affects porosity. Stated differently, the porosity of rocks is determined by the arrangement and spacing of their fractures. The size, shape, and degree of sorting and cementation of the grains determine the porosity of the rock in sediments. With these rocks, the packing or sorting order is crucial. Because well-rounded sorted sediments are nearly all the same size, their porosity is important.

Due to the tendency of the fine-grained particles to fill the void spaces, poorly sorted sediments typically have low porosity. Because the grains in well-rounded coarse-grained sediments don't fit together well, they typically have higher porosity than fine-grained sediments. Because of the surface tension that retains the water in them, very finely grained rocks don't make excellent aquifers. Up to 90% of some shales are made up of open space. The grain shape has an impact on the porosity of sediments. They can resemble disks, books, rods, or other shapes. Porosity is also influenced by the particle's fabric or orientation if they are not spheres. In highly cemented sedimentary rocks, the amount of cement already filling the void spaces have a lower porosity than the cemented void spaces. Within carbonate sedimentary rocks, like limestone, groundwater is found in crevices and fractures created by the sediments dissolving. The largest fractures may have flow rates that are comparable to surface water velocities.

The porosity of extremely dense crystalline rocks called igneous and metamorphic rocks is regulated by faults, fissures, and cracks. Other significant elements are the weathering and structure of the rocks. Because the minerals are often intergrown and leave little void space, porosity is low. Conversely, higher fractured metamorphic and igneous rocks may have a high secondary porosity. Because the Sierra Nevada contains carbonate rocks, this kind of groundwater extraction from fractured rocks occurs there and other mountainous areas of the United States. Additionally, there is a lot of groundwater in the coastal and peninsular ranges. Porosity is largely determined by how the soil's particle packing is arranged. Cubic packing is used when particles are stacked exactly on top of one another. It has a higher porosity than rhombohedral packing, which is the term for particles arranged in a pyramid shape on top of two other particles. A lower porosity would arise from the smaller particles filling in the empty spaces left by the larger particles. Particles can take on a wide range of shapes, and the way these shapes are packed can affect how porous they are. A heterogeneous mixture of grain sizes and shapes leads to reduced porosity. It's critical to realize that grain diameter size has no bearing on porosity. The ratio of void space to total volume is known as porosity. If the ball arrangement or packing is the same in both scenarios, a box containing basketballs and ping pong balls will have the same porosity. There are two types of porosity: primary and secondary. The voids found in the first-formed sediment or rock are known as primary porosity. Secondary porosity is the term used to describe voids that are created in the rock after it has weathered or fractured. It is the porosity of fractured rocks that is caused by distinct discontinuities in the rock mass, such as joints, fractures, and faults. The homogeneity of the rock's grain size (sorting) and shape determines sediment's porosity. Because they are more uniformly sized, finer-grained sediments have a higher porosity than coarser-grained sediments. Because of their higher degree of weathering or fracturing, crystalline rocks have more porosity. The primary porosity of alluvial sediments decreases with consolidation because of void cementation and compaction. Secondary porosity also increases when consolidated rock is prone to stresses that cause fracturing.

- Permeability: Permeability measures the ability of a rock to transfer water through its pore spaces, whereas porosity indicates the amount of water that a rock or soil can hold. The ability of water to seep through the spaces between them is a characteristic of the soil. All soils are permeable because voids exist in them, even in the hardest clay soil. The average daily progress is only a few centimeters, so it's a slow process. The liquid or gas will not flow in the absence of the voids or connected pore spaces. The primary factors influencing permeability are the size and arrangement of soil grains, as well as the interconnectivity of these spaces. When describing the characteristics of aquifers and groundwater, the term hydraulic conductivity is used. According to hydraulic conductivity, water is the fluid that passes through different types of rock or soil. Water can seep through rock fractures and pore spaces. The geological material is more permeable the larger the pore space. Because the smaller grains fill the spaces left by the larger grains, a sample size of mixed grain sizes that has been improperly sorted has less permeability. Sand, gravel, and soil are permeable and porous materials. The pore spaces in clay and shale are small, allowing water to move through them slowly even though they are porous and can hold a lot of water. Because of its small grain size, large surface area, and low pore space connectivity, clay has low permeability. Because of the numerous interconnected pore spaces between the grains, sandstone and conglomerate are extremely permeable materials. The velocity of groundwater flow increases when the ground's permeability is uniform and the water table's gradient slope increases. We refer to this as a hydraulic gradient.

## 2.3.6. Aquifer and Well Types

There are two categories of aquifers: confined and unconfined. The vadose zone above an unconfined aquifer is mostly made up of permeable material, and these aquifers are frequently shallow. The water table is located at the aquifer's surface. Seasons affect how high and low the water table is. It reaches its maximum during the rainy season because of reduced pumping rates and higher recharge (e.g., no irrigation), and lowest in the dry season due to increased use and insufficient recharge (e.g., a high irrigation rate).

Low-permeability strata (such as these) divide confined aquifers from the surface, which can be shallow or deep., geologic strata) that enclose subsurface groundwater. Groundwater in a confined aquifer is typically under pressure. Like the water table in an unconfined aquifer, this water pressure may fluctuate seasonally. Because of the pressure within a confined aquifer, groundwater in a well bore will rise above the level of the aquifer that the well has penetrated. Squeezing a milk or juice pouch with a straw punctured in the top will help you see this. The liquid will rise into the straw and above its level inside the pouch if the straw fits snugly into the squeezed pouch. Wells drilled into confined aquifers are typically described as artesian or flowing artesian wells. When groundwater in a well rises above the level of the aquifer it has penetrated, the well is called an artesian well. The water in an artesian well will rise to a level known as the hydrostatic level when the pressure of the water in the aquifer is equal to the pressure reflected by the well's elevation. A well is referred to as a flowing artesian well if groundwater uses its own pressure to rise to the surface.

## 2.3.7. Groundwater Movement

As with land, the water table (or the potentiometric surface of a confined aquifer) is not a flat surface; rather, it has high and low points, much like hills and valleys. Groundwater flows predominantly downward from higher-elevation water-table areas (potentiometric regions) to lower-elevation water-table areas (potentiometric regions), following the same general trend as surface water. Higher elevations at the land surface are typically, though not always, associated with higher water-table areas of unconstrained aquifers, and lower water-table areas are associated with lower areas. Because streams, lakes, and wetlands frequently occur in low points of the watershed, groundwater in unconfined aquifers tends to flow towards and discharge into these water bodies. Larger area-wide rivers often receive the discharge of groundwater from even small aquifers. Wells and springs are two more typical groundwater discharge locations. An area where groundwater can reach the surface of the earth is called a spring. Sometimes as precipitation seeps through the ground and descends toward the underlying aquifer, it encounters a layer of rock or sediment that is comparatively impermeable. The low-permeability layer prevents groundwater from passing through it, so it travels along its top until the layer is exposed to the ground's surface, at which point the water can emerge as a spring. The movement of the water in this typical "gravity spring," the most popular type of spring, is propelled by gravity. These types of springs typically appear at a bluff or canyon wall, or at the foot of a hill. In other situations, fractures permit the movement of groundwater from the aquifer to the surface. A spring's groundwater may discharge directly into a lake, stream, or ocean, or it may run off the surface of the ground. An additional link between groundwater and the surface of the earth is provided by a well. Usually, a pump is used to raise groundwater to the surface of the earth so that it can be utilized.

### 2.3.7.1. Flow Rate

Hydrologists typically use feet per second or meters per second to describe surface water velocity. Groundwater often moves only a few centimeters (inches) per day, which is far slower than the flow of water in streams. The steepness of the water table (or potentiometric surface) and the permeability of the aquifer regulate groundwater velocity. The potentiometric surface, or water table, has a steeper slope depending on how permeable the aquifer is the pressure gradient), groundwater moves more quickly. Groundwater can travel up to 10 meters (33 feet) a day in highly permeable gravels or fractures.

### 2.3.7.2. Recharge

An unconfined aquifer is replenished at the ground surface directly above it. This process is known as recharge. However, where the aquifer is no longer confined—that is, where the materials above it are permeable and permit rainfall to seep into the confined aquifer—recharge to a confined aquifer can happen many miles away, usually at a higher elevation. Groundwater flows down a gradient to the aquifer's confinement once it has been replenished.

# 2.4. ESSENTIALS OF TOXICOLOGY

Toxicology is the scientific study of the adverse effects of chemicals on living organisms and their environments. It is a multidisciplinary field that combines biology, chemistry, medicine, and environmental science to understand how various substances can cause harm, assess their risks, and develop measures to prevent or mitigate toxic exposure. Essential components of toxicology include the identification of toxic substances, understanding their mechanisms of action, determining the dose-response relationship, and evaluating the risk posed by exposure to these substances.

One of the core principles in toxicology is the concept of dose-response, which states that the magnitude of a toxic effect is related to the amount of exposure to the substance. This principle helps toxicologists establish threshold levels below which a substance is considered safe and above which it becomes harmful. The dose-response relationship is critical for setting regulatory limits and guidelines for exposure to chemicals in the workplace, environment, and consumer products.

Toxicology also involves studying the absorption, distribution, metabolism, and excretion (ADME) of toxicants. These processes determine the concentration of a toxic substance at its site of action and influence the severity and duration of its effects. For instance, a toxicant that is rapidly metabolized and excreted may pose less risk than one that accumulates in the body.

Risk assessment is a crucial aspect of toxicology, involving the evaluation of potential health risks posed by exposure to toxic substances. This process includes hazard identification, dose-response assessment, exposure assessment, and risk characterization. Risk assessment helps regulatory agencies develop policies and standards to protect public health and the environment.

Furthermore, toxicology addresses specific categories of toxicants, including environmental toxins (like pesticides and industrial chemicals), pharmaceuticals, natural toxins (such as those produced by plants and animals), and household chemicals. Each category requires specialized knowledge to understand its sources, effects, and ways to manage exposure.

Emerging areas in toxicology include molecular toxicology, which explores how toxicants affect cellular and molecular processes, and computational toxicology, which uses computer models to predict toxic effects and reduce the need for animal testing. These advancements enhance our ability to identify and mitigate toxic risks more effectively.

Overall, the essentials of toxicology provide a comprehensive framework for understanding the complex interactions between chemicals and biological systems, assessing risks, and protecting public health and the environment from harmful exposures.

## 2.4.1. Types of Toxic Effects

Toxic effects refer to the harmful consequences that result from exposure to toxic substances. These effects can vary widely depending on the type of toxicant, the dose, the duration of exposure, and the individual characteristics of the exposed organism. Here are the primary types of toxic effects:

### 2.4.1.1. Acute Toxic Effects

Acute toxic effects occur shortly after a single or short-term exposure to a toxicant, often within minutes to hours. These effects are typically severe and can be life-threatening. Examples include:

- Chemical burns: Caused by exposure to strong acids or bases.
- Respiratory Distress: Resulting from inhaling toxic gases or vapors.
- Organ Failure: Such as liver or kidney failure due to ingestion of a high dose of a toxic substance.

### 2.4.1.2. Chronic Toxic Effects

Chronic toxic effects result from prolonged or repeated exposure to a lower dose of a toxicant over an extended period, often months or years. These effects are usually less immediate but can lead to serious long-term health issues. Examples include:

- Carcinogenesis: Development of cancer due to prolonged exposure to carcinogens like asbestos or tobacco smoke.
- Neurotoxicity: Long-term damage to the nervous system from exposure to heavy metals like lead or mercury.
- Reproductive Toxicity: Impairment of reproductive functions or developmental defects due to substances like certain pesticides.

### 2.4.1.3. Subacute and Subchronic Toxic Effects

These effects are intermediate between acute and chronic. Subacute effects occur from repeated exposure over a few days to weeks, while subchronic effects result from exposures lasting up to several months. Examples include:

- Mild Liver Damage: From repeated low-dose exposure to alcohol or certain medications.
- Respiratory Problems: From ongoing exposure to air pollutants.

### 2.4.1.4. Local Toxic Effects

Local toxic effects occur at the site of contact with the toxicant. They often involve the skin, eyes, or mucous membranes. Examples include:

- Skin Irritation: Caused by contact with detergents or solvents.
- Eye Damage: From exposure to chemical splashes.

#### 2.4.1.5. Systemic Toxic Effects

Systemic toxic effects occur when a toxicant is absorbed into the bloodstream and distributed throughout the body, affecting various organs and systems. Examples include:

- Hematotoxicity: Effects on blood, such as anemia from exposure to benzene.
- Nephrotoxicity: Kidney damage from substances like certain antibiotics or heavy metals.

#### 2.4.1.6. Reversible and Irreversible Toxic Effects

- Reversible Toxic effects which subside once exposure to the toxicant ceases. Examples include temporary skin irritation or mild liver enzyme elevation.
- Irreversible Toxic effects result in permanent damage, such as scarring from chemical burns or chronic lung disease from long-term exposure to asbestos.

#### 2.4.1.7. Immediate and Delayed Toxic Effects

- Immediate Toxic effects appear rapidly after exposure, such as dizziness or nausea from inhaling fumes.
- Delayed Toxic effects may not become apparent until long after exposure has occurred, such as cancer developing years after exposure to radiation or certain chemicals.

#### 2.4.1.8. Specific and Non-specific Toxic Effects

- Specific Toxic effects targets organs or systems. For example, hepatotoxins specifically damage the liver.
- Non-Specific Toxic effects cause general harm to multiple systems or tissues. For example, cyanide poisoning affects cellular respiration across various tissues.

#### 2.4.1.9. Allergic and Idiosyncratic Reactions

- Allergic Reactions occur when the immune system overreacts to a toxicant, such as anaphylaxis from a bee sting or drug allergy.
- Idiosyncratic Reactions are unusual and unpredictable effects that occur in a small fraction of individuals, often due to genetic differences. An example is a severe reaction to a medication that is safe for most people.

### 2.4.2. Dose–Response Relationship

The dose-response relationship is a fundamental concept in toxicology that describes the correlation between the dose of a substance and the magnitude of its effect on a living organism. This relationship is crucial for understanding how different levels of exposure to a toxicant can produce varying degrees of biological response, ranging from no effect to harmful or even lethal outcomes. It typically follows a sigmoidal curve, where low doses may have little or no observable effect, moderate doses elicit a proportional increase in response, and high doses lead to maximum effect or toxicity. The dose-response relationship helps toxicologists to determine the threshold dose below which no adverse effects are expected (NOAEL: No Observed Adverse Effect Level) and the dose at which adverse effects begin to occur (LOAEL: Lowest Observed Adverse Effect Level).

Additionally, it aids in establishing the median lethal dose (LD50), the dose required to kill 50% of a test population, which is a standard measure of a substance's acute toxicity. This relationship is essential for risk assessment, regulatory decisions, and the development of safe exposure guidelines, ensuring that the use of chemicals and drugs is both effective and safe for human health and the environment.

## 2.4.3. Hypersensitivity and Hyposensitivity

Hypersensitivity and hyposensitivity refer to abnormal responses to stimuli, particularly in the context of immune and sensory reactions, and are important concepts in both medicine and physiology.

### *2.4.3.1. Hypersensitivity*

Hypersensitivity involves an exaggerated or inappropriate immune response to an antigen, often leading to tissue damage and disease. It is commonly classified into four types based on the mechanism and time course of the reaction:

- Type I (Immediate) Hypersensitivity: Also known as allergic or anaphylactic reactions, these occur within minutes of exposure to an allergen (such as pollen, pet dander, or certain foods). The reaction is mediated by Immunoglobulin E (IgE) antibodies, which triggers the release of histamines and other inflammatory mediators from mast cells and basophils. Symptoms can range from mild (hives, itching) to severe (anaphylaxis, which can be life-threatening).
- Type II (Cytotoxic) Hypersensitivity: This occurs when antibodies target and destroy specific cells in the body, often involving IgG or IgM antibodies. Examples include hemolytic anemia, where red blood cells are destroyed, and certain drug reactions where drugs bind to cells and trigger their destruction.
- Type III (Immune Complex-Mediated) Hypersensitivity: This involves the formation of immune complexes (antigen-antibody complexes) that deposit in tissues and cause inflammation. Examples include systemic lupus erythematosus (SLE) and rheumatoid arthritis, where immune complexes cause widespread tissue damage.
- Type IV (Delayed) Hypersensitivity: Unlike the other types, this response is mediated by T cells rather than antibodies and occurs 24-72 hours after exposure to the antigen. Examples include contact dermatitis (such as poison ivy reactions) and the tuberculin skin test for tuberculosis.

### *2.4.3.2. Hyposensitivity*

Hyposensitivity, on the other hand, refers to a diminished or absent response to stimuli, which can be immune-related or sensory:

- Immune Hyposensitivity (Immunodeficiency): This occurs when the immune system's response to antigens is inadequate, leading to increased susceptibility to infections. Primary immunodeficiency diseases, such as severe combined immunodeficiency (SCID), are genetic and present at birth. Secondary immunodeficiencies can result from factors like malnutrition, chemotherapy, or diseases like HIV/AIDS.

- Sensory Hyposensitivity: In this context, hyposensitivity refers to reduced sensitivity to sensory stimuli, such as touch, sound, or pain. This can be seen in conditions like autism spectrum disorder (ASD), where individuals may not respond to sensory input in typical ways. Hyposensitivity can affect the ability to perceive temperature, pressure, or even bodily pain, leading to potential risks and injuries.

### 2.4.4. Exposure Paths of Toxicants to Humans

There are three main ways that pollutants can enter the human body: through the skin, food, or inhalation. The term "dose" refers to the quantity of a particular pollutant that is administered. Organ dose precisely refers to the amount that reaches the human organ where the relevant effects can occur; for example, the dose will depend on the length and intensity of the exposure to the lungs. Exposure can frequently happen concurrently from several sources and through several pathways. Lead exposure pathways include, for instance, drinking water contamination from industrial and traffic emissions, food, tobacco smoke, dust, paint, and other industrially produced materials, and soil. Therefore, a comprehensive understanding of the geographic distribution of the pollutants of concern, the temporal fluctuations in pollution levels, and the exposure processes are typically necessary for valid exposure assessment. Individuals frequently experience multiple pollution exposures at once. These may be exposed to in a variety of settings (e.g., both at work and/or at home) and at various times. There is a considerable possibility that a wide range of factors will need to be investigated. The term "environmental pollutants" refers to a wide range of substances that are released into the environment, either continuously or sporadically, under controlled conditions (such as energy production, industry, and pesticide use), from a variety of sources (such as radioactivity, hazardous chemicals, dusts, and particulates) intentional releases) or unintentionally.

To effectively develop and execute a plan for enhancing local air quality, local government agencies must initiate the required monitoring initiatives to ascertain pollutant concentrations and pinpoint their origins. The amount of monitoring will depend on the resources available, but it's crucial to emphasize that even with relatively cheap equipment, adequate monitoring can be done. Local authorities can set priorities for improvement based on the resources at their disposal, the costs and benefits of different courses of action, and monitoring's ability to validate earlier assumptions that pollution levels are unacceptably high. Programs for long-term monitoring demonstrate the effectiveness of the policies implemented, allowing for the possible reallocation of resources when deemed necessary.

New technologies for sampling and analysis are being developed, such as remote sensing, automated samplers, and inexpensive sampling tools (e.g., passive evaluators). Better knowledge of the temporal and spatial fluctuations in pollution is gradually leading to better monitoring; for instance, sampling the microenvironment where exposure mostly takes place, such as indoor environments (e.g., in living rooms and kitchens when measuring radon and

$NO_2$. There are additional chances to enhance exposure estimates through the application of total exposure monitoring, which samples every potentially relevant microenvironment. Furthermore, an element of personal exposure monitoring is being added to environmental health assessments.

Local authorities should aim, as a minimum, to monitor the following pollutants, all of which can affect local air quality and human health: sulfur dioxide, particles, nitrogen dioxide, ground-level ozone, carbon monoxide, and airborne lead. (Lomonosov, 2012)

## SUMMARY

- Elements of the atmosphere are variably unstable and subject to the processes of weathering and erosion. Weathering and erosion break the original rock down into smaller fragments and carry away dissolved material.
- Transpiration is the evaporation of water from plants through stomata. Stomata are small openings found on the underside of leaves that are connected to vascular plant tissues.
- Rainwater gets absorbed into the ground through the process of infiltration. The level of absorption varies based on the material the water has seeped into. For instance, rocks will retain comparatively less water than soil.
- Tectonic plates can move because of the relative density of oceanic lithosphere and the relative weakness of the asthenosphere. Dissipation of heat from the mantle is the original source of the energy required to drive plate tectonics through convection or large-scale upwelling and doming.
- The vector of a plate's motion is a function of all the forces acting on the plate; however, therein lies the problem regarding the degree to which each process contributes to the overall motion of each tectonic plate.
- Hydrogeology is the study of the laws governing the movement of subterranean water, the mechanical, chemical, and thermal interaction of this water with the porous solid, and the transport of energy, chemical constituents, and particulate matter by flow.

## REVIEW QUESTIONS

1. Describe the three main types of rocks in the rock cycle.
2. What are the two main types of igneous rocks and how do they form?
3. Explain the difference between contact metamorphism and regional metamorphism.
4. What are the primary processes involved in the hydrologic cycle?
5. How do plate movements contribute to the occurrence of earthquakes?
6. Discuss the evidence supporting or refuting the impact of tidal forces on tectonic activity.
7. Compare the significance of mantle convection versus slab pull in driving plate movements.
8. Define the geochemical cycle and its importance in earth systems.
9. How does hydrogeology impact water resource management?
10. What are the different types of toxic effects that substances can have on organisms?

## REFERENCES


1. Coates, R. R. (1962). Magma type and crustal structure in the Aleutian arc. In The Crust of the Pacific Basin (Geophysical Monograph Series, Vol. 6, pp. 92–109). American Geophysical Union.
2. Doglioni, C. (1990). The global tectonic pattern. Journal of Geodynamics, 12(1), 21–38.
3. Grove, T. L., Till, C. B., & Krawczynski, M. J. (2012). The role of $H_2O$ in subduction zone magmatism. Annual Review of Earth and Planetary Sciences, 40(1), 413–439.
4. Hofmeister, A. M., Criss, R. E., & Criss, E. M. (2022). Links of planetary energetics to moon size, orbit, and planet spin: A new mechanism for plate tectonics. In G. R. Foulger, L. C. Hamilton, D. M. Jurdy, C. A. Stein, K. A. Howard, & S. Stein (Eds.), In the Footsteps of Warren B. Hamilton: New Ideas in Earth Science (Special Paper 553, 10 pp.). Geological Society of America.
5. Holmes, A. (1931). Radioactivity and Earth movements. Transactions of the Geological Society of Glasgow, 18(3), 559–606.
6. Lomonosov, M. (2012). On the strata of the earth. In S. M. Rowland & S. Korolev (Trans. & Comm.), The Geological Society of America, Special Paper 485.
7. Lovett, R. A. (2006). Moon is dragging continents west, scientist says. National Geographic News.
8. Lyell, C. (1991). Principles of Geology. Chicago, IL: University of Chicago Press.
9. Scoppola, B., Boccaletti, D., Bevis, M., Carminati, E., & Doglioni, C. (2006). The westward drift of the lithosphere: A rotational drag? Geological Society of America Bulletin, 118(1–2), 199–209.

CHAPTER

# 3 Environmental Laws

## LEARNING OBJECTIVES

- Understand united states environmental law
- Describe the rivers and harbors act
- Define atomic energy act
- Discuss on solid waste disposal act
- Know resource conservation and recovery act (RCRA)
- Describe national environmental policy act
- Deal with the occupational safety and health act
- Discuss about clean water and toxic substances control act

## KEY TERMS FROM THIS CHAPTER

Dilution prohibition
Environmental health
Environmental monitoring system
Material safety data sheets (MSDSs)
National environmental policy Act (NEPA)
Waste recycling
Environmental agencies
Environmental law
Human health
Microbial source tracking (MST) techniques
Sustainable material management (SMM)

# INTRODUCTION

Waste management laws control the movement, handling, disposal, and storage of various types of waste, including nuclear, hazardous, and municipal solid waste. Waste laws include measures to lessen waste generation and encourage or require waste recycling. Waste laws are intended to reduce or eliminate the uncontrolled dispersal of waste materials into the environment in a way that may cause ecological or biological harm. Waste types are identified and categorized, and procedures for transportation, treatment, storage, and disposal are required.

Determining whether a material qualifies as hazardous waste under the US Resource Conservation and Recovery Act is one example of a complicated question. Waste determination is the process by which a specific material is classified as a waste subject to regulation.

The way a material must be handled going forward may depend on whether it falls under a specific waste category. For instance, used motor oil is considered hazardous and cannot be disposed of in landfills; instead, it must meet stricter guidelines for handling, storing, treating, and disposing of it. On the other hand, non-hazardous municipal solid waste can be sent to a landfill in the United States. Technical guidelines for evaluating and categorizing waste are provided by the UK government in the form of the WM3 reference manual. It includes all waste classifications in the UK.

There might be several other wastes that have different definitions and demands for how they should be handled. A waste stream may be identified in each situation. Waste is created when an item that was once useful is discarded or abandoned. It may then pass through several defined areas for storage, recycling, and treatment before ending up at a final disposal site.

Disposal standards specify where, how, and when a certain waste can be disposed of. These regulations might be created to safeguard environmental values as well as people's comfort and health. There are numerous ways to regulate disposal.

A disposal ban can completely limit how waste is disposed of. The most popular and widely applied of these standards is the one against littering. Trash abandonment or deposition elsewhere may result in civil or criminal penalties if a jurisdiction has approved a particular location or system for garbage collection. Other, more restrictive disposal laws regulate the ultimate resting place of different wastes. Examples include the requirement that paint not be poured down drains and the designation of national repositories for radioactive waste. It might be necessary to separate certain wastes for recycling instead of disposal. All these restrictions are, in a way, conditional since they limit the disposal location rather than completely forbid the disposal of the material.

Restrictions on waste disposal can also be conditional, such as requiring a waste to be treated in a certain way before it can be disposed of at a specific location. Land Disposal Restrictions under the Resource Conservation and Recovery Act Subtitle C hazardous waste management program is one such program run by the US Environmental

Protection Agency. These regulations limit the land disposal of hazardous waste (mainly, dumping it in landfills) in the absence of previously authorized programs. Hazardous waste is prohibited from being disposed of on land until it has been treated to meet certain requirements (acceptable ignitability, corrosivity, reactivity, and toxicity), or until treated using specific approved treatment methods. The dilution prohibition prohibits addition of water, soil, or non-hazardous waste to avoid specific treatments. Instead of storing waste endlessly to evade treatment the storage prohibition permits waste to be stored only for treatment purposes.

Another option is the adoption of guidelines for the design and management of waste disposal facilities. For instance, landfills might have to install liner systems and collection systems to reduce groundwater contamination from leachate; adopt operation policies that minimize dust, and other nuisances; install methane removal or collection systems to eliminate landfill gas; be covered and otherwise closed upon decommissioning; and operate environmental monitoring systems to ensure compliance. Certain landfill locations may also have other requirements.

# 3.1. UNITED STATES ENVIRONMENTAL LAW

Environmental law in the United States deals with regulations to safeguard public health and enhance the country's natural environment. Despite facing criticism, both domestically and internationally, regarding protection, enforcement, and overregulation, the nation continues to be a valuable resource for environmental legal knowledge and experience.

## 3.1.1. Scope

Laws pertaining to the environment are known as environmental laws. The body of laws, agreements, rules, and common law that control how people interact with their surroundings is known as environmental law. This covers rules pertaining to the environment, the management of natural resources like fisheries, forests, and minerals, and related subjects like environmental impact assessments. Environmental law is defined as the body of laws pertaining to the protection of all living things, including humans, from the harm that human activity may inflict upon them or their species, either directly or indirectly, or to the media and the customs that they rely upon.

Federal statutes aimed at controlling and remediating pollution have been enacted by the US Congress. These include the Comprehensive Environmental Response, Compensation, and Liability Act (CERCLA, or Superfund), which addresses the cleanup of contaminated sites, the Clean Water Act, which addresses water pollution, and the Clean Air Act, which addresses air pollution. Additionally, federal laws like the National Forest Management Act, the Coastal Zone Management Act, and the Endangered Species Act that regulate biodiversity and the use of natural resources are heavily impacted by environmental principles. All these areas may be impacted by the National Environmental Policy Act, which regulates environmental impact reviews for actions taken or approved by the US federal government.

The national environmental laws of the United States have been influenced by federalism. Cooperative federalism is a mechanism used by many federal environmental laws; the US states collaborate with the federal government to administer numerous regulatory programs. In addition, states have typically passed their legislation to address matters not covered by federal law. This covers both areas where Congress has left regulation largely to the states (e.g., water resources law) and areas where Congress has acted in limited ways (e.g., state site cleanup laws to handle sites outside Superfund).

## 3.1.2. History

Environmental law in the United States has its origins in common law doctrines such as the public trust doctrine and the law of nuisance. The Rivers and Harbors Act of 1899 was the first environmental statute, but the Clean Water Act (CWA) has essentially replaced it. Nonetheless, majority of significant environmental laws in effect today, including those federal statutes mentioned above, were enacted between the late 1960s and the early 1980s. Most federal environmental laws were not nearly as comprehensive before these statutes were passed.

The environmental movement in the United States is often credited to Rachel Carson's 1962 book Silent Spring. The book provided information on how pesticides, particularly DDT, affect birds and other wildlife. The 1969 Santa Barbara oil spill was one of the biggest environmental catastrophes of the 1960s, causing a great deal of public indignation at a time when Congress was debating several important environmental laws.

## 3.1.3. Legal Sources

Environmental matters are covered by laws from all tiers of the US legal system. Although Congress has enacted several historic environmental regulations, numerous other federal laws are just as significant, if not more so. At the same time, many similar sets of laws have been passed by the legislatures of the fifty states. Administrative regulations at the state and federal levels are layered on top of one another. In the meantime, the administrative decisions made by the numerous agencies that handle environmental matters are reviewed by the US judicial system in addition to legislative enactments. The common law kicks in where statutes and regulations finish.

### *3.1.3.1. Federal Regulation*

US federal agencies promulgate regulations in the Code of Federal Regulations that complete the comprehensive programs enacted by Congress, by the federal statutes that they oversee. The Code of Federal Regulations Title 40, which contains the Environmental Protection Agency's (EPA) regulations, is the most important of these. Titles 10 (energy), 18 (Conservation of Power and Water Resources), 21 (Food and Drugs), 33 (Navigable Waters), 36 (Parks, Forests, and Public Property), 43 (Public Lands: Interior), and 50 (Wildlife and Fisheries) are additional significant sections of the CFR.

### *3.1.3.2. Judicial Decisions*

The federal and state judiciaries have played an important role in the development of environmental law in the United States, in many cases resolving significant controversy regarding the application of federal environmental laws in favor of environmental interests. The decisions of the Supreme Court in cases such as Calvert Cliffs Coordinating Committee vs US. Atomic Energy Commission (broadly reading the procedural requirements of NEPA), Tennessee Valley Authority vs Hill (broadly reading the Endangered Species Act), and, much more recently, Massachusetts vs EPA (requiring EPA to reconsider regulation of greenhouse gases under the Clean Air Act) have had policy impacts far beyond the facts of the particular case.

#### 3.1.3.3. Common Law

When it comes to settling environmental disputes outside the purview of regulated activity, the Torts law is a valuable tool. The doctrines of nuisance (public or private), trespass, negligence, and strict liability assigned harm and liability for actions causing pollution which may be subject to regulatory regimes, before the current proliferation of environmental regulations. These theories are still applicable, and plaintiffs attempting to hold parties accountable for the effects of climate change have most recently used them.

The common law is still a major influence on American water law, particularly about the ideas of prior appropriation and riparian rights.

## 3.1.4. Administration

The administration of environmental laws in the United States is the responsibility of multiple federal and state agencies, each with distinct, sometimes conflicting missions. The most well-known federal organization is the Environmental Protection Agency (EPA), which is in charge of many of the nation's waste, hazardous material, and air quality programs. While some federal agencies, like the US Forest Service and the Bureau of Land Management, prioritize the beneficial use of natural resources, other agencies, like the US Fish and Wildlife Service and the National Park Service, focus primarily on conservation.

Within the boundaries of their authority, federal agencies conduct their business. For instance, the Clean Water Act restricts the EPA's authority to American waters. Federal laws frequently permit states to impose stricter regulations and to transfer some federally mandated duties from the federal to the state level of governance. As a result, environmental agencies are consistently included in US state governments' administration of state laws enacted according to state police powers or federal law by delegation. Different jurisdictions have different levels of federal law inspiration or divergence from state environmental laws.

Thus, while a single state agency may be required to grant a permit for filling the non-federal wetlands, multiple federal and state agencies may need to approve larger and more complex projects, like building a coal-fired power plant.

#### 3.1.4.1. Enforcement

Environmental law infractions in the US are typically treated as civil offences, punishable by fines and possibly by civil sanctions like injunctions. Serious infractions of environmental laws may also result in criminal penalties. If the government fails to take enforcement action, a US citizen may also file a lawsuit against a violator of certain federal laws, such as the Clean Water Act.

A separate enforcement office is frequently included in environmental agencies. Its responsibilities include monitoing approved activities, conducting compliance inspections, issuing citations, and prosecuting violations (either civilly or criminally, depending on the offence). One such organization is the Office of Enforcement and Compliance Assurance of the EPA. Some, like the US Park Police, engage in more conventional forms of law enforcement.

Under the framework of administrative law, agencies frequently handle adjudicatory proceedings for environmental violations. Appeals are sometimes managed internally as well (for e.g. the Environmental Appeals

Board of the EPA). Final agency decisions can usually be challenged later by the relevant court.

Another method of enforcement is emerging as environmental law becomes more prevalent in the US. The next generation of enforcement strategies consists of financial penalties and rewards. On the other hand, some researchers think that enforcing non-environmental laws that can also benefit the environment is the best way to enforce environmental regulations. Bipartisan support for improved environmental performance has been observed by some researchers. (Wassersug, Stephen R.; Ercmann, Sevine; Khan, Rahmatullah; Rummel-Bulska, Iwona, 1990)

# 3.2. RIVERS AND HARBORS ACT

The term "Rivers and Harbors Act" may be used to describe any of the numerous laws and appropriations that the US Congress has passed since the first one was passed in 1824. Congress then appropriated $75,000 to remove sandbars, snags, and other obstructions to improve navigation on the Ohio and Mississippi rivers. Similar to when it was originally passed, the Secretary of War (or, more recently, the Secretary of the Army) and the Chief Engineer of the United States Army Corps of Engineers (USACE) were to oversee the legislation's administration.

The Supreme Court held in the historic Gibbons v. Ogden case that the Constitution's Commerce Clause gave the federal government jurisdiction over interstate commerce, including riverine navigation. This decision largely ended the bitter debate over transportation reforms between proponents of federalism and state rights activists. Soon after, the General Survey Act gave the president the power to commission surveys of potential routes for roads and canals that would be important for the country's economy or military, or that would be required for the delivery of public mail. The President gave the Corps of Engineers charge of conducting the surveys. Congress passed the first federal rivers and harbors legislation in May to expand the range of potential improvements, and the USACE was once again tasked with overseeing the project.

In 1826, new river and harbor laws gave the president the power to commission river surveys to deepen and clean up specific waterways and for other river and harbor enhancements. Congress also increased the workload for Army engineers in that year and approved the first survey for a canal between the Atlantic Ocean and the Gulf of Mexico. While the 1824 legislation aimed at enhancing the Mississippi and Ohio rivers is frequently referred to as the initial legislation on rivers and harbors, it was the 1826 act that pioneered the combination of authorizations for surveys and the projects themselves, setting a precedent that persists to this day.

## 3.2.1. Early Legislation

Authorizations for preliminary surveys of the navigation safety of rivers used for transportation were included in many of the early river and harbor laws; as these were developed, authorizations for specific improvements were added. The Erie Canal, which was entirely funded by the state of New York, was a great commercial success and its rapid growth in the use of steamboats on inland waters drove many of these improvements.

The Tennessee River was first surveyed by the Corps in 1828, and the first steam-powered snag boat was launched on the Ohio River in 1829 at New Albany, Indiana. Development was permitted along the Cumberland River in Tennessee and Kentucky in 1832 and along the Hudson River in New York in 1834. The federal waterway improvement program nearly collapsed in 1837 due to the panic of that year, while other initiatives were being carried out by the individual states. Development of the Tennessee River and the Illinois Waterway, which lay downstream of the Illinois and Michigan Canal built by the state, was approved in 1852. Although the Corps and the federal government focused on rivers that could be navigated, they also helped with the construction of canals, most of which were built by individual states. For instance, the engineers' report states that the River and Harbor Act of August 11, 1858, allotted $102,000 for the construction of a lock at Taylorsville and the renovation of the lock at Zanesville, which would improve the Muskingum River in Ohio. It also contains an early mention of waterpower that is connected to and dependent upon the use of the water for navigation. According to the act, the Secretary of War is (hereby) authorized and empowered to grant licenses or leases for the use of the Muskingum River's water-powers at any rate, under any circumstances, and for any length of time that he deems just, equitable, and convenient. He is also authorized to grant leases or licenses for the occupation of United States lands on the Muskingum River, as may be necessary for mill sites or other purposes not in conflict with the requirements of navigation. The only restriction on the leases or licenses is that they must be restricted to the use of excess water that is not needed for navigation. The United States Treasury will receive all funds received under these leases or licenses, and the Chief of Engineers' annual report will be accompanied by an itemized statement. However, nothing in this act may be interpreted to impair any lessee of water-power on said river's vested rights, if any.

### 3.2.2. Later Legislation

Federal funding for railroads began to flow after the Civil War, and funding for improvements to rivers and harbors increased significantly. The 1869 and 1930 Rivers and Harbors Acts provide an example of how appropriations for rivers and harbors increased from over $2 million to tens of millions of dollars annually by the end of the 19th century. Every one of these acts lists hundreds of projects that the Secretary of War will oversee and the Army Corps of Engineers will build. The pattern demonstrated the Corps' intimate relationship to the building of civil works projects.

Gifford Pinchot noted that the federal government oversees navigable rivers and their tributaries under the authority bestowed upon it by the US Constitution. However, Congress gave the states complete control over how water-power was regulated for a century following the adoption of the Constitution. When federal stream legislation first started, its focus was on preventing or eliminating obstacles to navigation. Later periods saw a change in this due to the building of locks and dams for improved navigation. Electricity demand increased quickly with the introduction of Edison's DC light bulb in the early 1880s and the later adoption of AC, which had the superior capacity to be transmitted over long distances. Around this same period, the proliferation of hydroelectric plants and other contemporary multipurpose improvements became a problem for river and harbor legislation. (McFarland, Charles. 1966)

Like previous appropriations acts, the Rivers and Harbors Appropriations Act of 1899 started with a lengthy list of navigational improvement projects (section 1), funded a survey program for said projects (section 2), and gave the President full authority to conduct a thorough investigation of the Panama Isthmus to  build the Panama Canal. The Act's Section 9 gave Congress control over causeways, bridges, dams, and dikes. Section 10 gave Congress control over unauthorized obstructions in navigable waters and mandated that any wharf, pier, dolphin, boom, weir, breakwater, bulkhead, jetty, or other structure be authorized. Section 13 forbids disposal of trash in navigable waters, with certain restrictions and licenses being an exception. The use or modification of various navigation-related public works anchored or submerged vessels, and surveys for harbor improvements are covered in other sections. Certain permit requirements of the US Army Corps of Engineers are based on this act.

# 3.3. ATOMIC ENERGY ACT

The United States' control and management of the nuclear technology it had developed in collaboration with the United Kingdom and Canada, its allies in World War II, were outlined in the Atomic Energy Act of 1946, also known as the Mahon Act. Most notably, the Act created the United States Atomic Energy Commission and declared that civilian, not military, authorities would oversee the development of nuclear weapons and nuclear power.

The United States Senate Special Committee on Atomic Energy was chaired by Senator Brien McMahon, a Democrat from Connecticut, who sponsored the bill. His hearings in late 1945 and early 1946 helped to refine and pass the Act. The Act was approved by the House of Representatives 265–79 and the Senate by a unanimous voice vote. President Harry S. Truman signed it into law on August 1, 1946, and it came into force on January 1, 1947, when the Manhattan Project's nuclear energy division was transferred to the Atomic Energy Commission. (Gott, Richard, 1963)

Later, in 1954, the Act was changed to encourage private nuclear energy development as part of the Eisenhower administration's Atoms for Peace initiative. By denying other nations access to nuclear data, it strained relations between the US and its allies, especially Canada and Britain, who had taken part in the Manhattan Project. As a result, complex command and control systems were established, and Britain began producing its nuclear weapons. In 1958, the Act was modified to permit sharing information between the US and its close allies.

## 3.3.1. Origins

The Manhattan Project, which was active during World War II, developed nuclear weapons. Leading scientists involved in the project expected their advancement to have far-reaching consequences. But the project manager, Major General Leslie R. Groves Jr. resisted using project funds for anything other than the necessities of winning the war. However, the Manhattan Project's governing body, the Military Policy Committee, commissioned a study by Richard Tolman that was comparable, and Arthur Compton of the Metallurgical Project in Chicago commissioned a report on post-war nuclear energy. A comprehensive nuclear energy program with military, scientific, and industrial components was recommended in both the reports and would be funded by the government.

In July 1944, James B. Vannevar Bush, Sr. Conant and Irvin Stewart suggested a plan for domestic nuclear energy regulation laws. Conant sent this to Henry L. Secrecy of War. Stimson in September 1944, and subsequently to the Interim Committee, which

President Harry S. Truman established in May 1945 to oversee, control, and regulate nuclear energy until Congress established a permanent body to do so. The Interim Committee questioned George L. in June 1945.

## 3.3.2. Creation of the Act

### *3.3.2.1. May–Johnson Bill*

Harrison brought in two experienced Harvard Law School-educated War Department lawyers, Brigadier General Kenneth Royall and William L. Marbury Jr., to draft the legislation. The Bush and Conant proposal served as the foundation for the legislation, and the Manhattan Project's current organizational structure served as the model for the proposed organization. Five civilians and four military members would have made up the nine-member commission that their draft bill called for. It gave the commission extensive authority over property acquisition, facility management, research, nuclear energy regulation, and the administration of its own security, administrative, and auditing systems.

Experts would oversee nuclear energy with little to no political intervention, according to Royall and Marbury. The President would have limited authority to remove the commissioners, who would be appointed for undefined terms. Four advisory boards—one each for military applications, industrial uses, research, and medicine—whose membership would be limited to individuals with technical qualifications, would provide support for them. An administrator and his deputy would oversee the organization's daily operations. At its meeting on July 19, the Interim Committee examined the Royall–Marbury Bill and made revisions based on their recommendations. Following the atomic bombings of Hiroshima and Nagasaki, the Manhattan Project's secrecy was lifted, allowing Royall and Marbury to confer with the Office of Scientific Research and Development, the Attorney General, and the Judge Advocate General. In August, the President received a draft for review and feedback from the relevant government agencies. The only party with concerns was the State Department, which objected on the premise that it was still engaged in efforts to reach a global nuclear energy agreement.

On October 3, 1945, the bill was introduced in the House of Representatives by Congressman Andrew J. May from Kentucky, the chairman of the House Military Affairs Committee, and in the Senate by Senator Edwin C. Johnson from Colorado, the ranking member of the Senate Military Affairs Committee. The bill was known as the May-Johnson Bill for its sponsors. May immediately referred the bill to the Military Affairs Committee, which held hearings on October 9. Bush, Conant and Groves all testified before the committee. But in the Senate Military Affairs Committee, the bill was held up by Senator Arthur H. Vandenberg.

Scientists, especially those at the Metallurgical Laboratory in Chicago, unleashed a torrent of criticism. Two particularly well-known critics were Leó Szilárd and Harold Urey. The bill clearly stated that the administrator and deputy administrator might be members of the armed forces and established a strong administrative team. They were thought to have the upper hand among the part-time commissioners. Suspicions that the administrator would be an Army officer and the deputy a Navy officer werefurther stoked by the bill's emphasis on the administrator's need to keep the deputy fully informed. Many scientists were also alarmed by the secrecy provisions, which carried harsh penalties for security breaches: up to $10,000 in fines and ten years in jail. The War Department was accused of trying to push the legislation through Congress, according to the headline in the Chicago Sun.

The situation that the legislators were placed in was peculiar and unsettling. The nature of nuclear energy was not well known, and nuclear weapons were terrifying. Legislators had no precedents or policies to guide them because it was so new, and traditional party alignments were nonexistent. The scientists who created the new technology never expressed their opinions, but they have done so now. Despite the great prestige the armed forces enjoyed after winning World War II, Americans' long-standing mistrust of standing armies and their custom of civilian control over the military persisted.

### *3.3.2.2. McMahon Bill*

The Senate Military Affairs Committee had drafted an alternative atomic energy bill, which Senator Brien McMahon introduced on December 20, 1945. The bill was soon dubbed the McMahon Bill. When it came to the regulation of scientific research, this bill was very liberal at first and had the support of many scientists. Although the May-Johnson Bill also allowed for civilian control, McMahon presented the debate as one between military and civilian authority over atomic energy. The contentious elements of the May-Johnson Bill were attempted to be addressed by the McMahon Bill. Five commissioners were appointed, all of whom would serve full-time. Officers in the military who were currently serving were not exempt. An amendment stipulated that their terms would be staggered by five years.

While the bill was being debated, the news broke on February 16, 1946, of the defection of Igor Gouzenko in Canada, and the subsequent arrest of 22 people. The members of Congress debating the bill feared that "atomic secrets" were being systematically stolen by Soviet atomic spies. McMahon convened an executive session at which Federal Bureau of Investigation Director J. Edgar Hoover, Secretary of State James F. Byrnes and Groves were called to appear. Groves revealed that the British physicist Alan Nunn May had passed information about the Manhattan Project to Soviet agents.

The more conservative elements in Congress then moved to toughen the act. Section 10, which was formerly titled "Dissemination of Information," then became "Control of Information." This new section contained the novel doctrine later described as "born secret" or "classified at birth." All information concerning the design, development and manufacture of nuclear weapons was "restricted data," and regardless of how it was derived or obtained, was considered classified unless it was specifically declassified. This restriction on free speech, covering an entire subject matter, is still enforced. The "wall of secrecy" set up by the Act meant that atomic energy research and development had to be conducted under the supervision of the Atomic Energy Commission.

The McMahon Bill's House sponsor, Representative Helen Gahagan Douglas, fiercely defended Section 10's dissemination provisions against objections. She refuted concerns that it would reveal the bomb's secret, arguing that while the bill might keep America at the forefront of scientific research, its advantage in nuclear weapons might only last temporarily. Section 2 of the Act also established a General Advisory Committee and a new Joint Committee on Atomic Energy to supervise the new organization. This significant addition, known as the Vandenberg Amendment, established a Military Liaison Committee to advise the commission on defence matters. On June 1, 1946, the Senate unanimously approved the Act by voice vote. It took a lot of political wrangling before the House passed it on July 20 by a vote of 265–79. On July 26, both houses then approved a compromise bill. On August 1, 1946, Truman signed the compromise bill into law, which became the Atomic Energy Act of 1946. On January 1, 1947, at midnight, the newly established Atomic Energy Commission took over the wartime Manhattan Project's nuclear energy responsibilities.

## 3.3.3. Amendment

### *3.3.3.1. Private Production of Nuclear Energy*

Since military applications of nuclear energy at the time outnumbered all other uses, it was significant that the Atomic Energy Act of 1946 did not mention non-governmental use. The act's restrictions on secrecy, control over fissile materials, patent ownership, and operation of production facilities placed several legal roadblocks in the way of private nuclear power plants.

This ran counter to the Atoms for Peace initiative of the Eisenhower administration and put pressure on federal officials to create a civilian nuclear power sector to support the government's large spending on the nuclear weapons program. The Joint Committee on Atomic Energy received a set of draft amendments from the Atomic Energy Commission in 1953 for review. This led to some discussion and the creation of the 1954 Atomic Energy Act.

The goals of the Atomic Energy Act of 1954, which was to promote privately constructed and operated nuclear reactors, were not met. Following a string of research reactor mishaps, including partial core meltdowns, private businesses became wary of using nuclear energy without liability protection. As a result, the Price–Anderson Nuclear Industries Indemnity Act of 1957 was passed, capping private liability for nuclear mishaps and offering suitable accident victims' compensation.

### *3.3.3.2. International Relations*

The McMahon Act's implementation caused a significant rift between the US and the UK. The British and Canadian governments had agreements with the United States regarding the post-war sharing of nuclear technology, but the new control over restricted data prevented the allies of the United States from receiving any information. This was true even before the governments contributed manpower and technology to the Manhattan Project. The Quebec Agreement of 1943 served as the formalization of those agreements. In the instance of the United Kingdom, these were expanded upon in the 1944 Hyde Park Agreement, which Franklin Roosevelt and Winston Churchill signed.

The Hyde Park Agreement was lost in Roosevelt's papers after his death, and until the American copy of the document was found American officials were puzzled when the British mentioned it. The Quebec Agreement was an executive agreement that only applied to the Roosevelt administration, and the Senate had not seen the document. McMahon told Churchill in 1952 - "If we had seen this Agreement, there would have been no McMahon Act." The McMahon Act fueled resentment from British scientists and Churchill and led to Britain developing its nuclear weapons.

In January 1958, Lewis Strauss, the chairman of the Atomic Energy Commission, suggested that the President should have access to nuclear information if allies were significantly and materially bolstering the country's defence and security. Since 1948, Britain has not only maintained its nuclear arsenal but also welcomed American Strategic Air Command nuclear bombers. In June 1958, the Atomic Energy Act of 1954 was amended by Congress, and as a result, the United States and the United Kingdom resumed sharing nuclear research under the 1958 US-UK Mutual Defense Agreement.

The Act's provisions generated a great deal of controversy when it came to discussions about NATO's military command structure. Striking Fleet Atlantic and the US Sixth Fleet have never been permitted to be positioned anywhere other than directly under the Supreme Allied Commander Atlantic and Commander-in-Chief, Allied Forces Southern Europe, respectively, due to the prevailing legal interpretation of the McMahon Act, which holds that non-US commanders are unable to control nuclear striking forces. For this reason, in October–November 1952, Striking Fleet Atlantic was established as a separate organization rather than being operationally under the command of the UK Admiral who was the Commander-in-Chief of the Eastern Atlantic. This was also the reason why when the European commands were agreed upon simultaneously, the Sixth Fleet—then known as Naval Striking and Support Force, South—was placed under American command instead of Allied Forces Mediterranean.

# 3.4. SOLID WASTE DISPOSAL ACT

Congress of the United States passed the Solid Waste Disposal Act (SWDA) in 1965. The Act is the first federal attempt to advance waste disposal technology, according to the US Environmental Protection Agency. Following the Second Industrial Revolution, all economic sectors produced more solid waste because of the country's growing industrial and commercial activity as well as rising consumer demand for goods and services. The act created a framework that states could use to better regulate the disposal of solid waste and set minimum safety standards for landfills. To manage the country's waste, Congress deemed the SWDA's provisions inadequate in 1976 and passed the Resource Conservation and Recovery Act (RCRA). The Hazardous and Solid Waste Amendments of 1984 (HSWA) were significant additions to SWDA that Congress passed.

## 3.4.1. Background

The nation's economic output increased significantly after the second industrial revolution in the late 19th and early 20th centuries, which also significantly increased the amount of waste produced. To get rid of this waste, private businesses and local and regional governments created a wide range of often dangerous and unhygienic disposal methods. The most popular methods of disposal were burning trash and open dumps. Waste management received more attention in the United States as a public health emergency as the germ theory of disease evolved in the late 19th and early 20th centuries. The SWDA was passed by Congress in 1965 due tothe growing demand for improved waste management throughout the 1960s. (Rushton, Lesley, 2003)

## 3.4.2. Overview

The purpose of SWDA was to promote improved municipal waste disposal technology, minimize pollution, and safeguard the health of people and the environment while reducing waste. It prescribes how large volumes of industrial and municipal waste are to be disposed of. Additionally, it establishes waste management standards, encourages the development of waste management technology, and defines solid waste as a local responsibility.

## 3.4.3. Amendments

### *3.4.3.1. Resource Recovery Act*

The first modification to the SWDA was the Resource Recovery Act of 1970 (RRA). The amendment introduced criteria for the disposal of hazardous waste, encouraged waste reduction as well as recycling technology, and called for increased government involvement in waste management.

### *3.4.3.2. Hazardous and Solid Waste Amendments*

By regulating underground storage tanks, closing substandard landfills, and addressing small-scale producers of hazardous waste, the Hazardous and Solid Waste Amendments of 1984 (HSWA) reinforced the existing legislation.

### *3.4.3.3. Federal Facilities Compliance Act*

Passed in 1992, the Federal Facilities Compliance Act (FFCA) makes federal facilities, including military bases, liable for all fines and penalties specified in the SWDA and its amendments. This amendment was made in response to the Department of Energy v. Supreme Court decision of 1992. Ohio. The Court decided that because the US Department of Energy possessed sovereign immunity, it was not liable for managing a uranium processing facility.

# 3.5. RESOURCE CONSERVATION AND RECOVERY ACT (RCRA)

Enacted in 1976, the Resource Conservation and Recovery Act (RCRA) is the primary federal law in the United States that regulates the disposal of hazardous and solid waste.

## 3.5.1. Definition and General Aspects of RCRA

The framework for handling both hazardous and non-hazardous solid wastes is provided by the Resource Conservation and Recovery Act (RCRA), a legislation. Given the rising volume of solid waste produced regularly, this law is crucial because it promotes effective cleanup innovations. To support economic growth, RCRA is also committed to incorporating resource conservation into the US economy.

## 3.5.2. History and Goals

To address the growing issues the country was facing as a result of its expanding volume of industrial and municipal waste, Congress passed the RCRA. The Solid Waste Disposal Act of 1965 was amended by RCRA.

The act set national goals for:

- Protecting human health and the natural environment from the potential hazards of waste disposal.
- Energy conservation and natural resources.
- Reducing the amount of waste generated, through source reduction and recycling
- Maintaining environmental health standards.
- Ensuring the management of waste in an environmentally sound manner.

The US Environmental Protection Agency (EPA) sets the fundamental requirements for the RCRA program, which is a combined federal-state initiative that states adopt, modify, and implement. The rules issued under the RCRA that establish guidelines for the handling, storage, and disposal of hazardous waste in the US are currently the most well-known aspects of the law.

Parts 239 through 282 of Title 40 of the Code of Federal Regulations contain the waste management regulations published by the EPA. Part 260 contains the regulations pertaining to the management of hazardous waste. States are entrusted with developing state implementation plans for solid waste management and are allowed to run their hazardous waste programs, which must adhere to federal standards at least as strictly. The primary agency in California responsible for enforcing the RCRA regulations and the 1972 California Hazardous Waste Control Law (HWCL) is the Department of Toxic Substances Control (DTSC).

## 3.5.3. Why is RCRA Important?

Figure 3.1 outlines the various aspects of society that are impacted by the RCRA, including resource conservation, economic growth, human health and the environment, and the cleanup of contaminated sites. It is among the most significant pieces of legislation enacted in the US that mandates cleanups and encourages collaborations to create jobs, especially in the waste recycling industry, as well as resource conservation.

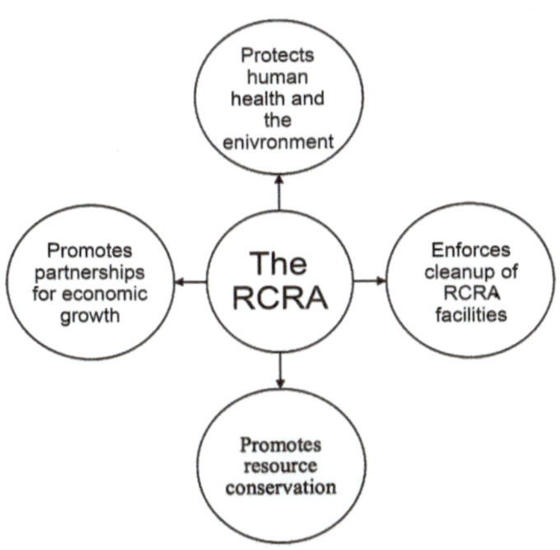


Figure 3.1. The importance of RCRA.

Source: https://www.wwdmag.com/what-is-articles/article/21169925/what-is-the-resource-conservation-recovery-act-rcra.

### *3.5.3.1. Protecting Human Health and the Environment*

Under the RCRA subtitle C program, hazardous waste is managed under the regulatory framework of the RCRA. This extensive program requires waste generators (who possess the necessary RCRA permits) to certify on an annual basis that they have a program in place to reduce the volume and toxicity of the waste, in addition to tracking hazardous wastes across the country.

Long-term stewardship for waste disposal is required, in addition to the RCRA hazardous waste permitting program that oversees performance to ensure adherence to the permits. This is the essence of stewardship: waste management facilities carry out post-closure activities, such as tank demolition or landfill final capping, once they have completed their management duties under the contract. Following closure, since the waste is left in situ (such as in a landfill), long-term stewardship involving both public and private stakeholders is required. Long-term stewardship also involves local government, states, tribes, and the EPA. In addition to providing waste management expertise for disasters, the EPA, which enforces the RCRA, emphasises on recycling and reusing the debris created by disasters.

### *3.5.3.2. Enforcing Cleanup: The RCRA Corrective Program*

Facilities that handle hazardous wastes are obligated to pay for the cleanup of any hazardous constituent leaks under the RCRA Corrective Action Program. Old waste must be cleaned up by approved waste generators as part of this program.

The RCRA Corrective Action Program is implemented in two stages: the corrective measure and the stabilization measures.

Policies are implemented in the first stage to stop both short- and long-term exposure to contaminated sites. The first step is advanced in the second step. This stage develops and puts into practice solutions, like restricting the amount of land used, to shield people from potentially harmful substances. These two actions are carried out by the EPA and the states through permitting programs and legislation. To shield the public from exposure to contaminated groundwater, the RCRA Corrective Action Program applied to 76% of the priority sites in 2014 (referred to as the RCRA Corrective Action sites).

#### *3.5.3.3. The Sustainable Material Management Program*

The Sustainable Material Management (SMM) Program aims to reduce the environmental impact of the material's hazardous constituents and it is implemented by the EPA and certain states. It considers the material's entire lifecycle for its management. The retirement of mercury-containing devices and product reformulation to create lead-free PVC products are two examples of SMM program practices.

#### *3.5.3.4. Partnering for Economic Growth*

Involving local communities and stakeholders in decisions about hazardous waste management, including cleanup efforts and emergency preparedness and response, is one of the main goals of the RCRA cleanup program. Permit decisions are also open to public participation.

To protect underprivileged communities from the effects of hazardous wastes, the RCRA promotes dialogues aimed at understanding these communities' needs and hearing their opinions to make decisions that are sustainable and effective. Furthermore, the Integrated Waste Management Plans and other tribal solid waste programs are developed and implemented with assistance from the RCRA program in partnership with the tribes.

The Organization for Economic Development (OECD) is one of the international partners in the RCRA. The RCRA manifest, which includes details on the kind and volume of waste transported as well as instructions for handling the waste, is applicable to shipments from OECD member nations that are listed by the EPA.

# 3.6. NATIONAL ENVIRONMENTAL POLICY ACT

The goal of the National Environmental Policy Act (NEPA), a US environmental law, is to encourage environmental improvement. It established the President's Council on Environmental Quality (CEQ) and new laws mandating US federal government agencies to assess the environmental effects of their decisions and actions. The US Congress passed the Act in December 1969, and President Richard Nixon signed it into law on January 1st, 1970. National environmental laws based on NEPA have been passed by more than 100 countries worldwide to date.

The most important result of NEPA was the requirement that all executive Federal agencies prepare environmental impact statements (EISs) and environmental assessments (EAs). NEPA requires Federal agencies to assess the environmental effects of their actions. The possible environmental effects of suggested Federal agency actions are stated in these reports. Furthermore, as stewards for future generations, each person should protect and improve the environment, as acknowledged by the US Congress. Since the President, Congress, and the Federal courts are not by definition Federal agencies, they are exempt from NEPA's procedural requirements. NEPA's procedural requirements may apply to a Federal agency acting as a final agency action under the President's authority.

## 3.6.1. History

NEPA originated from the growing public awareness and concern for the environment in the United States during the 1960s, a time of increased industrialization, suburban and urban growth, and pollution. Around this time, the 1964 Wilderness Act and other laws (such as the 1970 Clean Air Act and 1972 Clean Water Act) were supported by environmental interest groups and the public after Rachel Carson's 1962 book Silent Spring. The NEPA laws were being drafted in Congress at the same time as the public outcry following the early 1969 Santa Barbara oil spill. Shortly after the Senate's unanimous vote, a Time magazine article revealed the existence of a Cuyahoga River fire. The 1960s highway revolts, a wave of demonstrations across numerous American cities, were sparked by the Interstate Highway System's demolition of many communities and ecosystems. These protests served as a major impetus for the legislation of NEPA. Furthermore, as demonstrated by the work of political scientist Lynton Caldwell from Indiana, NEPA was also motivated by worries about the environmental irresponsibility of foreign development initiatives in the 1950s and 1960s.

In 2015 a United States District Court provided a documented concise background of NEPA being created to protect the environment from actions involving the Federal government as follows:

After almost a century of fast population growth, industrialization, urbanization, and economic expansion, it was evident by the late 1960s that American progress came at an environmental cost. Many indications of egregious mismanagement of the nation's resources and environment, particularly at the hands of the federal government, were found during a congressional investigation. Consequently, both lawmakers and the public demanded an immediate and comprehensive policy for environmental protection.

In response, Congress passed NEPA, which has been our fundamental national environmental protection statute since 1970. With the passage of NEPA, Congress required federal agencies to carefully consider the effects their actions would have on the environment and to take all reasonable precautions to avoid causing harm to the environment. Additionally, to address the general mistrust of federal agencies, Congress included action-forcing provisions in NEPA that mandate agencies adhere to certain protocols to complete any federal project.

Since it was passed, any significant projects involving federal funding, federally performed work, or permits from federal agencies—whether at the state, municipal, or federal levels—were required to comply with NEPA. According to court rulings, actions where permits issued by federal agencies are mandatory regardless of whether federal funds are used to act, as well as actions which are fully funded and managed by private-sector entities, necessitate NEPA-related environmental studies. This legal interpretation is predicated on the reasoning that obtaining a permit from a federal agency necessitates the processing and approval of a permit application by one or more federal employees (or contractors in certain cases), which inevitably results in the expenditure of federal funds to support the proposed action, even in cases where no federal funds are specifically designated to finance the specific action.

## 3.6.2. Contents

The preamble to NEPA reads:

To declare a national policy which will encourage productive and enjoyable harmony between man and his environment; to promote efforts which will prevent or eliminate damage to the environment and biosphere and stimulate the health and welfare of man; to enrich the understanding of the ecological systems and natural resources important to the Nation; and to establish a Council on Environmental Quality.

The three sections of NEPA are as follows: the first sets forth national environmental policies and goals; the second provides guidelines for federal agencies to follow; and the third creates the CEQ within the Executive Office of the President.

NEPA's goals are to create a national environmental policy and guarantee that environmental considerations are given the same weight as other considerations in the decision-making processes used by federal agencies. The act also encourages the CEQ to counsel the President in drafting a yearly report on the federal agencies' progress toward NEPA implementation. To counsel the president on environmental matters and policy, it also established the CEQ.

By mandating that federal agencies prepare an environmental impact statement to go along with reports and recommendations for Congressional funding, NEPA established the national environmental policy. An EIS is the name given to this impact statement. Since NEPA is an action-forcing law, all NEPA enforcement must take place through the legal system because the act does not carry any criminal or civil penalties. When a federal agency finances any part of a project, it must comply with NEPA regulations. On the other hand, a federal employee's review of a project may be considered a federal action, necessitating NEPA-compliant analysis performance.

NEPA applies to a wide range of federal agency actions, but it does not apply to state actions in which there is no federal funding or influence at all. NEPA's regulations also contain exclusions and exemptions, such as EPA exemptions and particular federal projects that are outlined in legislation. Exemptions also apply in cases where an impact analysis akin to the one required by NEPA is necessary to comply with other environmental laws. The Federal Insecticide, Fungicide, and Rodenticide Act, the Safe Drinking Water Act, the Resource Conservation and Recovery Act, and the Clean Air Act are a few examples of such laws.

### 3.6.3. Council on Environmental Quality

The Council of Economic Advisers, an executive branch body established by the Employment Act of 1946 to counsel the President on economic issues, served as the model for the Council on Environmental Quality (CEQ). With Executive Order 11514, President Richard Nixon broadened the CEQ's jurisdiction shortly after NEPA's creation. He gave the CEQ instructions on how to compile and organize federal programs pertaining to environmental quality, as well as how to properly prepare an EIS. Under the Executive Office of the President of the United States, the council was established. It consists of three members that the Senate confirms after the President appoints them.

An important role in the development of the EIS process has been played by the CEQ. According to its original rules, which were published in 1971, every federal department and agency had to adopt rules that were in line with those set by CEQ. Although these guidelines lacked the official status of agency regulations, they were frequently accorded the same respect in legal proceedings. By Executive Order 11991, President Jimmy Carter gave the CEQ permission to adopt regulations in 1977 as opposed to merely providing basic guidelines for EIS preparation.

Among other advisories, the CEQ has created advisory documentation in the last few years that explains the nature of cumulative impacts and the general framework of NEPA. As part of its ongoing responsibilities, the CEQ also maintains a website called the Citizen's Guide to NEPA. According to CEQ regulations, agencies and other federal actors must incorporate NEPA requirements with other planning obligations as early in the process as possible. This is to guarantee that environmental values are informed and are reflected in all decisions. Additionally, it avoids any conflicts or delays that might result from imposing these requirements. Section 102(2)(C), which is NEPA's action-forcing provision, stipulates that with every recommendation or report on proposals for legislation and other major Federal actions significantly affecting the quality of the human environment, all federal agencies shall include "a detailed statement by the responsible official ... on the environmental impact of the proposed action." This statement must specify all the details required for the EIS.

### 3.6.4. Consequences for Violating the National Environmental Policy Act

Reputational damage resulting from NEPA violations may put agencies' access to future funding opportunities at risk. Alleged violations of the guidelines must often go through the legal system because the CEQ lacks the authority to enforce them. These lawsuits have the potential to cause more harm to the agencies by putting years of project delays and financial demands on the federal government. Such delays have resulted in the cancellation or shutdown of several major projects, including the proposed Keystone XL Pipeline. The expansion was abandoned by TC Energy in 2021 after delays including the rejection of its review under NEPA, in the case Indigenous Environmental Network v. US Department of State. The most common consequence imposed by courts for violations of NEPA is a mandated repeat of whichever environmental review or report was found to have been insufficient.

### 3.6.5. National Environmental Policy Act and Environmental Justice

By mandating federal agencies to incorporate low-income and minority populations in their NEPA-mandated environmental analyses, the National Environmental Policy Act advances environmental justice. Executive Order 12898 mandates that federal agencies that follow NEPA consider the effects of the environment on human health as well as the economic and social implications.

The Promising Practice Report, released by the Environmental Justice Interagency Working Group, lists standard procedures for incorporating environmental justice into NEPA.

# 3.7. OCCUPATIONAL SAFETY AND HEALTH ACT

The federal law of occupational health and safety in the US federal government and private sectors is governed by the Occupational Safety and Health Act of 1970, a labor law. Congress passed it into law in 1970, and on December 29, 1970, President Richard Nixon signed it. Its major objective is to guarantee that employers give workers a safe working environment free from known risks, like exposure to hazardous chemicals, loud noises, mechanical risks, heat stress, or unhygienic conditions. The National Institute for Occupational Safety and Health (NIOSH) and the Occupational Safety and Health Administration (OSHA) were established by the Act.

## 3.7.1. History of Federal Workplace Safety Legislation

Before OSHA was passed, there were not many federally funded workplace health and safety safeguards. The mass production system in the United States promoted the use of machinery, but the legal framework did little to ensure worker safety. For most employers, replacing a deceased or injured worker was less expensive than implementing safety precautions. However, survivors of deceased workers or injured employees had few options for obtaining justice under tort law. Following the Civil War, advancements in life insurance coverage were made more widely available, state railroad and factory commissions were established, and new technologies—like the railway air brake—were adopted. However, the overall effect of these advancements was insignificant. (Cuming, Rocco, McEachern, 2008)

The Progressive Era saw the passage of the first federal safety laws. The first federal law requiring safety equipment in the workplace was the Safety Appliance Act, passed by Congress in 1893 (the law applied only to railroad equipment, however). Congress established the United States Bureau of Mines in 1910 to carry out research into mine safety in response to a string of well-publicized and fatal mine explosions and collapses (although the Bureau had no authority to regulate mine safety).

Many states also passed workers' compensation laws, supported by trade unions, discouraging employers from allowing hazardous workplaces. For a while, there was a noticeable decline in worker accidents because of these laws, the rising influence of labor unions, and public indignation over inadequate workplace safety.

During World War II, American industrial production rose dramatically, and the number of industrial accidents also skyrocketed. However, after the war ended, workplace accident rates remained high and even started to rise. Winning the war took precedence over workplace health and safety, and most labor unions were more concerned with maintaining wages in the face of severe inflation than with these issues. 14,000 workers died and another 2 million were injured or disabled annually as a result of workplace hazards in the two years before OSHA was passed. The chemical revolution also brought a wide range of novel chemical compounds into the manufacturing sector. There was a lack of knowledge regarding the health effects of these chemicals, and workers were not adequately protected against high or prolonged exposure. While some states, like California and New York, have passed legislation pertaining to both workplace health and safety, many states have not updated their workplace protection laws since the year 2000.

### 3.7.2. Passage

Growing knowledge of the harm many chemicals cause to the environment by the middle of the 1960s gave rise to a politically significant environmental movement. Seizing upon the public's mounting concerns about chemicals in the environment, some labor leaders contended that the effects of these substances on worker health were even more detrimental than the low-level exposure that plants and animals experienced in the wild. The 23rd of January 1968 saw President Lyndon B. Johnson sent Congress a comprehensive bill pertaining to occupational health and safety. Business opposition to the legislation was strong, with the National Association of Manufacturers and the US Chamber of Commerce leading the charge. Numerous labor leaders, such as the AFL-CIO leadership, testified in favor of the legislation during congressional hearings. The bill was killed in committee.

President Richard Nixon presented two bills to Congress on April 14, 1969, that would have safeguarded the health and safety of workers. Compared to the Johnson bill, the Nixon legislation was far less prescriptive and only required advice on workplace health and safety. But Senator Harrison Williams and Representative James G. O'Hara presented a far harsher bill modelled after the previous year's Johnson legislation.

An all-purpose general duty clause was also imposed on the enforcing agency by companion legislation that was introduced in the House. Republicans introduced a new, competing bill in response to the Democratic bill's stricter approach, which appeared to be supported by both chambers and unions. Nixon's preferred independent research and standards-setting board was established by the compromise bill, which also established a new enforcement agency. Additionally, as in the Democratic bill, the compromise bill authorized the Department of Labor to bring legal action on behalf of the enforcement agency. Both houses acted in November 1970: the Senate passed the more stringent Democratic bill (which now included the general duty clause) and the House passed the Republican compromise bill.

Early in December 1970, a conference committee reviewed the final bill. Members of the conference committee were under pressure from union leaders to transfer the standard-setting authority from an independent board to the Department of Labor. In exchange, unions consented to remove a clause in the legislation that would have allowed the Secretary of Labor to close factories or halt production processes that put workers in immediate danger of harm. They also agreed to grant an independent review commission veto power over enforcement actions. Republicans offered to create an independent agency for research on occupational health and safety; Democrats countered by gaining the right for union representatives to accompany federal inspectors on inspections and incorporating the general duty clause. The conference committee bill was signed into law by President Nixon on December 29, 1970, after passing both chambers on December 17, 1970. Tony Mazzocchi, an environmentalist and labor activist, is credited with spearheading the legislation, according to the New York Times.

### 3.7.3. Description

Congress expressed its intention to protect human resources and to ensure safe and healthy working conditions for all working men and women in the country when it passed the Act.

The Department of Labor's Occupational Safety and Health Administration (OSHA) was established by the Act. The power to establish and implement workplace health and safety regulations was granted to OSHA. The Autonomous Occupational Safety and Health Review Commission was also established by the Act to evaluate enforcement actions, priorities, and cases.

Additionally, the Act created the National Institute for Occupational Safety and Health (NIOSH), an independent research organization currently housed within the Centers for Disease Control and Prevention but formerly part of the Department of Health, Education, and Welfare.

The definition of an employer under the Act is any individual with employees who are involved in a business that affects commerce; the United States, any state, or any political subdivision of a state are not included in this definition. Employers of all stripes, including manufacturers, construction companies, law firms, hospitals, nonprofit organizations, labor unions, and private schools, are covered by the Act.

If a church or other religious organization hires people for nonreligious reasons, they are protected. The self-employed, family farms, and businesses subject to other federal laws (such as mining, the production of nuclear weapons, railroads, and airlines), as well as state and local governments (unless authorized by state law), are not included in the Act. The United States Postal Service and federal agencies are covered by the Act.

Section 5 of the Act contains the "general duty clause." The "general duty clause" requires employers to

- Maintain conditions or adopt practices reasonably necessary and appropriate to protect workers on the job.
- Be familiar with and comply with standards applicable to their establishments.
- Ensure that employees have and use personal protective equipment when required for safety and health. OSHA has established regulations for when it may act under the "general duty clause."

The four criteria are

- There must be a hazard;
- The hazard must be a recognized hazard (e.g., the employer knew or should have known about the hazard, the hazard is obvious, or the hazard is a recognized one within the industry);
- The hazard could cause or is likely to cause serious harm or death;
- The hazard must be correctable (OSHA recognizes not all hazards are correctable).

It is challenging to satisfy all four requirements, even though in theory, it is an effective tool against workplace hazards. To fulfil its legal responsibilities, OSHA has therefore taken a leading role in the creation of regulatory rules.

Because of the complexity of the Administrative Procedures Act-governed rule-making process, OSHA has prioritized fundamental mechanical and chemical hazards over procedures. Currently, its standards cover the following major areas: hazardous waste, infectious diseases, dangerous atmospheres, machine hazards, confined spaces, toxic substances, harmful physical agents, electrical hazards, fall hazards, and hazards related to trenches and digging.

The Act's Section 8 addresses reporting obligations. If an employee passes away due to a work-related incident or if three or more employees are hospitalized due to a work-related incident, all employers are required to report these events to OSHA within eight hours. Furthermore, all fatal heart attacks that occur while working must be reported. Section 8 of the Act authorizes OSHA inspectors to enter, examine, and investigate any workplace covered by the Act during regular business hours. Employers are also required to notify workers about potential hazards at work. Employers are required by OSHA regulations to maintain a record of all non-consumer chemical products used in the workplace. Employees must have access to and be able to read comprehensive technical bulletins known as material safety data sheets (MSDSs) to prevent chemical hazards. Additionally, OSHA mandates that employers record all work-related illnesses and injuries that need medical attention (other than first aid) on OSHA Form 300, Log of Work-Related Illnesses and Injuries. Records must be retained for a minimum of five years, and an annual summary must be posted for three months.

According to Section 11(c) of the Act, an employer is not allowed to fire, take disciplinary action against, or treat a worker unfairly just because they have exercised their legal rights. These rights include filing a complaint with OSHA, requesting an inspection, taking part in an inspection, and offering testimony in any proceedings pertaining to an OSHA inspection.

If state regulations and enforcement are or will be at least as successful in ensuring safe and healthy employment as the federal OSH Act, Section 18 of the Act encourages and permits states to adopt their own occupational safety and health plans. Twenty-two states and territories had fully functional plans in place as of 2007, while four more only had plans that addressed the public sector.

# 3.8. CLEAN WATER ACT

The main federal statute in the US addressing water pollution is the Clean Water Act (CWA). Its goals are to preserve the integrity of wetlands, acknowledge the states' obligations to address pollution and support them by funding publicly owned treatment facilities that enhance wastewater treatment. It also aims to restore and maintain the chemical, physical, and biological integrity of the country's waters.

The Federal Water Pollution Control Act is the official name of the law. The Federal Water Pollution Control Act was first passed in 1948, but it wasn't until 1972 that it underwent a comprehensive revision that resulted in the 1972 Federal Water Pollution Control Act Amendments. The Clean Water Act of 1977 and the Water Quality Act (WQA) of 1987 are two examples of amendatory legislation that has since been introduced and contains significant changes.

Contamination of groundwater is not specifically addressed by the Clean Water Act. The Safe Drinking Water Act, the Resource Conservation and Recovery Act, and the Superfund Act all contain provisions for groundwater protection.

## 3.8.1. Background

### *3.8.1.1. Health Implications of Water Pollution*

Contamination can happen in the source water, or in the distribution system that supplies drinking water. Water contamination can arise from various sources such as naturally occurring minerals and chemicals (like arsenic, radon, and uranium), local land use practices (like pesticides, fertilizers, and concentrated feeding operations), manufacturing processes, and wastewater releases or sewer overflows. Infertility issues, neurological disorders, and gastrointestinal illnesses are a few health effects of contaminated water. Certain contaminants may be dangerous for young children, pregnant women, the elderly, and those with weakened immune systems from AIDS, chemotherapy, or transplant drugs.

### *3.8.1.2. Gastrointestinal Illness*

Constipation, irritable bowel syndrome, haemorrhoids, anal fissures, perianal abscesses, anal fistulas, diverticular diseases, colitis, colon polyps, and cancer are examples of digestive disorders. The elderly and young are generally most vulnerable to gastrointestinal disorders. An interquartile range increase in turbidity was linked to a 9 increase in hospital admissions for gastrointestinal illness in a study about the relationship between drinking water quality and gastrointestinal illness in elderly Philadelphia residents. The association was stronger in the over-75 population than in the 65–74 age group. Water quality nine to eleven days before the visit was found to be negatively associated with hospital admissions for gastrointestinal illness. To some extent, this illustration represents the impact of current water treatment methods which still put Americans at risk for gastrointestinal illnesses caused by contaminated water.

#### 3.8.1.3. Reproductive Problems

Any disease affecting the reproductive system is referred to as a reproductive problem. The connection between increasing male fertility issues and water pollution is reinforced by new research from Brunel University and the University of Exeter. A study found a class of compounds in contaminated water that function as anti-androgens, blocking the action of the male hormone testosterone and lowering male fertility.

#### 3.8.1.4. Neurological Disorders

Diseases of the brain, spine, and the nerves that connect them are referred to as neurological disorders. According to a recent study involving over 700 participants in the Central Valley of California, there was a higher incidence of Parkinson's disease in those who had probably ingested tainted private well water. Those with private wells in proximity of fields sprayed with commonly used pesticides faced a ninety percent increased risk. Private wells are not subject to the same regulations as public water sources in big cities, and they are also not checked for pollution. Some of the crop chemicals used to kill weeds and pests can seep into groundwater, and many of them are found at shallow depths of less than 20 yards. Consequently, pesticides that target developing brains (in the womb or during infancy) are likely present in private wells and can result in neurological diseases later in life. According to a study conducted by Beate Ritz, a professor of epidemiology at UCLA, individuals with Parkinson's disease were found to have consumed private well water more frequently and to have done so for an average of 4 and a half years longer than those without the illness.

### 3.8.2. Waters Protected

As per the Supreme Court's 2023 ruling, the CWA applies to all bodies of water, including rivers, lakes, streams, and oceans with a continuous surface connection to navigable waters.

The 1972 statute frequently uses the term "navigable waters" but also defines the term as "waters of the United States, including the territorial seas." Regulations interpreting the 1972 law have included water features such as intermittent streams, playa lakes, prairie potholes, sloughs and wetlands as "waters of the United States." In 2006, in Rapanos v. United States, a plurality of the US Supreme Court authored by Justice Antonin Scalia held that the term "waters of the United States" "includes only those relatively permanent, standing or continuously flowing bodies of water 'forming geographic features' that are described in ordinary parlance as 'streams [,]... oceans, rivers, [and] lakes.'" The concurrent written opinion of Justice Anthony Kennedy defined the term more broadly, including wetlands with a "significant nexus" to traditionally defined navigable waters. Since Rapanos, the EPA and the U.S. Army Corps of Engineers have attempted to define protected waters in the context of Rapanos through the 2015 Clean Water Rule, but this has been highly controversial. The agencies considered the CWA to cover bodies of water with a "significant nexus" with traditional navigable waters, according to Justice Kennedy's definition.

In 2023, the Supreme Court rejected the "significant nexus" test in Sackett v. EPA and established the current definition.

## 3.8.3 Pollution Control Strategy

### 3.8.3.1. Point Sources

The National Pollutant Discharge Elimination System (NPDES), a permit-based system for controlling pollution point sources, was established by the Clean Air Act (CWA).

Point sources include:

- Industrial facilities (including manufacturing, mining, shipping activities, oil and gas extraction and service industries).
- Municipal governments (particularly sewage treatment plants) and other government facilities (such as military bases), and
- Some agricultural facilities, such as animal feedlots.

Pollutants from point sources are prohibited from entering surface waters without an NPDES permit. The EPA oversees the system in collaboration with state environmental agencies. The EPA has permitted 47 states the ability to the discharging facilities directly. Tribes may also grant permits under the CWA, but the EPA has not given any tribes permission for the same. An EPA regional office issues the permits in the remaining states and territories.

Congress permitted states to create water quality standards in legislation passed before 1972. These standards would have limited facility discharges based on the unique qualities of each body of water. However, those standards were only going to be developed for interstate waters, and the methodology and data needed to support that process were still in the early stages of development scientifically. There was no permit system in place to enforce the requirements, and that system was ineffective. Congress inserted the permit system and the need for technology-based effluent limitations in the 1972 CWA.

In the 2020 Supreme Court case County of Maui v. Hawaii Wildlife Fund, the Court also validated that some discharges may not be point sources but are the "functional equivalent of a direct discharge" to navigable waters, such as in this case, the injection of wastewater into groundwater injection wells. The EPA had not yet developed regulations in this area at the time of the case's ruling, and the Court directed the EPA to collaborate with the courts to define such functional equivalents. The Court stated that this would probably depend most on how far the pollutants traveled and how long it took them to reach navigable waters, considering the materials they passed through, any chemical or physical interactions they had with elements of the ground, and the amount of pollutant that reached the navigable water. After the Supreme Court's ruling in July 2021, the Hawaii District Court ruled that the groundwater injection of sewage by the Maui County sewage treatment plant was the functional equivalent of a direct discharge, necessitating the plant to obtain an NPDES permit.

- Technology-based Standards

A new requirement for technology-based standards for point source discharges was established by the CWA in 1972. The performance of pollution control technologies is the basis on which the EPA develops these standards for different categories of dischargers, without considering the specific conditions of a receiving water body. Congress aimed to level the playing field by using Best Available Technology to establish a basic national discharge standard for all facilities within a category. The standard turns into a permit's minimal regulatory requirement. Water quality standards may be used if the federal standard is deemed insufficiently protective at a specific location. In such a case, the permit authority (state or EPA) will incorporate effluent limitations based on water quality into the permit.

- Water Quality Standards

The 1972 act allowed the technology-based standards to be used in tandem with the water quality-based approach going forward. If water quality for a specific water body is still compromised after applying technology-based standards to a permit, the permitting agency may add water quality-based limitations to that permit. The permittees would have to install extra controls, and the additional limitations would be stricter than the technology-based restrictions. Water quality standards consist of four basic elements:

- Designated uses;
- Water quality criteria;
- Antidegradation policy and
- General policies.
  - Designated Uses: States and federally recognized tribes and nations are required by the water quality standards regulations to designate suitable uses for water bodies under their jurisdiction. The value and use of public water supplies, fish and wildlife protection, recreational waterways, and agricultural, industrial, and navigational waterways are all considered when determining appropriate water uses. States and tribes/nations evaluate a water body's suitability based on its physical, chemical, and biological characteristics. States, tribes, and nations assess the appropriateness of designated uses for water bodies by considering economic factors, geographical settings, and scenic qualities. States or tribes must update standards to reflect the uses that are being attained if those standards show that designated uses are less than those that are currently attained. For any body of water with designated uses that do not include "fishable/swimmable" target use that is identified in section 101(a)(2) of CWA, a "Use Attainability Analysis" must be conducted. Every three years, such bodies of water must be re-examined to verify if new information is available that demands a revision of the standard. If new information is available that specifies "fishable/swimmable" uses can be attained, the use must be designated.
  - Water Quality Criteria: The EPA publishes water quality criteria under CWA section 304(a), which states and federally recognized Indigenous Nations adopt to protect their designated areas. They can also modify the criteria to account for site-specific conditions or adopt criteria based

on other methods that are supported by science. Water quality standards can be numerical standards known to cause toxicity to protect against pollutants. Water quality standards, which form the foundation for limiting the toxicity of waste discharges to aquatic species, are narrative criteria. A water body's aquatic community, which characterizes the variety and quantity of species present, serves as the foundation for a biological criterion. A sediment criterion describes the conditions of contaminated and uncontaminated sediments to prevent undesirable effects, whereas a nutrient criterion only guards against nutrient over-enrichment.

* Anti-degradation Policy: A three-tiered anti-degradation program must be established by states and tribes per the anti-degradation policy included in the water quality regulations. Anti-degradation protocols specify the actions and inquiries that must be made when certain activities have an impact on the quality of the water. "Tier 1" requirements apply to all surface waters. These requirements maintain and protect current uses and the water quality conditions to support existing uses. Current uses are identified by showing that fishing, swimming, and other water uses have occurred and are suitable since November 28, 1975. "Tier 2" requirements maintain and protect water bodies with existing conditions that are better to support "fishable/swimmable" uses according to CWA section 101(a)(2). "Tier 3" requirements maintain and protect water quality in "outstanding national resource waters" (ONRWs), which are the highest quality waters in the US with ecological significance.
* General Policies: The EPA reviews and approves general policies that states and Native American tribes adopt regarding water quality standards. These regulations on water quality standards cover low flow policies, variances, and mixing zones. The area around a point source discharge where sewage is diluted by water is known as the "mixing zone policy." The location, dimensions, form, and quality of mixing zones are determined by the methodology of the mixing zone procedure. An alternative to eliminating a designated use is to implement a variance policy, which temporarily relaxes water quality standards. Variance may be a part of a states or tribe's water quality standard. Variances must be reviewed by the public every three years to support efforts to improve the quality of the water. The Low-Flow policy describes the methods used to identify critical low-flow conditions and is relevant to water quality standards for states and tribes.
* Laboratory test methods

Most NPDES permit holders must gather wastewater sample collection and use the test procedures outlined in their permits for sample analysis. The permittees use analytical methods published by the EPA. As mandated by the act, the procedures identify the chemical compounds and microbiological components of wastewater. Chemical element detection, including the identification of metals known to cause cancer, is one of the test procedures for chemical compounds. In certain microbiological test procedures, biological and chemical trends that could justify new regulatory limits on pollutants are calculated and identified using microbial source tracking (MST) techniques.

#### 3.8.3.2. Nonpoint Sources

Congress was unsure of the status of some sources of water pollution and excluded some from the 1972 CWA's point source definition. As a result, these sources were regarded as nonpoint sources exempt from the permit program.

Permit requirements were specifically waived for irrigation return flows and agricultural stormwater discharges. On the other hand, the US Department of Agriculture's research, technical, and financial support programs to enhance farm runoff management practices were supported by Congress.

The 1972 law did not specifically address stormwater runoff from industrial sources, municipal storm drains, or other sources. Following the agency's initial NPDES program implementation, which excluded urban runoff and industrial stormwater discharges, an environmental group sued the EPA. In 1977, the D.C. Circuit Court of Appeals ruled that stormwater discharges must be covered by the permit program.

Studies carried out beginning in the late 1970s and early 1980s revealed that in many US regions, stormwater runoff was a major contributor to the degradation of water quality. The Nationwide Urban Runoff Program (NURP) was implemented by the EPA in the early 1980s to map out the scope of the urban stormwater issue. The agency started drafting stormwater permit coverage regulations but faced pushback from the business community and local governments. This resulted in more legal battles. While Congress was debating additional CWA amendments in 1986, the litigation was still ongoing.

In response to the stormwater issue, Congress passed the Water Quality Act of 1987, which designated municipal separate storm sewer systems (MS4) and industrial stormwater dischargers as point sources and mandated that they obtain NPDES permits by a certain date. Agriculture discharges were still exempt from permits, but Congress established several grants and programs, such as the EPA's demonstration grant program, to increase the study and creation of non-point controls and management strategies.

#### 3.8.3.3. Financing of Pollution Controls

With the 1972 CWA, Congress established a significant public works financing program for municipal sewage treatment. Title II authorized and provided funding for a system of grants for the construction of municipal sewage treatment plants. Under the original program, the state would finance the remaining portion of each grant, with the federal portion amounting to up to 75% of the capital cost of the facility. Congress changed the grants' federal percentage in later amendments, and in 1987 the WQA switched to a revolving loan program under Title VI. According to the polluter pays principle, industrial and other private facilities must fund their treatment upgrades.

# 3.9. TOXIC SUBSTANCES CONTROL ACT

Enacted by the 94th US Congress in 1976, the Toxic Substances Control Act (TSCA) is a federal law that governs substances not covered by other US federal statutes, including substances that are already in use in commerce and newly introduced substances. It is enforced by the US Environmental Protection Agency (EPA). All chemicals that were already in use were grandfathered in under the TSCA because they were thought to be safe. Its three primary goals are to evaluate and control newly developed commercial chemicals before they hit the market, to control substances that were already on the market in 1976 but presented an unreasonable risk of harming human health or the environment, like PCBs, lead, mercury, and radon, and to control the use and distribution of these substances.

In contrast to what the name suggests, TSCA does not classify chemicals as toxic or non-toxic. Instead, it outlaws the production of chemicals or their importation if they are not included in the TSCA Inventory or fall under one of several exemptions. Chemicals that are not listed on the TSCA inventory are referred to as new chemicals, and those that are listed are known as existing chemicals. The term "chemical substance" is defined by the TSCA as any element or uncombined radical, as well as any organic or inorganic substance of a specific molecular identity, including any combination of these substances that arises in whole or in part because of a chemical reaction or occurs in nature. However, chemicals regulated by other federal statutes are excluded from the definition of a chemical substance.

Generally, before producing or importing new chemicals for sale, manufacturers should notify the EPA of their pre-manufacturing process. The Federal Food, Drug, and Cosmetic Act regulates foods, and food additives; the Federal Insecticide, Fungicide, and Rodenticide Act regulates pesticides; the Bureau of Alcohol, Tobacco, Firearms and Explosives regulates tobacco and tobacco products; Section 5(h)(3) governs substances used only in small quantities for research and development; and the Nuclear Regulatory Commission regulates radioactive materials and wastes.

The EPA examines new chemical notifications and may regulate a substance by restricting its uses or production volume or outright outlawing it if it determines an unreasonable risk of harm to human health or the environment. The Frank R. Dot Lautenberg Chemical Safety for the 21st Century Act underwent its first significant revision in several years in 2016.

### 3.9.1. Overview

The Toxic Substances Control Act of 1976 required the EPA to regulate the production, distribution, use, sale, and disposal of chemicals to safeguard the public from unjustified risks to health or the environment. The Resource Conservation and Recovery Act, the Clean Water Act, and the Clean Air Act regulate pollution; this act does not address it. TSCA, on the other hand, regulates commercial products, much like FIFRA (the Federal Insecticide, Fungicide, and Rodenticide Act). The act required the EPA to compile a list of all chemicals currently in use and the industry to notify the EPA of any new chemicals being produced. It also gave the EPA the power to regulate the production and use of chemicals. The act also gave the EPA the authority to gather information on and require manufacturers (and importers) to test products. Methylene chloride, for instance, is no longer permitted in consumer paint and coating removal products by the EPA.

Existing and novel chemicals are the two main categories into which the act divides chemical types. The act regulates the two categories of chemicals differently, which is why the distinction is made. Any chemical presently listed under TSCA section 8(b) of the TSCA Inventory is considered an existing chemical. Any chemical substance not listed in the list of chemical substances created and released under section 8 is considered a new chemical (b). Approximately 8,800 chemicals that were produced or imported in quantities greater than 10,000 pounds were included in this list, which comprised all chemicals produced or imported into the United States before December 1979, or 99 of the EPA's mandate as outlined in the bill.

### 3.9.2. Regulation of Existing Chemicals

The TSCA Inventory contains a list of all chemicals currently on the market. Approximately 62,000 chemicals were never tested by the EPA despite being tasked with safeguarding the public from hazardous and possibly cancer-causing substances because they were grandfathered in and legally did not pose an unreasonable risk. Without testing, TSCA grandfathered the use of these chemicals into TSCA's list of "existing" (as opposed to "new") chemicals. For existing chemicals, manufacturers need to generate and report data on risk, manufacturing and processing, adverse health effects, published and unpublished health and safety studies, and "substantial risks" to the EPA. Also, if a substance is subject to Significant New Use Rule (SNUR), a company would have to submit and receive approval of a Significant New Use Notice (SNUN) before undertaking the significant new use(s) defined in the SNUR.

The Environmental Protection Agency (EPA) may mandate that manufacturers test chemicals that are already on the market under section 4. However, the EPA must make multiple formal findings in each case, the first of which is subject to judicial review under the substantial evidence standard. Specifically, the requirement that the EPA establish whether the chemical poses an unreasonable risk creates a regulatory Catch-22 because the EPA cannot access or order the generation of much of the data needed to make this determination without first making the determination itself. During the TSCA reform talks, this burden was one of the factors that prompted modifications to Section 4.

Nearly all products containing asbestos were prohibited from being manufactured, imported, or processed in the United

States by a final rule issued by the EPA in 1989 under section 6. It was, however, only partially successful in enforcing the chemicals tested and found to be harmful to public health using the authority granted under TSCA section 6. There have been heated discussions regarding the EPA's legal responsibility for chemical bans because of the agency's inadequate regulation of these substances. In its 38-year history, the EPA has been successful in limiting five chemicals (PCBs, dioxin, asbestos, hexavalent chromium, and chlorofluorocarbons) by section 6 authority; however, the asbestos ban was partially overturned in 1991. A few current chemicals are subject to EPA restrictions through Section 5 SNURs.

Only 250 of the more than 60,000 chemicals in use today have undergone direct testing by the EPA as of 2015. Of those, 60 underwent testing following the manufacturer's voluntary consent, and 140 underwent testing as required by law.

Furthermore, 3,000 high production volume (HPV) chemicals are produced or imported in amounts greater than one million pounds annually. Although HPV chemicals make up only one-third of all chemicals currently in use, their large volumes raise concerns about the dearth of fundamental hazard information.

Numerous environmental organizations, including the Natural Resources Defense Council, express dissatisfaction over the EPA's apparent lack of ability to regulate hazardous chemicals, including those that are known to cause cancer or other grave health consequences. More than 2,200 chemicals have had new data on them or previously collected data made available thanks to the HPV program.

### 3.9.3. Regulation of New Chemicals

Companies must submit a PMN to the EPA at least ninety days before they manufacture or import a new chemical for a non-exempt purpose to notify the agency of their intention to manufacture (or import) a chemical that is not on the TSCA Inventory. Since the TSCA's Section 5 does not mandate toxicity testing before PMN submission, the EPA is forced to rely on computer modeling to assess whether a new chemical may pose an unreasonable risk. Safety information is also not required to be included in the PMN. To regulate new chemicals, the EPA must determine that the chemical ..". may present an unreasonable risk of injury to health or the environment..." or show that the chemical ..". is or will be produced in substantial quantities, and such substance either enters or may reasonably be anticipated to enter the environment in substantial quantities or there is or may be significant human exposure to the substance."

The EPA cannot take action to stop the production of the new chemical if these conditions are not met. After receiving a PMN, the EPA has ninety days to act so that the new chemical can be legally marketed and used in products. As a result, only 40% of acute toxicity and mutagenicity tests are ever finished, and even fewer data are ever produced regarding long-term consequences or endpoints, such as sub chronic, neurotoxicological, developmental, reproductive, and chronic endpoints. Furthermore, fewer than 5% of data on aquatic organism toxicity are submitted with a PMN. Between 1979 and 1994, more than 24,000 PMNs were submitted to the EPA's new chemical program, which attempted to postpone manufacturing and request more information for just 5,000 of those submissions. Nevertheless, half of these eventually made their way onto the market and are still there.

According to an EPA evaluation, there are deficiencies in 85% of PMN data regarding health effects and 67% of PMNs have data on any kind of environmental or health effects. Section 5 of the TSCA established The Structure Activity Team (SAT) and structural-activity relations (SARs) to review PMN chemicals to make up for this lack of data. A group of knowledgeable scientists and specialists called SAT assess the possible environmental fate of novel chemicals as well as their potential health and environmental risks.

Hazard assessments for chemicals primarily rely on models, SARs based on similar chemicals, or, in certain situations, data on the subject chemical obtained from public databases or reference material like Beilstein, as there is little to no data received with the PMNs. On the other hand, some contend that the risks connected to novel chemicals are not sufficiently assessed by SARs or the SAT review procedure. For example, there is no minimum data set other than what they have on hand when they submit the PMN. Only substances that meet the requirements of the TSCA are subject to EPA regulation, and the agency is also responsible for demonstrating the safety of chemicals that are already on the market.

## 3.9.4. TSCA and Environmental Justice

In communities where minorities and low-income residents are disproportionately exposed to toxic chemicals, TSCA can advance environmental justice by lowering the risk of chronic illnesses and conditions like obesity, learning disabilities, asthma, prostate cancer, and infertility.

Communities battling environmental racism linked to industrial facilities located close to low-income neighborhoods can be transformed and strengthened by public policy. Minority communities are disproportionately impacted by environmental threats, which range from toxic contamination, locally unwanted land use (LULUs), unsafe and substandard housing, and the extraction of natural resources. These communities cannot afford to wait years for bureaucratic procedures to show that living under these conditions has put their health at risk. Research has indicated that residing close to an industrial facility that releases pollutants is a significantly greater likelihood for individuals with lower incomes than for those with higher incomes.

The industry's desire to locate its facilities in areas with inexpensive land, access to industrial labor pools, and readily available materials triggers changes in the market dynamics. Due to noise, pollution releases, and concerns about potential health effects, industrial facilities may eventually cause the value of surrounding properties to decrease and affect the value of housing. Rich neighborhoods aggressively oppose the location of industrial facilities close to them, but low-income and communities of color are more easily targeted because they lack the resources and organization to effectively oppose the government and industry.

### *3.9.4.1. Populations Vulnerable to Toxic and Chemical Substances Exposure*

Environmental justice organizations can influence local, state, and federal policies by involving the community. By promoting the production of safe chemicals and restricting the market for dangerous chemicals, TSCA regulations can safeguard public health.

The elderly, young children, pregnant women, and some occupational workers are among the vulnerable populations that are more susceptible to illnesses brought on by exposure to harmful chemicals. Older people are more susceptible to chemical exposure, which can weaken their immune systems and lead to heart disease and other health problems from drug interactions. Since children's immune systems are still developing and their bodies cannot handle the amount of chemicals compared to their size, they are more susceptible to the negative health effects of environmental injustice. Better policies could be implemented by the federal, state, and local governments to safeguard vulnerable groups against the growing number of chemical exposure incidents that occur daily. The Frank R. Lautenberg Chemical Safety for the 21st Century Act takes this concern into account by requiring the EPA to assess risk to susceptible and potentially exposed subpopulations.

- Children's Exposure to Toxic Chemicals

Chemical exposure may play a role in pediatric disease development, as children are more likely to experience chronic illness patterns. The three primary health issues linked to exposure to environmental chemical pollutants are obesity, lead poisoning, and asthma. The yearly expenses associated with lead poisoning-related diseases in children in America are estimated to be $433.4 billion. The health of newborns, young children, teenagers, and developing fetuses is at risk due to toxic chemicals. Children should be shielded from chemical exposures because their bodies are not as capable as adults of breaking down and excreting harmful substances. Newly developed chemicals that are used in consumer goods expose children. These chemicals can also be found in food, water, air, houses, schools, and communities. Families may be disproportionately affected by the health issues that hotspot communities bring because of social and environmental stressors. Consequently, knowledge about possible toxicity can assist parents in choosing products that are safer for their kids to use. For approximately two-thirds of the 3000 high production volume (HPV) chemicals, toxicity data is available. Children's exposure to hazardous materials can be decreased by laws and regulations designed to protect vulnerable populations.

To prevent children from being exposed to lead in toys and bisphenol A, or BPA, in baby bottles, for example, the state of Maine passed the Kid-Safe Act in 2008. In January 2016, the Center for Science in the Public Interest released Seeing Red: Time for Action on Food Dyes, a report that criticized the continued use of artificial food coloring in the United States. Citing data from the Centers for Disease Control and Prevention, the report estimated that over half a million children in the United States experience adverse behavioral reactions because of ingesting food dyes, with an estimated cost exceeding $5 billion annually. The Food and Drug Administration is urged by the report to take action to prohibit or limit the use of these dyes. Readers are reminded that the Federal Food, Drug, and Cosmetic Act regulates artificial food coloring and baby bottles, not the Toxic Substances Act. TSCA regulation, however, extends to other hazardous materials (like bisphenol A) that may be present in baby food.

- Pregnant Women's Exposure to Chemicals

Exposure to toxic chemicals frequently during pregnancy can affect a woman's reproductive and developmental health during critical developmental windows. This can increase the risk of birth defects, childhood illnesses, and disabilities that can last a lifetime. Women who are pregnant or intend to become pregnant can receive information from health professionals about how to minimize risks and exposure to harmful chemicals in the environment. Consuming organic food is one way to lessen the effects of harmful chemicals in your diet. Breastfeeding mothers may be exposing their infants to harmful substances in their milk. Children's bodies contain fewer pesticides when they switch from a conventional to an organic diet, but low-income families must prioritize their needs, and purchasing organic food may not be feasible due to financial constraints. It is important to remind readers that the Federal Insecticide, Fungicide, and Rodenticide Act governs pesticide regulations instead of the TSCA. On the other hand, the TSCA regulates other hazardous substances that are transferable through breast milk, such as per- and polyfluoroalkyl substances (PFAS).

- Occupational Worker Exposure to Toxic Chemicals

Workers have the right to a workplace that is safe and does not present a risk of serious harm, according to the Occupational Safety and Health Administration (OSHA). Chemical exposure at work can occur by direct skin contact, ingestion, inhalation, or eye contact. Certain jobs expose workers more frequently to hazardous chemicals, which may have long-term detrimental effects on their health. Exposure rates that are too high for the body to process and get rid of can cause accumulation and even damage to the body. Toxic chemical exposure can affect both male and female fertility. The Frank R. Lautenberg Chemical Safety for the 21st Century Act specifically includes workers in its definition of potentially exposed and susceptible subpopulations, indicating concerns about exposure to workers.

- Corporate Support for Mitigation of Toxic Chemical Exposure

Workers have the right to a workplace that is safe and does not present a risk of serious harm, according to the Occupational Safety and Health Administration (OSHA). Chemical exposure at work can occur by direct skin contact, ingestion, inhalation, or eye contact. Certain jobs expose workers more frequently to hazardous chemicals, which may have long-term detrimental effects on their health. Exposure rates that are too high for the body to process and get rid of can cause accumulation and even damage to the body. Toxic chemical exposure can affect both male and female fertility. The Frank R. Lautenberg Chemical Safety for the 21st Century Act specifically includes workers in its definition of potentially exposed and susceptible subpopulations, indicating concerns about exposure to workers.

Businesses can demonstrate their social responsibility by taking products off the market and replacing them with ones that are similar and helping vulnerable populations. For example, a major buyer of medical supplies, Kaiser Permanente, has a policy to stay away from chemicals linked to cancer, reproductive issues, and genetic mutations. Corporate social responsibility, or CSR, is the moral duty of the business to build success in a way that respects communities, upholds moral principles, and fosters sustainability and a positive reputation. Businesses can innovate and enhance their reputation by addressing the growing consumer demand for environmentally friendly chemicals as a means of lowering their exposure to harmful substances. (Rushton, 2001)

# CASE STUDY

## SOLID WASTE MANAGEMENT IN MALAYSIA: ISSUES, CHALLENGES AND ITS BUSINESS MANAGEMENT PRACTICES

Waste implies unwanted resources, which create additional costs and environmental issues if they are not well managed. There is a need to help Malaysia, as a developing country, to implement a good waste management system, especially in managing the issues and challenges related to the collection, transportation, handling of solid waste and the issues of public cleanliness. A good waste management system will ensure both sustainable and environmentally friendly operations by solid waste management companies. Hence, the study proposes that the primary purpose of having business management practices, which serve as an integral part of strategic and operation management, is to help organizations make appropriate decisions while dealing with risks and challenges.

To ensure sustainability and remain competitive, organizations must ensure that their resources are managed effectively, which can be achieved by adopting appropriate business management techniques. It is recognized that business management practices are important to the success and sustainability of an organization. In Malaysia, almost all profit-oriented companies are currently facing challenging times, especially with the increase in cost due to the current global recession. Hence, there is a need to use sound business management practices in tackling issues and challenges for any type of organization in a developing country such as Malaysia.

### *Issues and Challenges in Solid Waste Management Companies*

Waste management has become a very interesting and controversial issue in the past decade or so in Malaysia. The Malaysian Government has been taking many initiatives to enhance the waste management system in Malaysia to protect the environment and sustain the quality of life experienced by its citizens. Nevertheless, most of the research conducted thus far emphasizes the general issues of waste management in Southeast Asia and the methods in use by different countries, both developed and developing countries, to improve their waste management systems.

The Malaysian Government, in 2007, approved a Bill on Solid Waste and Public Cleansing Management (Act 672) which is the guiding framework for the solid waste concessionaires to provide their services. However, the provoking question is how effective and how efficient the concessionaires have been in managing their services. What are the problems and difficulties they face and how are those problems handled by the concessionaires' management. To answer these questions, we need to investigate the business management techniques used in providing the proposed solutions to the relevant management issues that lead to the sustainability of solid waste management companies.

The management of waste in Malaysia is the responsibility of the Ministry of Housing and Local Government (MHLG). However, to handle municipal solid waste (MSW) through an integrated management system, Malaysia initiated privatization in 1993 and granted concessions to four concessionaires. These organizations oversee the collection and transportation of waste from residential and commercial sites to final disposal centers. The main mode of municipal waste disposal and treatment in the country is landfilling. Nevertheless, these landfills are quickly filled up and rapid urbanization inhibits land acquisition for new landfill sites. Recently, new problems from both active and closed landfills due to water source pollution have forced the government to form a special Cabinet Committee to propose a more comprehensive waste management structure for the country, especially in densely populated regions. One of the recommendations is incineration as an alternative to landfill for MSW treatment.

Municipal solid wastes (MSW) is often described as the waste that is produced from residential and industrial (non-process wastes), commercial and institutional sources except hazardous and universal wastes, construction and demolition wastes, and liquid wastes (water, wastewater, industrial processes). Waste management involves the collection, processing or disposal, treatment, recovery, management and monitoring of waste materials. Public Cleaning involves cleaning the public streets, public places, public toilets, drains, public markets/hawking centers, illegally dumped waste, beaches, grass in public places, and animal carcasses. The increase in human population has increased the amount of waste being generated. If solid waste is not managed properly, the citizens will be exposed to environmental risks such as flash floods, blocked drainages, polluted water from waste dumps and disposal sites, and polluted water due to leachate flowing from waste dumps and disposal sites. The nature and composition of solid waste varies between areas, and this has implications for handling and disposal.

Among the main and current issues in waste minimization is the increased amount of solid waste and the increased cost of SWM, lack of basic data on solid waste management (SWM) and recycling, lack of awareness on waste minimization, lack of policies to promote waste minimization and limited information and linkage among the stakeholders. Other than that, Malaysia also encounters problems in terms of technology, manpower, land scarcity and insufficient facilities to cope with the increased rate of waste generation. The main component of Malaysian MSW is organic waste, which contributes approximately 40% of the total waste stream. This has implications for the selection of containers, vehicles and systems that are needed to be able to cope with such high-density waste.

### *Fresh Air Company Background*

Fresh Air was appointed by the State Government to manage solid waste and public cleansing in December 2009. Later, on 19th September 2011, the Federal Government signed the privatization agreement with all concessionaire companies including Fresh Air, to manage waste collection and public cleansing. The concession is governed by Act 672 Solid Waste Management and Public Cleanliness Act 2007, and this has been in force since 1st September 2011. Following this, two institutions were established, namely the Department of National Solid Waste Management and the Solid Waste and Public Cleansing Management Corporation.

Owned by two parent companies, Fresh Air operates with nine departments and is headed by the Chief Executive Officer. The various departments include Operation & Planning, Fleet Engineering Management and Safety, Health & Environment, Corporate Communication, Quality Improvement & Technical Audit, Finance & Account, Human Resource & Administration, Management Information Systems, Strategic Planning, Business Development and Risk Management. Each department is headed by their respective Heads of Department (HODs) except for Operation & Planning, Fleet Engineering Management and Safety, Health & Environment departments, which are headed by a Senior Manager. The Operation & Planning Department is further divided into 12 Operation Service Units headed by 12 Service Area Managers. These twelve areas were clustered into two categories: the large and small areas. For the large areas, the locations are FSA1A, FSA2E, FSA3J, FSA3M, FSA3L and FSAIB whereas for small areas, the locations are FSA1C, FSA1D, FSA2F, FSA3H and FSA3G.

Generally, Fresh Air's services cover three categories or schemes. In terms of waste collection, the schemes are further categorized into premises (housing or residential areas), commercial buildings, institutions, industrial areas, governmental and places of worship. Each area or scheme is defined based on the number of premises (ranging from 20,000 to 25,000 premises) or the density of the population. In terms of public cleansing, the scheme is further categorized into types of drains (based on size and rates) and public washrooms. In addition, Fresh Air also offers special services to cater for waste collection and cleansing after special events such as carnivals, feasts, night or Ramadhan markets. Fresh Air also has six sub-contractors serving them, which translates to about 10% of their activities. The concession agreement allows concessionaires to subcontract their work not exceeding 30% to other parties.

Fresh Air is also involved with other concession teams throughout Malaysia to promote public awareness of their existence, together with the Solid Waste and Public Cleansing Corporation or Perbadanan Pengurusan Sisa Pepejal dan Pembersihan Awam (PPSPPA). They have conducted roadshows and programs to promote public awareness, especially on the 3R program (recycle, reuse and reduce). Under Act 672, FreshAir is expected to help preserve Malaysia's environment by having a proper and uniform waste management system that will eventually bring benefits to all. For example, a new ruling starting in September 2015 on solid waste separation was seen as a positive step in creating a more efficient and effective waste management system in Malaysia. This new rule is a way of educating and creating awareness among Malaysians to protect the environment.

### *Business Management Techniques Used in FreshAir*

Given the limited accessibility of data, the study could only rely on the data from the interviews with three key personnel and the initial meeting with the Chief Executive Officer (CEO) and the Heads of Departments (HODs). The aim is to understand the use of business management techniques within the few areas selected for discussion in this paper, i.e., strategic management, general and operational management, as well as quality management. With the understanding of business management practices and their challenges, the study would be a very good resource for the writing of teaching cases and could also lead to a better understanding of effective ways to adapt and improve the implementation of business management theories, particularly on cost management (cost savings) which was raised by the management team of FreshAir during the initial meeting.

### *Discussion and Conclusion*

The aim of the privatization of solid waste management in Malaysia is to resolve the issues of solid waste management faced by local authorities such as finance and cost management, lack of expertise in integrated, effective, efficient and technologically advanced systems, illegal dumping and lack of management skills on disposal and landfill system. The Solid Waste and Public Cleansing Management Act 2007 (Act 672) adapted from best management practices in solid waste management from Japan, Denmark, Switzerland, Germany and the United States is expected to improve and ensure high-quality services in solid waste management in Malaysia. Act 672 also authorized the appointed concessionaires to take over the solid waste management from state government and local authorities to provide recommendations and implementation of policies and strategies pertaining to solid waste management services, as well as promoting participation and awareness among the public. Inadequate budget, low economic status of households and low attitude toward solid waste workers and their jobs are identified as hindering factors of appropriate SWM service.

One of the biggest challenges raised by FreshAir's CEO in the first meet-up with the management team is the increasing costs of operations in public cleansing and waste collection. Disposal of waste through landfilling is becoming the biggest issue in waste management as most of the municipal landfills have been filled up at a very fast rate. To add a new landfill will be difficult as the capital expenditure for a new landfill might reach up to millions of ringgits on average whereas the operating expenditure of landfills could be high due to the increase in population. Zainu, Wan Mohamad Ali and Songip (2015) reported that most solid waste companies or local authorities could not cope with the capital expenditure which could reach RM30 million for a new landfill to be set up or even the operating expenditure of a landfill around RM30-40 per ton on average. Hence, FreshAir would need to work out ways of managing the operations of waste collection and disposal most efficiently and cost-effectively. As described in the previous section, the business development unit of FreshAir has been cost-effective by designing the transmittal station, intermediate treatment facilities and underground bin system, which led to a reduction in the collection and disposal costs. The recommendation of Agamuthu et al., indeed makes sense in practice as FreshAir has designed adapted or localized facilities, vehicles and systems that could cope with high-density wastes.

FreshAir would also need to work together with other appointed concessionaires around Malaysia to improve the design of landfills to include the latest technology in waste treatment, such as leachate treatment systems, gas ventilation systems and waste reduction facilities or treatment before disposal. This is consistent with finding alternatives to landfills for waste treatment as recommended by Sharifah et al.

Delays in collection due to improper ways of dumping waste by households add further complications to the waste collection task. Many households in Malaysia have insufficient knowledge about waste separation practices making the collection task quite problematic. A study carried out by the Solid Waste And Public Cleansing Management Corporation in 2006 found that public awareness on recycling was relatively high. This is consistent with our findings on public awareness. Unfortunately, we believe that though the public is aware of the 3R practices, they are not practising them.

Implementation of a recycling campaign is a way to reduce the waste disposal problem. Recycling is cheaper and a more environmentally friendly alternative than seeking a new landfill site, and capable of extending the lifespan of existing landfills. Besides, the program is more economical by substituting raw materials with used materials, conserves energy, and creates jobs. To ensure successful recycling, people must know how to recycle and be motivated to recycle. For FreshAir to sustain its campaign it would need to balance three elements suggested by the three-legged stool of sustainability, economic, social and environmental. It would need to support and strengthen the coordination between FreshAir and household/public, their employees as well as the concessionaires, local authorities and the federal government. The use of a balanced scorecard is recommended to equip FreshAir with appropriate techniques for managing performance, particularly in cost management. The focus of a balanced scorecard on four perspectives, namely financial, customer, internal business processes and learning and growth would provide a better measurement of how FreshAir manages its activities. The concept of a strategy map in relating the measures for each perspective would trigger immediate action to manage any issues pertaining to its sustainability.

## CLASS ACTIVITY

Ask students the following questions to generate discussion and determine how much they already know about environmental law. Write student responses on the board. Seek at least five responses to write down per question.

1. What does the term "environmental law" mean to you?
2. Can you think of some examples of environmental laws?
3. Why do we need environmental laws? What do they accomplish?
4. Who regulates all these activities and enforces environmental laws?

## SUMMARY

- Waste management laws govern the transport, treatment, storage, and disposal of all manner of waste, including municipal solid waste, hazardous waste, and nuclear waste, among many other types.
- Waste determination is the process by which a particular material is classified as a "waste" subject to regulation.
- United States environmental law concerns legal standards to protect human health and improve the natural environment of the United States.
- In the United States, responsibilities for the administration of environmental laws are divided between numerous federal and state agencies with varying, overlapping, and sometimes conflicting missions.
- Rivers and Harbors Act may refer to one of many pieces of legislation and appropriations passed by the United States Congress since the first such legislation in 1824.
- The Solid Waste Disposal Act (SWDA) is an act passed by the United States Congress in 1965. The United States Environmental Protection Agency described the Act as "the first federal effort to improve waste disposal technology."
- The Resource Conservation & Recovery Act (RCRA) is a law, and it is the framework for managing hazardous and non-hazardous solid wastes. It is one of the most important laws considering the increasing amount of solid waste generated regularly and it also encourages efficient cleanup innovations.
- The National Environmental Policy Act (NEPA) is a United States environmental law designed to promote the enhancement of the environment. It created new laws requiring U.S. federal government agencies to evaluate the environmental impacts of their actions and decisions, and it established the President's Council on Environmental Quality (CEQ).
- The Occupational Safety and Health Act of 1970 is a US labor law governing the federal law of occupational health and safety in the private sector and federal government in the United States.
- The objective of the Clean Water Act (CWA) is to restore and maintain the chemical, physical, and biological integrity of the nation's waters; recognizing the responsibilities of the states in addressing pollution and aiding states to do so, including funding for public owned treatment works for the improvement of wastewater treatment; and maintaining the integrity of wetlands.
- The Toxic Substances Control Act (TSCA) is a United States law, passed by the 94th United States Congress in 1976 and administered by the United States Environmental Protection Agency (EPA), that regulates chemicals not regulated by other U.S.

## REVIEW QUESTIONS

1. What is the US Environmental Policy Act?
2. What is the purpose of the Rivers and Harbors Act?
3. What authority does the Rivers and Harbors Act grant to the federal government?
4. What are the key provisions of the Resource Conservation and Recovery Act (RCRA), which amended the Solid Waste Disposal Act?
5. How does the RCRA define "solid waste"?
6. How does the OSH Act define the responsibilities of employers and employees regarding workplace safety?
7. What are the rights and protections afforded to workers under the OSH Act?
8. How does the Clean Water Act address point source and non-point source pollution?
9. How does TSCA define "chemical substances" and "new chemicals"?
10. What authority does the Environmental Protection Agency (EPA) have under TSCA?

## REFERENCES


1. Cointreau, S. (2006). Occupational and environmental health issues of solid waste management: Special emphasis on middle and lower-income countries. World Bank.
2. Cuming, R., Rocco, T. S., & McEachern, A. G. (2008). Improving compliance with Occupational Safety and Health Administration standards. AORN Journal, 87(2), 347–356; quiz 357–360.
3. Ferronato, N., & Torretta, V. (2019). Waste mismanagement in developing countries: A review of global issues. International Journal of Environmental Research and Public Health, 16(1060). https://doi.org/10.3390/ijerph16061060.
4. Gott, R. (1963). The evolution of the independent British deterrent. International Affairs, 39(2), 238–252. https://doi.org/10.2307/2611300.
5. McFarland, C. K. (1966). The federal government and water power, 1901–1913: A legislative study in the nascence of regulation. Land Economics, 42(4), 441–452. https://doi.org/10.2307/3145402.
6. Michaels, D., & Barab, J. (2020). The Occupational Safety and Health Administration at 50: Protecting workers in a changing economy. American Journal of Public Health, 110(5), 631–635.
7. Neitzel, R. L., Gershon, R. R., McAlexander, T. P., Magda, L. A., & Pearson, J. M. (2012). Exposures to transit and other sources of noise among New York City residents. Environmental Science & Technology, 46, 500–508.
8. Pearson, C., Littlewood, E., Douglas, P., Robertson, S., Gant, T. W., & Hansell, A. L. (2015). Exposures and health outcomes in relation to bioaerosol emissions from composting facilities: A systematic review of occupational and community studies.

Journal of Toxicology and Environmental Health, Part B: Critical Reviews, 18, 43–69. https://doi.org/10.1080/10937404.2015.1009961.

9. Raltz, S., Kozarek, R. A., Kim-Deobald, J., Pethigal, P., & Moorhouse, M. A. (1994). The impact of OSHA regulations on nursing care cost and compliance. Gastroenterology Nursing, 17(3), 106–109.
10. Rushton, L. (2003). Health hazards and waste management. British Medical Bulletin, 68(1), 183–197. https://doi.org/10.1093/bmb/ldg034.
11. Wassersug, S. R., Ercmann, S., Khan, R., & Rummel-Bulska, I. (1990). XIII. Special reports. Yearbook of International Environmental Law, 1(1), 353–386. https://doi.org/10.1093/yiel/1.1.353.
12. Ziraba, A. K., Haregu, T. N., & Mberu, B. (2016). A review and framework for understanding the potential impact of poor solid waste management on health in developing countries. Archives of Public Health, 74(55). https://doi.org/10.1186/s13690-016-0166-4.

CHAPTER 4

# Municipal Solid Waste

## LEARNING OBJECTIVES

- Discuss historical perspective
- Understand environmental and health impacts from mismanagement of municipal solid waste
- Know US laws regulating solid waste management
- Focus on source, composition, and quantity of MSW
- Explain collection and disposal of MSW
- Identify physical and chemical properties of MSW
- Describe landfill
- Focus on bioreactor landfill
- Perform waste audits

## KEY TERMS FROM THIS CHAPTER

Bioreactor landfills
Cholera
Dengue
Incineration
Landfills
Malaria
Carbohydrates
Composting
Environmental protection agency
Land disposal
Lipids
Matter

# INTRODUCTION

State solid is one of the four common states of matter. All waste materials that are thrown away because they are unneeded, or undesirable are collectively referred to as solid waste. In a particular area, it is produced by commercial, industrial, and residential activity. A considerable ability to dispose of various waste types is necessary for this variation.

In a crowded place, each product creates some waste. This is currently one of the most important environmental issues. The World Bank estimates that 2.01 billion tons of municipal solid waste are generated annually worldwide; therefore, addressing the solid waste problem is the responsibility of all of humanity. In the United States alone, the amount of municipal waste generated in 2018 was 292 million tons. First and foremost, there needs to be a shift from consumerism to more responsible habits and a rise in personal waste awareness.

However, managing such a massive amount of waste calls for more than just community involvement and conventional solid waste management. Local governments and municipalities bear primary responsibility for nearly all aspects of solid waste management.

# 4.1. HISTORICAL PERSPECTIVE

## 4.1.1. Early Waste Disposal

Waste was thrown onto unpaved streets and roadways in ancient cities, where they accumulated. The first known law against this practice was not established until 320 BCE in Athens. At that point, Greece and the eastern Mediterranean cities dominated by Greeks started to develop a system for disposing of waste. Property owners oversaw keeping the streets in front of their buildings clean in ancient Rome. However, organized waste collection was limited to events sponsored by the state, like parades. The most basic form of disposal was through open pits just outside the city walls. There was an attempt to move waste outside of cities as the population grew.

During the Middle Ages, there was a downturn in waste collection and municipal sanitation following the fall of Rome. Scavengers were assigned the duty of transporting waste to dumps outside of city walls towards the end of the 14th century. In smaller towns, however, this was not the case, many people continued to throw trash on the streets. Every English city was not obliged to have an official scavenger until 1714. In America, municipal garbage collection started in Boston, New York City, and Philadelphia towards the end of the 18th century. However, waste disposal techniques remained incredibly primitive. For instance, trash gathered in Philadelphia was merely thrown into the Delaware River downstream of the city.

## 4.1.2. Developments in Waste Management

In the latter half of the 1800s, a technological approach to solid-waste management started to take shape. In the US, stronger vehicles were employed to collect and transport waste, and watertight garbage cans were first introduced. The first refuse incinerator was built in England in 1874, marking a major advancement in solid-waste treatment and disposal methods.

At the start of the 1900s, 15% of the largest American cities were burning solid waste. But even then, most of the biggest cities continued to use antiquated disposal techniques like open dumping on land or in water.

In the first part of the 20th century, technological advancements persisted, leading to the creation of pneumatic collection systems, compaction trucks, and garbage grinders. However, by the middle of the 20th century, it was clear that improper incineration and open dumping of solid waste were endangering public health and causing pollution issues. Sanitary landfills were consequently created to take the place of open dumping and lessen the dependency on waste incineration.

Many nations have separate regulations for the disposal of hazardous and non-hazardous waste, which are classified into two categories. To reduce risks to the environment and public health, landfills were constructed and run. To meet strict air quality standards, new refuse incinerators were built with extensive air pollution control devices and designed to recover heat energy from the waste. Instead of focusing on incineration and land disposal, many developed countries' modern solid-waste management facilities now prioritize recycling and waste reduction at the source.

# 4.2. ENVIRONMENTAL AND HEALTH IMPACTS FROM MISMANAGEMENT OF MUNICIPAL SOLID WASTE

Inadequate handling of solid waste, which includes waste from both human and animal sources, leads to the following problems: diseases and infections spread by rodents and other animals; pollution from chemicals emitted from landfills and greenhouse gases; plastic waste; and respiratory ailments.

## 4.2.1. A Spread of Infections and Diseases

With over two billion tons of municipal solid waste produced annually, it is critical to assure that it is disposed of appropriately to reduce negative effects on the general public's health. This is because of the high levels of contamination in the soil, water, and air, which have impact the lives of adults and children alike. Hazardous waste or unsafe waste treatment methods, such as open burning, can have negative effects on workers and other people who are involved in the burning process. Additionally, children and vulnerable adults in the community are more vulnerable to health risks. Inadequate waste collection has been linked to negative environmental effects, marine pollution, and obstruction of water drains. This can lead to floods cholera and other vector-borne illnesses like dengue and malaria can spread as a result. In a 2019 survey, the World Health Organization estimated that the annual production of e-waste was around 54 million tons. They comprised computers, phones, and televisions. It is anticipated that this will increase to 75 million tons by 2030.

Inappropriate organic waste disposal can lead to a fermentation process, which creates the perfect environment for microbial pathogens to "survive and thrive" (Kenekar A. 2021). This created major health risks because people who come into close contact with this kind of solid waste can get infections and chronic illnesses. The consequences of leaving unattended solid waste on roadsides can be dangerous because the waste eventually becomes a breeding ground for rat, mosquito, and cockroach infestations. It is commonly known that rodents can cause malaria, dengue fever, and food poisoning. Therefore, it is essential to dispose of solid waste properly to reduce the number of pests that spread disease and their negative effects on the general public's health.

The people most at risk when solid waste is not disposed of properly using scientific methods are rag pickers and municipal employees. This also applies to situations in which people have been exposed to toxins and pollutants, as these can potentially cause skin irritation and blood infections. A plan for facilitating the sustainable management of organic waste is shown in Figure 4.1. It involves the production of waste, which needs to be properly managed, sorted, stored, and processed. Following collection, processing, recovery, transfer, and transportation, the waste needs to be properly disposed of.

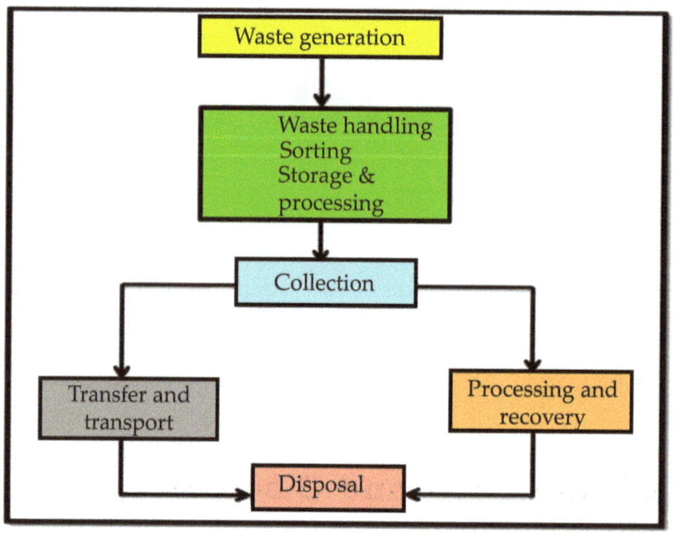


Figure 4.1. Sustainable strategy of organic waste management facilitation.

Source: Somani, P. (2023). Health Impacts of Poor Solid Waste Management in the 21st Century. IntechOpen. doi: 10.5772/intechopen.1002812.

## 4.2.2. Pollution from Chemicals

Research indicates that waste composition is influenced by the local economy. Rich people are probably the ones who use packaged goods, such as paper, plastic, glass, and metal. Therefore, disparities in waste composition have an impact on and reflect waste management practices (Sridevi P, Modi M, Lakshmi M, Kesavarao L. 2012). Certain wastes are classified as hazardous, such as batteries, pesticides, and medications, and they are combined with regular municipal waste. Furthermore, biomedical waste—such as blood-stained clothing, sanitary products, and disposable syringes—should not be combined with municipal solid waste due to the risk of contamination and elevated infection rates. In contrast, fruits and vegetables are classified as organic waste (Bhalla B, Saini M, Jha MK. 2013). However, household use generates most of the organic waste; in contrast, inert waste is produced by road sweeping and demolition produce. However, different cities may have different municipal solid waste compositions (Rawat M, Ramanathan AL, Kuriakose T. 2013).

The negative effects of junkyards on the environment and public health are substantial. Methane is produced in open dumps when anaerobic conditions are created by the breakdown of biodegradable waste. On the other hand, methane gas plays a significant role in causing global warming and enabling explosions and fires. The formation of leaches that build up inside the water reservoirs and the production of unpleasant odors are additional issues that affect health adversely. This is especially noticeable in developing nations like India, where temperatures can soar to 45°C. Furthermore, the uncontrolled burning of waste without proper controls has increased the incidence of respiratory illnesses. As a result, fine particles have formed, which has led to smog and respiratory illnesses. Inadequate waste management has had a major impact on public health, contributing to an increase in infections. This may include bacterial infections, inflammation of the throat and nose, asthma, allergies, dyspnea, and lowered immunity.

Significant volumes of waste are being illegally dumped, even in developed nations like the United Kingdom, according to research. This affects environmental elements and poses a risk to human health. This is because dangerous chemicals have the potential to contaminate the air, soil, and water, harming wildlife and marine life as well as human health. The British Medical Bulletin conducted a study with residents living close to a dumpsite, and the findings showed that both residents living close to and far from the dumpsite were impacted by poor waste management. They mentioned experiencing diarrhea, chest pains, and cholera. While efforts are being made to make nations greener and cleaner, it is imperative to guarantee that improved solid waste management procedures are in place. Fly-tipping, which is thought to have several negative effects on the environment and the public, and unconventional waste disposal companies are still used by some people and businesses (Wade T. 2018). According to an analysis conducted by the local government association in the United Kingdom, costs have increased by 13% over the previous year. The landowner and taxpayer have been impacted for fly-tipping to occur. Furthermore, people still fly-tipp hazardous waste because they are ignorant of the harmful consequences of waste materials. It was discovered that over forty tons of electrical waste were produced annually and unlawfully disposed of. The combination of chemicals meant that both people and wildlife were negatively impacted by exposure.

### 4.2.3. Plastic Waste and Respiratory Diseases

It has come to light that in suburban areas, barely 10% of waste generated is properly collected and disposed of thereby increasing the risks to the environment and public health. This has been connected to acute respiratory infections, which are more common in kids who live close to landfills. Those who experienced increased diarrhea were among the other symptoms mentioned.

For valuable resources to be extracted from solid waste using standard process, resource abstraction needs to be well-organized. Additionally, the position can help those with low incomes. It is possible to find and recycle new materials, energy, and nutrients using this method. Investment in solid waste management is necessary to accomplish this, as it will enable the creation and implementation of innovative R&D projects and assist in the identification of recyclable materials. There are many developing nations, such as India, that lack the structure and systematic procedures necessary to separate waste for household or communal use. The management and disposal of solid waste is the responsibility of the waste producer, or the unorganized sectors. Their lack of procedures causes ineffective sorting and segregation, which breeds instability and produces subpar resource abstraction procedures. This is because the higher financial gain leads to the extraction of only valuable waste products.

Global societies should be made aware of the need to apply suitable techniques for managing solid waste. According to the 2016 MoUHA, landfills are the least preferred solid waste management method because they produce excessive air pollution. People who live close to landfills are at a risk of inhaling harmful pollutants like hydrogen sulfide and methane, which can have a serious negative impact on their health. Additionally, "high toxic emission levels that can put human health at risk" have been revealed using incinerators in solid waste management. On the other hand, the best way to guarantee that solid waste management is carried out efficiently is through composting, which businesses and individuals can do locally. This is thought to lighten the load on local governments, their communities, and governing bodies. People ought to follow the rules for properly sorting their waste. This will enable the organic waste to be separated, processed, and used as fertilizer to go through the composting process. This can be made from leftover food, waste that has accumulated from fruits and vegetables, and paper that is used around the house. Organic waste can be broken down using several different treatment techniques. One of them is available in spray form, allowing waste to be broken down quickly by bacteria and producing rich compost. It is the responsibility of every member of society to safeguard the environment and one another from the damaging effects of solid waste management.

It's crucial to keep in mind that improper disposal of solid waste can increase the risk of cancer, death rates, and problems with reproductive health. For this reason, solid waste management is essential in both urban and rural settings. Legislators, organization leaders, and the public must determine creative, eco-friendly, economic and responsible ways to manage solid waste. Waste can be divided into categories such as municipal solid waste, e-waste, or health waste depending on where it is generated. Solid waste really refers to any materials or trash that has been discarded. In 2019, approximately 17% of e-waste had accurate documentation. The significant increase in e-waste and inappropriate e-waste disposal can be detrimental to development and health, especially in young children. Therefore, it is now essential to dispose all solid waste safely and to categorize it for easier management.

Raising awareness in international societies about the problems associated with solid waste management and the ensuing negative health effects is imperative. People will gain a better understanding of the long-term detrimental effects of solid waste management as well as strategies for building a progressive, sustainable, and safer future for the twenty-first century and beyond through this.

# 4.3. US LAWS REGULATING SOLID WASTE MANAGEMENT

The Environmental Protection Agency (EPA) oversees waste disposal laws in the US. These rules specify how residents and companies are required to dispose their waste. Businesses that do not adhere to the enforced regulations may face steep fines and other consequences. The EPA regulates two primary categories of commercial waste: hazardous and non-hazardous. "Waste that is dangerous or potentially harmful to human health or the environment" is the definition of hazardous waste. The remainder is classified as non-hazardous waste.

The primary federal statute governing the disposal of solid and hazardous waste in the US is the Resource Conservation and Recovery Act (RCRA). Guidelines for the appropriate management of waste, from generation to disposal, are provided by the RCRA. It also specifies specifications for facilities that handle, store, and dispose of waste.

Businesses must make sure that their commercial waste is appropriately managed according to the RCRA. This includes making certain that wastes are transported to approved waste facilities for disposal with the appropriate labels applied. Companies that produce hazardous and solid wastes are also required to maintain documentation of their waste management practices, which they must provide to EPA inspectors upon request.

Different facets of waste disposal in the US are governed by several federal laws in addition to the RCRA. The Comprehensive Environmental Response, Compensation, and Liability Act (CERCLA), for instance, creates liability for those who unlawfully dispose of hazardous materials. The Superfund program, which offers financing for remediation efforts at hazardous waste sites, was also established by the CERCLA.

Wastes and other hazardous materials are subject to transportation regulations under the Hazardous Materials Transportation Act (HMTA). Businesses that ship hazardous materials are required by the HMTA to correctly label their shipments and give their staff training on safe handling techniques.

There are numerous additional state-level laws and rules pertaining to the waste disposal. While some states concentrate entirely on other facets of solid waste management, others have implemented programs that are stricter than the federal requirements. For instance, California's comprehensive solid waste management strategy includes landfill bans, reporting requirements for businesses, and recycling laws.

Over time, there have been significant modifications to the EPA's waste disposal regulations. The Resource Conservation and Recovery Act (RCRA), passed by Congress in 1976, gave the EPA the power to control the disposal of hazardous waste. Then, in 1980, Congress added regulation of nonhazardous industrial solid waste to the RCRA through an amendment. Ultimately, the Hazardous and Solid Waste Amendments (HSWA), passed by Congress in 1984, added more regulations to the disposal of hazardous waste.

Businesses now find it much harder to properly dispose of their waste because of these amendments. For example, companies can no longer dispose of any hazardous waste without first obtaining an EPA permit. They also must store and transport this kind of waste according to tight regulations. And lastly, they must constantly monitor all of their hazardous waste.

The rules set forth by the EPA are essential for safeguarding both the environment and public health. Medical waste is one type of hazardous waste that can contain harmful chemicals and other materials that harm our soil, contaminate our water supplies, and pollute our air. In addition, mishandling the transportation and storage of these wastes may result in mishaps that endanger both humans and animals. For this reason, companies must abide by EPA regulations to guarantee the appropriate disposal of their hazardous waste.

States oversee carrying out their unique laws and regulations regarding the disposal of hazardous waste, even though the EPA establishes broad guidelines in this area. For instance, businesses may or may not need a permit in some states before disposing of hazardous waste. Furthermore, some states might have more lax regulations on the kinds of hazardous wastes that can be disposed of, while others might have stricter regulations. For example, waste disposal laws are extremely strict in some states, like California. Others with far more lenient laws are Texas. Some people, however, fall in the middle.

The reasons for these variations can be traced back to the various approaches taken by the states in disposing of their waste. For instance, California is working on its waste reduction plans to reduce the quantity of garbage that ends up in landfills because the state has set a "zero waste" goal. California has implemented several stringent waste disposal regulations to accomplish this goal.

Texas, on the other hand, approaches garbage disposal in a more relaxed manner. There are no state laws governing the disposal of waste in the state. Rather, the decision about what rules, if any, to implement regarding waste disposal rests with each city or county. There are notable variations in waste management practices between states because of this disparity in strategy. For instance, California recycles far more than Texas does. However, Texas also has a far higher rate of landfill waste disposal.

### 4.3.1. Challenges Businesses Face When Complying with Waste Disposal Regulations

Let's examine each of the difficulties that companies encounter when navigating regulations in more detail.

- Laws are subject to frequent changes. The fact that waste disposal laws are always changing presents one of the main obstacles that businesses must overcome. For businesses, this can make it challenging to stay current and make sure they are compliant. For instance, the Environmental Protection Agency (EPA) published a new rule in 2015 requiring companies to monitor and report waste disposal. Compared to the prior rules, which exempted businesses from tracking or reporting their waste disposal, this represented a significant change.
- There are numerous distinct regulations. Depending on the kind of waste being disposed of businesses must adhere to many different waste disposal regulations, which presents another challenge. For instance, the rules governing solid waste and hazardous waste are not the same. Furthermore, the rules may change based on the state in which a company is situated. Because of this, it may be challenging for businesses to stay on top of all the rules and monitor their compliace.

It can be difficult for businesses to comply with all the various federal, state, and local laws and regulations on the disposal of hazardous waste. Moreover, it can be costly and time-consuming to keep track of all this data. Businesses that violate these laws and regulations risk severe consequences, such as large fines, the cancellation of their business permits, and even jail time.

- The price of conformance. It can be expensive to follow waste disposal regulations. Companies might have to upgrade their facilities or spend money on new equipment. It might also be necessary for them to recruit more employees to abide by the rules. These expenses can mount up and pose a challenge to small enterprises.
- Penalties for non-compliance can be severe. Lastly, companies that disregard waste disposal laws at the risk of paying steep fines. For example, businesses that violate hazardous waste regulations may face fines from the EPA of up to $50,000 per day. These penalties can strain a company's finances as they quickly mount up.

It can be difficult for businesses to follow waste disposal regulations, but it's crucial to stay informed about any updates and assure compliance. It's advisable to avoid non-compliance at all costs as fines can be steep.

# 4.4. SOURCE, COMPOSITION, AND QUANTITY OF MSW

Municipal solid waste (MSW) represents a significant environmental challenge. Waste management is typically the responsibility of the municipality. They must offer the occupants a productive and efficient system. However, they frequently deal with several issues that go beyond what the municipal authority can manage when it comes to MSW. Financial resources, a lack of organization, and complexity are the main causes behind such issues.

There are notable differences in the MSW composition between municipalities and between countries. This variance is primarily influenced by lifestyle, economic conditions, waste management laws, and industrial structure. Determining the proper handling and management of municipal solid waste depends heavily on both the quantity and composition of the waste. Such data is necessary and helpful to establish the municipality's solid waste to energy conversion facility. Engineers and scientists can determine the suitability of MSW as fuel based on its elemental composition and calorific value. In the interim, this data will be useful in forecasting the composition of gaseous emissions. After that, energy conversion technologies like gasification and incineration are applied to this MSW. However, care should be taken to consider any potentially dangerous materials present in the ash. In this regard, the waste's composition will offer important insight into the material's suitability for composting or the biological conversion of waste into gas for biogas production.

In the meantime, the composition of MSW is greatly influenced by time. One important factor influencing the amount of recyclable material, especially the organic contents, is the MSW's biodegradation over time. According to EPA estimates, 254 million tons of MSW were generated in the US in 2013. (M. Sujauddin, Huda, Rafiqul., 2008)

The composition and classification by material of such MSW is given in Figure 4.2.

Wastes from households or cities are typically produced by a variety of human activities occurring in various locations. According to several studies, households account for 55–80% of the municipal solid waste generated in developing nations, with markets and commercial areas accounting for the remaining 10-15%. The latter is made up of erratic amounts produced by streets, institutions, businesses, and numerous other sources (C. Valkenburg, C.W. Walton, B.L. Thompson, M.A. Gerber, S. Jones, D.J. Stevens, 2008). Solid waste originating from these sources is typically very diverse.

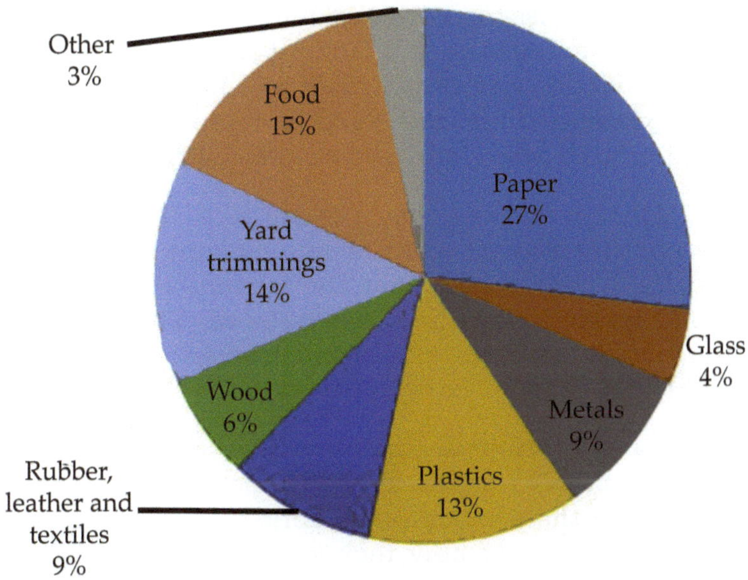


Figure 4.2. Composition and classification (by material) of MSW generated by the United States in 2013.

Source: U.S. Environmental Protection Agency, Wastes – Non-Hazardous Waste – Municipal Solid Waste, 1200 Pennsylvania Ave., N. W. Washington, DC 20460, U.S.A. (2013).

As a result, their physical and chemical makeup varies according to where they came from. Yard trash, food waste, plastics, wood, metals, papers, rubber, leather, batteries, inert materials, textiles, paint containers, demolition and building materials, and numerous other hard-to-classify items make up their composition. The main obstacle to sorting and using such generated solid waste as material is its heterogeneity. Thus, before beginning any significant treatment procedure, these wastes must be properly sorted and fractionated. One of the most significant and time-tested techniques in solid waste management is the sorting and separation of these wastes to provide information on the quality of the separated fractions for any prospective uses. Nevertheless, the success of any design for solid waste segregation depends mainly on public awareness and the active participation of such waste generators in the different communities (i.e., how they follow the fundamental principles of waste sorting and separation).

Solid waste generation (SWG) is a global problem that should worry everyone, but it is especially so in urban areas. Due to the significant environmental pollution caused by the large amounts of SWG, most developing countries consider this type of SWG to be one of their most difficult problems. Urban cities and their growing solid waste generation have a significant impact on basic services like water supply, transportation infrastructure, waste management, and sanitation-related issues.

Numerous studies revealed that one of the main issues facing urban cities and areas is the collection, storage, transportation, and ultimate disposal of solid waste. Cities in many developing nations, as well as in East and North Africa, are dealing with the same grave SWG-related issues.

The primary cause of these issues is the underdeveloped economy in these regions, which also explains the low success rate in solid waste management. Because of their limited resources and conflicting priorities, most of these developing nations struggle with solid waste management. Thus, the SWG is undoubtedly one of the significant issues that many cities around the world are dealing with.

In the meantime, other socioeconomic factors like the average family size, the number of rooms, the monthly income, and the employment status affected the SWG and composition. Additionally, it was stated that the community's social activities and the composition of the solid waste were directly correlated. The composition and quantity of solid waste in households are also influenced by various factors, such as shifts in source-sorting behavior and product consumption.

Factors regarding society, culture, economy, law, politics, the environment, and available resources are the primary concerns that impact MSW management worldwide. The impact and influence on the sociocultural fabric and local economy should thus be considered before implementing any new technology for MSW management and SWG.

The quantity and makeup of MSW have also changed due to shift in people's consumption patterns and the quick development of technology. The amount of MSW produced annually per person in 32 European countries between 2001 and 2010 was examined by the European Environmental Agency, which discovered that waste production rose in 21 of those countries and fell in 11 others. The study also looked at how much waste was produced in 26 different nations between 2001 and 2008; it discovered that six of those countries had less waste than that year. Accordingly, the quantities and types of waste differed between nations, as well as between regions and even within a single city, depending on the previously mentioned variables, such as people's usage patterns.

## 4.4.1. Food Solid Waste

The vast amounts of waste generated globally can provide a sustainable and significant source for certain industrial chemicals. Food waste and residues, including kitchen scraps, trash, and swill, are generally referred to as solid food wastes and byproducts. Food processing, cooking, distribution, production, and consumption all result in the production of these wastes. Nonetheless, there are significant differences in the definition of food waste between different cities and nations. "Raw or cooked of any food substances that are discarded, or intended or required to be discarded" is the definition of food waste in the European Union. On the other hand, the U.S. Environmental Protection Agency defines food waste as "Un-eaten food and food preparation wastes from residences and commercial establishments including restaurants, grocery stores, and produce stands, institutional cafeterias and kitchens, as well as industrial sources such as employee lunchrooms." Moreover, the United Nations recognizes "food loss" and "food waste" in different ways. Food losses are defined as a decline in food quantity and/or quality. In contrast, food losses resulting from the actions of retailers and/or customers are referred to as "food waste." Food waste, on the other hand, also comprises of wasted food items and uncooked raw materials, as well as edible items from the wet market or supermarket.

# 4.5. COLLECTION AND DISPOSAL OF MSW

One of the mandatory responsibilities of municipal authority is the collection, transportation, and disposal of solid waste.

## 4.5.1. Collecting and Transporting

Maintaining the environment's quality, safety, and public health depends on proper solid-waste collection. It is a labor-intensive process that makes up about 75% of the overall cost of solid waste management. The task is typically assigned to public employees, but occasionally it is more cost-effective for private businesses to perform it under contract with the municipality or for private collectors to be paid by individual homeowners. Every collection vehicle has a driver and one or two loaders serving it. Usually, these are enclosed trucks that can carry up to 30 cubic meters (40 cubic yards) of material. One can load from the side, back, or front. The amount of waste in the truck is reduced by compaction to less than half of its loose volume.

Choosing the best collection route is a challenging task, particularly in large, densely populated cities. A route that maximizes labor and equipment efficiency is called an optimal route, and choosing one necessitates the use of computer analyses that consider each of the numerous design variables present in a big, intricate network. Climate, haulage distance, service type, and frequency of collection are examples of variables. Due to low population densities and high unit costs, collecting trash in rural areas can pose unique challenges.

Food waste decomposes quickly, so refuse collection usually occurs at least once a week. Garbage disposals, or garbage grinders, can reduce the quantity of waste that ends up in a home's refuse. Sewerage systems are burdened more by ground garbage, although this is usually manageable. Nowadays, a lot of towns run source separation and recycling programs, where businesses and homeowners' separate recyclables from trash and put them in different containers to be collected. Furthermore, certain localities have recycling drop-off locations where citizens can bring their recyclables.

### *4.5.1.1. Transfer Stations*

One or more transfer stations might be required if the waste's final location is far from the community where it is generated. A transfer station is a central location where waste is merged into bigger vehicles, like tractor-trailer units, from several collection vehicles. Approximately 76 cubic meters (100 cubic yards) of uncompacted waste can be transported by open-top trailers to a local processing or disposal facility. Trailers like closed compactors are also available, but they need to have ejector mechanisms installed. Many collection trucks empty into the transport vehicle directly in a direct discharge station. At a storage discharge type of station, waste is first dumped into a platform or storage pit, and then the solid waste is pushed or hoisted into the transport vehicle using machinery. Over 500 tons of waste can be handled daily by large transfer stations.

## 4.5.2. Solid-waste Treatment and Disposal

Municipal solid waste can be treated after it is collected to lessen the overall amount and weight of waste that needs to be disposed of eventually. Waste undergoes treatment, which modifies its form and facilitates handling. In addition, it can be used to recover specific materials and thermal energy for recycling or other uses.

### *4.5.2.1. Incineration*

- Furnace Operation

Although it emits greenhouse gases, burning is a very efficient way to reduce the weight and volume of solid waste. In contemporary incinerators, the waste is burned under strictly regulated conditions inside a furnace that was built to specification. When oxygen and the combustible part of the waste mix, the main gases released are carbon dioxide, water vapor, and heat. An inert residue of ash, glass, metal, and other solid materials known as bottom ash is left behind after incinerating waste, which can reduce the volume of uncompacted waste by more than 90%. The incinerator airstream carries fly ash, a finely divided particulate matter, and the gaseous byproducts of incomplete combustion. Soot, dust, and cinders make up fly ash. Fly ash and other gaseous byproducts need to be eliminated from modern incinerators before they are released into the atmosphere, which calls for the installation of sophisticated emission control systems. These gadgets include electrostatic precipitators, acid gas scrubbers, and fabric baghouse filters. Typically, fly ash and bottom ash are mixed and dumped in a landfill. If toxic metals are discovered in the ash, it needs to be treated as hazardous waste.

Municipal solid-waste incinerators are built to hold and burn waste indefinitely. Enough room is provided for storing waste for approximately a day in a deep refuse storage pit, also known as a tipping area. A crane fitted with a bucket or grapple device lifts the trash out of the pit. After that, it is discharged onto a charging grate or stoker after being placed into a hopper and chute above the furnace. By shaking and moving waste through the furnace, the grate lets air flow around the material that is burning. Although rotary kiln furnaces and vertical circular furnaces are also available, modern incinerators are typically constructed with rectangular furnaces. Refractory bricks, which can tolerate high combustion temperatures, are used in the construction of furnaces.

In a furnace, there are two stages of combustion: primary and secondary. Primary combustion involves driving off moisture before igniting and volatilizing the waste. Odors are eliminated and the quantity of fly ash in the exhaust is decreased through the oxidation of the residual unburned gases and particulates during secondary combustion. Sometimes fuel oil or auxiliary gas is burned to initiate the primary combustion when the waste is extremely moist.

Air needs to be well mixed with the burning waste to supply enough oxygen for both primary and secondary combustion. Air is either admitted to the area above or supplied through openings beneath the grates. To achieve good combustion efficiency, the plant operator must determine the relative amounts of this under-fire air and overfire air. Mechanical forced-draft fans or a natural draft from a tall chimney can both keep the air flowing continuously.

- Energy Recovery

Depending on the amount of paper in it, the energy value of refuse can be as high as one-third that of coal. The heat released during incineration can be recovered by using a furnace with a refractory lining that is connected to a boiler. Reusing the energy content of the waste is made possible by boilers, which transform the heat of combustion into steam or hot water. This type of incinerator recycling heat energy is known as a waste-to-energy plant. For energy recovery, a water-tube wall furnace may also be utilized instead of a separate furnace and boiler. Vertical steel tubes are placed close enough to form continuous wall sections lining the furnace. To lessen heat loss, the exterior of the walls is insulated. In addition to helping to regulate combustion temperatures without the need for excessive air, the water circulating through the tubes absorbs heat to produce steam and lowers the costs associated with controlling air pollution.

Waste-to-energy facilities can run on refuse-derived fuel or mass-burn fuel. With a mass burn system, all the waste is burned without any prior preparation or treatment. Before burning, combustible wastes like glass and metal are separated from non combustibles by a refuse-derived fuel system. Cogeneration, the process of producing both steam and electricity at a plant, is possible if a turbine is installed there.

Because waste-to-energy systems require specialized equipment and controls, highly skilled technical personnel, and auxiliary fuel systems, they are more expensive to build and run than standard incinerators. However, most of the additional costs are offset by the sale of the generated steam or electricity, and recovering heat energy from waste is a practical solid-waste management solution from both an engineering and an economic standpoint. In the US, waste-to-energy plants constitute about 80% of municipal waste incinerators.

#### *4.5.2.2. Composting*

Composting, a biological process in which the organic component of waste is allowed to break down under carefully monitored conditions, is an additional method of handling municipal solid waste. The organic waste material is metabolized by microbes, which can cause a 50% reduction in volume. The stabilized product is referred to as humus or compost. It can be used as a mulch or soil conditioner and has a texture and smell like potting soil.

One way to process and recycle waste and sewage sludge simultaneously is through composting. Composting is expected to become more common as landfill options and solid-waste incineration are restricted by stricter environmental regulations and siting issues. Sorting and separation, size reduction, and digesting of the waste are the steps in the process.

- Sorting and Shredding

Sorting and separating processes are used to keep the decomposable materials in trash separate from glass, metal, and other inorganic objects. These are done mechanically, taking advantage of variations in the refuse's size, density, and magnetic qualities, among other physical attributes. By reducing the size of the waste materials, shredding or pulverizing produces a homogenous mass of material. Rotary shredders and hammer mills are used to accomplish it.

- Digesting and Processing

Composting of ground-up waste can be done in an enclosed mechanical facility or an open window. Long, low piles of trash are known as windrows. Every few days, they are mixed or turned to give the microbes that are breaking down the organic matter some air. The full breakdown of the waste may take five to eight weeks, depending on the moisture levels. An active compost pile reaches temperatures of approximately 65°C (150°F) due to the metabolic activity of aerobic bacteria, which kills any pathogenic organisms present in the waste material.

Composting in open windows requires a sizable amount of land. The amount of land needed for enclosed mechanical composting facilities can be decreased by roughly 85%. One or more closed tanks or digesters with rotating vanes that mix and aerate the shredded waste are used in mechanical composting systems. It takes roughly a week for the waste to digest completely.

It is necessary to process digested compost before using it as a soil conditioner or mulch. Processing entails granulating or pelletizing, screening, and drying. The most significant obstacle to the success of composting as a waste management option is the increase in the market value of compost that results from these actions. Due to transportation costs and competition from inorganic chemical fertilizers, the demand for digested compost in agriculture is typically low.

### *4.5.2.3. Sanitary Landfill*

The most popular method of managing municipal solid waste is land disposal. Sanitary landfills are disposal sites that are carefully chosen, designed, built, and run to protect the environment and public health. Refuse can be safely dumped there. The fact that the buried waste is never in contact with groundwater or surface water is one of the most crucial aspects of landfilling. The minimum distance required by engineering design is the distance between the seasonally high groundwater table and the landfill's bottom. A network of groundwater monitoring wells and an impermeable liner or barrier at the bottom are necessities for most newly constructed landfills. To prevent precipitation or surface runoff from getting near the buried waste, completed landfill sections need to be covered with an impermeable material. Layers of clay soil, flexible plastic membranes, or a combination of the two can be used for bottom and cap liners.

#### 4.5.2.4. Constructing the Landfill

The refuse cell is the fundamental component of a sanitary landfill. This is a restricted area of the property where waste is layered thinly and compacted. It is possible to compact multiple layers on top of one another down to a maximum depth of approximately 3 meters (10 feet). The compacted waste makes up roughly 25% of the loose volume that it originally had. To get rid of smells, insects, and rodent issues, the waste is covered with a layer of soil after each workday. Thus, the daily volume of compacted waste and soil cover is contained in one refuse cell. A lift consists of several nearby waste cells, and eventually, a landfill may have two or more lifts stacked on top of each other. It is also possible to cover the final cap of a finished landfill with a topsoil layer that encourages to growth of vegetation.

Daily cover soil can be brought in and piled from off-site sources, or it can be available on the spot. Several kinds of heavy equipment, like dozers with rubber tires or crawler tractors, are employed to spread and compact the dirt and trash. To achieve high-density compaction of the waste, large steel-wheeled compactors may also be used.

A new landfill's area and depth are precisely marked out, and the ground is ready for the construction of any necessary leachate collection systems and liners. To shield a plastic liner from landfill vehicles, at least 30 cm (12 inches) of sand must be carefully spread over it. The trench method of construction can be used at locations where excavations below grade are possible. When topography or groundwater conditions make this impractical, the area method can be used, creating a mound or hill that rises above the original ground. Soil typically needs to be transported to the site from another location because the area method does not involve any ground excavation. Variations on the area method can be used in situations where a landfill site is situated in a ravine, valley, or on sloping terrain. Eventually, the finished landfill becomes part of the surroundings.

- Controlling By-products

In a landfill, organic material breaks down due to anaerobic microbial activity. In most cases, complete decomposition takes more than 20 years. Methane gas is one of the byproducts of this breakdown. Methane is a strong greenhouse gas that is toxic and explosive when it gets diluted in the atmosphere. It can also travel great distances through porous soil layers, and hazardous situations can occur if it is allowed to build up in basements or other small spaces. Methane movement in contemporary landfills is regulated by gas-venting systems and impermeable barriers. Methane gas from landfills is sometimes extracted and used as fuel, either alone or in combination with other gases to create biogas.

In sanitary landfills, another byproduct of decomposition is leachate, a highly contaminated liquid. The majority of leachate is produced when runoff enters the refuse cells and meets the decaying trash. Sewage leakage onto the ground surface or into groundwater can cause major pollution issues for the environment, even to the point where drinking water supplies may get contaminated. Leachate can be managed by intercepting surface runoff to keep it out of the landfill and by putting impermeable liners or barriers between the waste and the groundwater. Leachate collection and treatment systems as well as groundwater monitoring wells should be installed at new landfill locations.

- Importance in Waste Management

Sanitary landfills are typically the most affordable option for disposing of non-recyclable waste in communities where suitable sites are available. Finding locations with sufficient capacity, accessibility, and environmental conditions is getting harder and harder, though. However, landfills will always be essential to the management of solid waste. Solid waste cannot have every component recycled; residuals from incineration and other treatment procedures will always need to be disposed of underground. Furthermore, low-quality land can be improved by landfills. Some towns turn legally closed landfills into parks, playgrounds, or golf courses for recreational purposes.

#### *4.5.2.5. Recycling*

Recycling includes separating, recovering, and reusing parts of solid waste that might still be valuable economically. The recovery and reuse of heat energy is one sort of recycling; this technique is covered separately in the context of incineration. Since composting recovers the organic components of solid waste for use as mulch or soil conditioner, it is also regarded as a recycling process. There is yet more potential for reuse with these waste materials. This section discusses the recovery of these materials, which include paper, metal, glass, plastic, and rubber.

- Separation

All materials need to be sorted and separated from raw waste before they can be recycled. Either a central processing facility or the waste's source can carry out separation. Individual citizens perform source separation, or curbside separation, by gathering newspapers, bottles, cans, and trash separately and setting them out for collection at the curb. "Commingling" of recyclables other than paper (glass, metal, and plastic) is permitted in many communities. In any scenario, the cost of municipal source-separated waste collection exceeds that of regular refuse collection.

Recyclable materials can be separated from trash at centralized mechanical processing plants in place of source separation. Experience has shown that the presence of moist trash and broken glass contaminates recovered recyclables from such facilities, lowering their quality. It is now accepted that the best course of action is for residents to sort their trash into a small number of categories, such as garbage and other nonrecyclables, mixed metals, glass, and plastics, newspapers, magazines, and other wastepaper. Separated from the other waste, newspapers, other paper wastes, and mixed recyclables are gathered and processed at a centralized material recycling facility, also known as an MRF (pronounced "murf" in waste-management terminology). Approximately 300 tons of recyclable waste can be processed daily by a modern MRF.

Mixed recyclables are placed onto a conveyor at a typical MRF. An electromagnetic separator removes steel cans (also known as "tin" cans, which are steel with a thin coating of tin), and the residual material is then passed over a vibrating screen to remove broken glass. The conveyor then travels through an air classifier, which divides lighter glass containers from the heavier aluminum and plastic containers. Glass is manually sorted by color, and an eddy-current separator separates aluminum cans from plastics by pushing the aluminum away from the conveyor belt.

- Reuse

Crushed recovered broken glass can be mixed with asphalt to create pavement. Crushed color-sorted glass is sold to glass makers as a cullet, a necessary component for glassmaking. Aluminum is compacted or baled so that smelters can reuse it, and steel cans are shipped in bales to steel mills as scrap. Aluminum has the highest value of any recyclable material, despite being one of the smallest components of municipal solid waste. Plastic recycling is difficult, primarily due to the wide variety of polymeric materials used in their manufacture. The only goods that can be made with mixed thermoplastics are inferior goods like "plastic lumber."

Old newspapers are manually sorted on a conveyor belt within the paper stream to remove mixed and corrugated papers. After that, they are loosely loaded or baled and transported to paper mills where they are recycled into more newspapers. To be sold to tissue mills, mixed paper is separated from corrugated paper. The market for recycled paper has expanded as more processing facilities have been established, even though the procedures of pulping, de-inking, and screening wastepaper are typically more costly than producing paper from virgin wood fibers.

In a process known as revulcanization, rubber is occasionally recovered from solid waste and shredded, reformed, and remolded; however, the resulting material is typically not as strong as the original. In addition to being marketed as an outdoor mulch, shredded rubber can be added to artificial turf and asphalt pavements. Tyre playgrounds are places where kids can use discarded tyres as swings and other play structures.

Identifying appropriate markets and applications for any solid waste material to be recycled presents the biggest challenge. Recycling alone won't address the escalating issue of disposing of and managing solid waste. Some solid residue that is completely worthless and unusable will always need to be disposed of in the end.

# 4.6. PHYSICAL AND CHEMICAL PROPERTIES OF MSW

Depending on their sources and types, solid wastes have different physical and chemical compositions. Because of the relative amounts of degradable and non-degradable components, the moisture content, and the unique characteristics of the biodegradable element, the type of waste that is deposited in a landfill will have an impact on the production and composition of gas and leachate. The bulk gases and the trace components will be impacted by the waste composition.

## 4.6.1. Physical Characteristics

For equipment selection, operation, analysis, and disposal facility design, information and data regarding the physical properties of solid wastes are crucial. The following are examples of the necessary data and information.

- Density
- Moisture Content
- Particle size & distribution
- Field Capacity
- Permeability of compacted wastes

### *4.6.1.1. Density*

Solid waste density, i.e., its mass per unit volume (kg/m$^3$), is an important consideration in the SWM system's design e.g., sanitary landfill construction, storage, kinds of vehicles used for collection and transportation, etc. In other words, optimal density compaction of wastes is necessary for a landfill to operate efficiently. Any typical compaction equipment can reduce waste volume by 75%, increasing the initial density from 100 kg/m3 to 400 kg/m$^3$. Put another way, a waste collection truck can transport four times as much waste when it is compacted as when it is not. A high initial waste density makes it impossible to achieve a high compaction ratio; the highest compaction ratio that can be attained is 1.5:1.

Significant density changes happen naturally as waste travels from the point of origin to the disposal. These changes can be attributed to decomposition, vibration in the collection vehicle, handling, scavenging, and weather-induced wetting and drying. Keep in mind that:

- One of the key physical characteristics of solid waste is soil cover, which helps to contain the waste.
- There is a limit to the density, and the conservative estimate of in-place density for waste in a sanitary landfill is between 600 and 1200 kg/m3.
- Increasing the moisture content of the waste has a negative effect because dry density decreases at higher moisture levels.

#### *4.6.1.2. Moisture Content of Solid Waste*

The weight of water (wet weight – dry weight) divided by the total weight of the wet waste is called as moisture content. It is among the crucial physical attributes of solid waste.

#### *4.6.1.3. Particle Size of Solid Waste*

Because it is crucial to the design of mechanical separators and shredders, the size distribution of the particles in the waste stream must be measured. The size distribution analysis results are typically expressed in the same way as the soil particle analysis results. They are represented as a plot of the percentage of particles (mm) that are smaller than a specified value. When using mechanical tools like magnetic separators and trammel screens, the size and distribution of the waste's component are crucial for material recovery. For instance, large, heavy ferrous objects might be too heavy to be separated using a magnetic belt or drum system.

#### *4.6.1.4. Field Capacity of Solid Waste*

The total amount of moisture that a waste sample can hold against the force of gravity is known as the field capacity of MSW. It is a crucial physical feature of solid waste since leachate, which can cause serious issues in landfills, is formed when water flows beyond the field's capacity. The amount of pressure applied and the wastes' level of decomposition both affect field capacity.

### 4.6.2. Chemical Characteristics

Chemical property information is crucial for assessing different options for processing and recovery. Waste is usually understood to consist of both combustible and non-combustible materials. We need to know the chemical makeup of solid waste if it will be used as fuel or for any other purpose.

#### 4.6.2.1. Lipids

They fall under the category of fats, oils, and grease. The main sources of lipids in the trash are fats and cooking oil. Because lipids have high heating values—roughly 38,000 KJ/Kg, or waste with a high lipid content can be used to recover energy. Lipids contribute to the liquid content during waste decomposition because they turn into liquids at temperatures that are just a little bit above ambient. Although they are biodegradable, their rate of biodegradation is sluggish due to their low water solubility.

#### 4.6.2.2. Carbohydrates

First and foremost, these come from foods high in cellulose and starch. These easily break down into methane, water, and carbon dioxide. Carbohydrate decomposition draws flies and rats, so it shouldn't be left out for extended period of time.

#### 4.6.2.3. Proteins

These are the substances that have organic acids with amino groups along with carbon, hydrogen, nitrogen, and oxygen. They are mostly found in food and garden wastes, but when they partially break down, amines are produced, which give off bad odors.

#### 4.6.2.4. Natural Fibers

These natural products have lignin and cellulose, which are comparatively resistant to biodegradation. They can be discovered in food waste, paper goods, and yard waste. Paper is nearly 100% cellulose, cotton is over 95%, and wood products are over 40–50%. These are extremely flammable goods that are best burned. Oven-dried paper products have a calorific value between 12,000 and 18,000 KJ/kg.

#### 4.6.2.5. Synthetic Organic Materials

Plastics now make up 1–10% of solid waste, which is a significant amount in recent years. Their presence in the waste is objectionable due to their high resistance to biodegradation. Reducing this component at disposal sites is currently a top priority. Plastics are ideal for incineration because of their high heating value of roughly 32000 KJ/kg. Nevertheless, among the plastics, burning Polyvinyl chloride (PVC) releases acid gas and dioxin. It has been established that the trace gases released when plastic burns are carcinogenic.

# 4.7. LANDFILL

Waste materials are disposed of at a landfill site, sometimes referred to as a tip, dump, rubbish dump, garbage dump, trash dump, or dumping ground. Although the systematic burial of waste with daily, intermediate, and final covers only started in the 1940s, landfills are the oldest and most popular place to dispose of waste. Historically, trash was merely piled up or dumped into pits; this is referred to as a midden in archaeology.

Certain landfills are utilized for waste management activities, including short-term storage, consolidation, and transportation, as well as different phases of waste material processing, including recycling, sorting, and treatment. Landfills may experience intense shaking or ground liquefaction during an earthquake if they are not stabilized. The space above a landfill can be reclaimed for different purposes once it is filled.

## 4.7.1. Operations

Operators of well-run landfills for non-hazardous waste meet predefined specifications by applying techniques to:

- Confine waste to as small as possible
- Compact waste to reduce volume

Additionally, they can cover waste (usually daily) with soil layers or other materials like woodchips and fine particles.

When waste collection vehicles arrive at the landfill, staff members may weigh them using a scale or weighbridge and inspect loads to make sure the wastes meet the landfill's waste-acceptance standards. After that, the waste collection trucks travel to the tipping face or working front to unload their contents by using the current road network. Compactors and bulldozers can spread and compact the waste on the working face after loads are deposited. The waste collection vehicles may pass through a wheel-cleaning facility before they exit the landfill boundaries. They return to the weighbridge, if needed, to reweigh without their load. Databases can keep track of the daily incoming waste tonnage statistics that are compiled by the weighing process. Certain landfills may have railroad container handling equipment in addition to trucks. Landfills can be situated at more remote locations thanks to "rail-haul," which eliminates the issues brought on by frequent truck trips.

The compacted waste is usually covered with soil or other materials every day in the working face. Chipped wood or other "green waste," various foam products sprayed on, chemically "fixed" bio-solids, and temporary blankets are examples of alternative waste-cover materials. Before disposing of waste, blankets can be lifted into position at night and taken out the next day. A daily cell is the area that the cover material and the compacted waste occupy daily. Compaction of waste is essential for prolonging the landfill's life. Waste densities are influenced by variables such as waste compressibility, the thickness of the waste layer, and the number of passes the compactor makes over the waste.

## 4.7.2. Sanitary Landfill Life Cycle

Usually, the term "landfill" refers to a sanitary or municipal landfill. Although these facilities were initially established in the early 1900s, their widespread use began in the 1960s and 1970s as a means of doing away with open landfills and other "unsanitary" methods of disposing of waste. Waste is separated and contained in an engineered facility called a sanitary landfill. Sanitary landfills are designed to function as biological reactors, or bioreactors, where microorganisms will gradually convert complex organic waste into less hazardous, simpler compounds. Regulations governing standards and procedures must be followed in the design and operation of these reactors (Figure 4.3).

Figure 4.3. Sanitary landfill diagram.

Source: ByFinonaL-Ownwork,CCBY-SA4.0,https://commons.wikimedia.org/w/index.php?curid=132496173.

In a landfill, aerobic decomposition is typically the initial phase of waste breakdown. Four stages of anaerobic degradation occur after them. Larger organic molecules typically break down into smaller molecules quickly when solid organic material is in the solid phase. The smaller organic molecules start to dissolve and transition into the liquid phase. Subsequently, these molecules undergo hydrolysis, which transforms and volatilizes them into carbon dioxide ($CO_2$) and methane ($CH_4$). The remaining waste remains in both the liquid and solid phases.

Early on, not much material volume reaches the leachate because the waste's biodegradable organic matter is rapidly losing volume. As the concentrations of the more recalcitrant compounds in the leachate rise relative to the more reactive ones, the leachate's chemical oxygen demand also rises. The efficacious conversion and stabilization of the waste are contingent upon the degree to which microbial populations perform in syntrophy, i.e., a collaboration between various populations to meet one another's nutritional needs (Letcher, T.M.; Vallero, D.A., eds.(2019)). A municipal landfill goes through five different stages in its life cycle:

#### 4.7.2.1. Initial Adjustment (Phase I)

There is a lot of molecular oxygen ($O_2$) in the void spaces created by the waste as it is disposed of in the landfill. The $O_2$ content of the landfill bioreactor strata gradually drops with the addition and compacting of wastes. Microbial densities rise along with their populations. Most biodegradation is aerobic, i.e., $O_2$ is the main electron acceptor.

#### 4.7.2.2. Transition (Phase II)

The microbial populations that are already present quickly break down the $O_2$. Conditions in the layers become more anaerobic and less aerobic as $O_2$ levels drop. Since $CO_2$ quickly displaces $O_2$ in the effluent gas, nitrates and sulfates are the main electron acceptors during transition.

#### 4.7.2.3. Acid Formation (Phase III)

Volatile fatty acids (VFAs) rapidly accumulate in the leachate because of the hydrolysis of the biodegradable fraction of the solid waste, which starts in the acid formation phase. As the amount of organic acid increases, the pH of the leachate drops from about 7.5 to 5.6. The decomposition intermediates, such as volatile fuel acids (VFAs), significantly contribute to the chemical oxygen demand (COD) during this stage. Acetic acid ($C_2H_4O_2$), $CO_2$, and hydrogen gas ($H_2$) are the products of the conversion of long-chain volatile organic acids (VOAs). Elevated levels of volatile fruit acids (VOAs) lead to elevated levels of both VOD and biochemical oxygen demand (BOD), which trigger the fermentation process and promote the development of $H_2$-oxidizing bacteria. Due to its completion after the acid formation phase, the $H_2$ generation phase is comparatively brief. The rate at which waste materials decompose and absorb nutrients increases when the biomass of acidogenic bacteria rises. During this phase, metals, which are generally more water soluble at lower pH values, may become more mobile, resulting in an increase in the concentration of metals in the leachate.

#### 4.7.2.4. Methane Fermentation (Phase IV)

The intermediate products of the acid formation phase (e.g., butyric, propionic, and acetic acids) are broken down by methanogenic microbes into $CH_4$ and $CO_2$. The pH of the landfill water returns to neutral as the methanogens break down the volatile organic compounds. As $CH_4$ and $CO_2$ gas production rises, the leachate's organic strength, measured in oxygen demand, falls off quickly. The longest stage of decomposition is this one.

#### *4.7.2.5. Final Maturation and Stabilization (Phase V)*

During the final stage of waste breakdown, the rate of microbiological activity decreases because the availability of nutrients restricts the chemical reactions that occur, e.g., as the amount of bioavailable phosphorus decreases. $O_2$ and oxidized species progressively reappear in the gas wells as $O_2$ percolates downward from the troposphere, causing $CH_4$ production to almost entirely cease. As a result, the leachate's oxidation-reduction potential (ORP) is changed to favor oxidative processes. As organic matter decomposes, the leftover organic materials may gradually transition to the gas phase; i.e., the organic matter is changed into substances that resemble humus.

### 4.7.3. Landfill Gas

In landfills, anaerobic digestion by microbes produces gases. This gas is collected and put to use in a landfill that is properly managed. It can be used for anything from straightforward flaring to producing electricity and using landfill gas. When hazardous gas buildup occurs, workers are notified by landfill gas monitoring. Landfill gas recovery is widespread in certain nations; in the US, for instance, there are currently over 850 landfills with operational systems for recovering landfill gas. (Powell et al., 2015)

# 4.8. BIOREACTOR LANDFILL

In many regions of the world, including the United States and Canada, landfills are the principal location for disposing of waste. It is anticipated that bioreactor landfills will decrease the quantity and expenses related to leachate management, boost the rate of methane (natural gas) production for commercial use, and require less land for landfills. Landfills with bioreactors are subject to monitoring and adjustments of moisture and oxygen levels to accelerate the pace of microbial decomposition.

## 4.8.1. Traditional Landfills and Associated Problems

According to Tammemagi, Hans (1999), landfills are the oldest known method of disposing of waste. Unless there are naturally occurring places available, waste is covered and buried in sizable pits that have been dug out. Bacteria and archaea decompose the waste over several decades producing several by-products of importance, including methane gas (natural gas), leachate, and volatile organic compounds (such as hydrogen sulfide ($H_2S$), $N_2O_2$, etc.).

If released from the cell, methane gas, a potent greenhouse gas, can accumulate inside the landfill and cause an explosion. Leachates are fluid metabolic byproducts of decomposition that include dissolved metallic ions and a variety of toxins. Plants and animals may experience health issues if leachate spills into groundwater. The production of acid rain and smog is linked to volatile organic compounds (VOCs). It is getting harder to locate suitable locations to store waste safely due to the rising volume of waste produced.

## 4.8.2. Working of a Bioreactor Landfill

A bioreactor can be classified as either anaerobic, aerobic, or hybrid if it combines both anaerobic and aerobic processes. To keep the landfill's moisture levels stable, collected leachate that has been supplemented with water is reintroduced in all three mechanisms. To reduce harmful emissions, the microorganisms in charge of decomposition are therefore encouraged to break down at a faster pace.

Air is pumped into the landfill in aerobic bioreactors using a horizontal or vertical pipe system. Methane, leachate toxicity, and the amount of volatile organic compounds (VOCs) decompose more quickly in an aerobic environment. Methane is produced by the landfill far more quickly and early in anaerobic bioreactors with leachate circulation than in conventional landfills. Because of its high concentration and abundance, methane can be used more profitably and requires less time to monitor methane production in the landfill. Hybrid bioreactors allow the landfill's upper sections to undergo aerobic-anaerobic cycles, speeding up the decomposition process, while the landfill's lower sections produce methane. Except for H2S, bioreactor landfills emit fewer VOCs than conventional landfills. Higher amounts of H2S are produced in bioreactor landfills. It is unclear exactly which biochemical pathway controls this rise.

## 4.8.3. Advantages of Bioreactor Landfills

Decomposition is accelerated in bioreactor landfills. More room is available for the disposal of waste as decomposition proceeds and the mass of biodegradable components in the landfill decreases. It is anticipated that bioreactor landfills will accelerate the rate of decomposition and reduce the amount of land required for landfills by up to 30%. Given the limited quantity of landfill space available and the rising volume of solid waste generated annually, bioreactor landfills can offer a substantial means of optimizing landfill space. This is not only more economical, but it's also better for the environment because it requires less land for landfills.

To make sure that no leachate or landfill gases leak into the neighborhood surrounding the landfill site, many landfills are also observed for at least three to four decades. On the other hand, bioreactor landfills should decompose in less than ten years to the point where they don't need to be monitored. Hence, depending on the location later, the landfill land may be utilized for parks or other uses like reforestation. Furthermore, the leachate is filtered by reusing it to moisten the landfill. As a result, processing the leachate takes less time and energy, improving process efficiency.

## 4.8.4. Disadvantages of Bioreactor Landfills

Landfills with bioreactors are a relatively new technology. The initial costs of monitoring bioreactor landfills in development are higher to assure that everything is found and controlled appropriately. This includes smells, gases, and leachate seeping into the ground. Due to an increase in pore water pressure within the waste mass, the bioreactor landfill's increased moisture content may weaken the landfill's structural stability. The increased moisture content of the waste can have an impact on gas collection systems because bioreactor landfills aim to maintain a high moisture content.

# 4.9. WASTE AUDIT

Waste audits offer a thorough examination of the composition of waste, evaluate the levels of contamination, and pinpoint possible areas for cost- and resource-saving improvements in waste management. According to earlier audits, general waste or recyclable materials make up 60% of clinical waste, and 45% of general waste is recyclable.

Using the department's waste audit guidelines ensures that audit data is detailed enough to:

- assess contamination in waste streams
- characterise and quantify waste streams
- identify waste diversion opportunities
- identify source reduction opportunities
- assess the effectiveness of waste management systems and identify ways to improve their efficiency.

Waste audits give you a starting point for reporting trends and comparing your facility to others. Waste audits need an upfront financial outlay, but most of the time, the opportunities for improvement will pay for themselves. Although visual waste assessments offer a broad picture of waste, they do not produce the data needed to create a benchmark, which makes it challenging to determine whether waste management system improvements have been made. As a result, the department suggests conducting audits. Waste audits can be carried out by your waste contractor, or an independent waste auditor hired directly by the health service. They are most beneficial if done every two years.

## 4.9.1. Benefits of a Waste Audit

The advantages of finishing a waste audit will differ based on the objectives of your business, but some advantages are as follows:

- Gathering specific data for waste reporting.
- Evaluating the effectiveness of the system already in place.
- Finding ways to save money on waste costs.
- Establishing the need for additional waste streams and staff education if contamination is found.
- Figuring out whether purchasing certain products that might be wasting money needs to be reduced. For instance, you can be able to modify your supply if you observe that milk is spoiling before it is consumed.
- Meeting certification and ESG standards. (EEA, 2013)

# CASE STUDY

## WASTE TO ENERGY – LANDFILL SITE

### Background

The first engineered landfill site in Ireland, North Kerry Landfill, began accepting waste in 1994. It was also Ireland's first landfill site to be run under an EPA Waste Licence (W0001). It took non-hazardous waste that contained some biodegradable material. To control the landfill gas being extracted from the gas wells, a flare was lit on the property. The breakdown of the waste also produced oxygen, carbon dioxide, and methane. Gas wells were dug into the waste body and linked to a flare system, which caused the gas to burn at temperatures higher than 1,000°C.

### Solution

Kerry County Council determined early in the new millennium that a well-managed gas utilization plant needed to be installed at the site to mitigate against any fugitive emissions and odors from the site and to use the gas produced as an energy resource where possible (Figure 1.0). A landfill gas management system's main goal was to gather all the gas generated by the waste, handle it to reduce emissions and odors from the landfill, and, where practical, provide energy recovery. Kerry County Council intended to meet EPA regulations and realize the financial value of landfill gas by generating electricity from landfills.

The gas utilization plant was identified along with the flare as the key gas infrastructure to:

- minimize the impact on air quality and the effect of greenhouse gases on the climate;
- minimize the risk of migration of landfill gas beyond the perimeter of the site
- minimize the risk of migration of landfill gas into services and buildings on site;
- avoid unnecessary ingress of air into the landfill and thereby minimize the risk of landfill fires;
- minimize the damage to soils and vegetation within the restored landfill area;
- permit effective control of gas emissions; and,
- where practicable, permit energy recovery.

To comply with the terms of the EPA Waste Licence, Kerry County Council also emphasized the necessity of hiring specialists in landfill gas management to oversee the production and use of gas. To use landfill gas from the landfill site to generate electricity, the contractor was in charge of building the generation facility as well as installing, running, and maintaining the electricity generation machinery.

### Benefits of Solution

#### *Environmental*

"[M]ethane is estimated to be 20–30 times more damaging (per molecule) to the global climate due to its greenhouse effect," per the EPA Landfill Manuals. Very small fugitive emissions have come from the North Kerry Landfill site due to the conversion of gas into electricity and the flaring of landfill gas produced at temperatures above 1,000°C. In line with EPA Air Guidance Note 5, this has been verified through questionnaires on volatile organic compounds. The gas utilization plant has consistently run 80–90% of the time at full capacity (330kW) over the years. There has been a decrease in gas production since 2018, but this decrease has not had any negative effects on the environment.

#### *Economic*

Kerry County Council receives a royalty payment each month from the landfill gas produced on the property. This is computed using actual and base energy prices in addition to the electricity generated from the landfill gas intake. The average annual electricity production between 2011 and 2018 was about 1.6 megawatt hours.

#### *Social*

Between 2012 and 2018, the number of complaints from the public about the landfill significantly decreased. Landfill gas management is improved with the installation of a gas utilization plant, with a particular emphasis on any fugitive emissions from the site. Although the site is not close to any residential areas, Kerry County Council has made odor control a top priority. There have been no complaints about smells from the North Kerry Landfill since it was closed to the public.

# CLASS ACTIVITY

## DISAPPEARING TRASH

### *Objective*

Students will learn which types of waste are biodegradable.

### ***Materials***

Garden soil or compost; Markers, crayons or stickers; Containers: yogurt, dixie cups, clear cups or egg cartons.

### ***Test Materials***

Tissue, fabric scrap, cotton, newspaper strips, fruit or vegetable scarp, toothpick, foil, plastic wrap, cardboard, leaves, candy wrappers, rubber and a marble.

### ***Procedure***

- Illustrate the amount of trash each student produces in a day by passing around something that weighs about five pounds. Or what a week's worth of student's trash would look like by having a student step into a trash bag and hold it open.
- Discuss that materials in landfills will decay slowly, if at all. Refer to Tracking Trash.
- Decaying Experiment: Organize students into pairs or teams. Each team will label 4 cups with the materials that they will be testing. Test materials should be placed in labeled cups and covered with soil. Keep soil moist during the experiment.
- Create a student handout or poster, so students can record: the team's testing materials, materials students predict will decay, and have students predict the order in which materials will decay.
- Observations can be recorded once a week or at the end of the month.

### ***Time***

One class period; month to two-month observation

### Conclusion

Compare what the students predicted to what happened with the materials. Did the materials stay the same? Change? Or complete decay? Discuss why certain materials did not decay. Why is it important when we use and throw away materials in the trash? Share ideas about substitutes for non-biodegradable products such as bringing your lunch in a cloth bag, using a thermos or water bottle, and storing leftovers in a reusable container.

### Extension

Demonstrate how much packaging comes with certain products such as Lunchables. Unwrap a product, show or weigh the amount of packaging. Discuss that students should look for products that can be reused, recycled or have minimal packaging.

## SUMMARY


- Poor solid waste management comprising of waste generated by human beings and animal activity, results in the following: a spread of infections and diseases through attracting rodents and other creatures, pollution from chemicals released through landfills and greenhouse gasses, plastic waste and respiratory diseases.
- Waste deemed as hazardous, or the implementation of waste treatment that is unsafe like open burning, can create a harmful impact on employees and other individuals involved in the burning processes.
- The local area economy has an impact on the amount of waste composition, according to research. The use of packaged goods is likely to be utilized by high earners, for example, plastic, glass, metals and paper. Hence, waste management practices are impacted and reflected through waste composition discrepancies.
- Solid waste management requires a well-organized method of resource abstraction so that valuable resources can be found in the waste through a systematic process.
- The waste disposal regulations in the United States are governed by the Environmental Protection Agency (EPA). These regulations dictate how citizens and businesses must dispose of their waste. Failure to comply with the regulations enforced can result in huge fines and other penalties for non-compliant companies.
- The composition of MSW varies significantly from one municipality to another and from country to country. Such variation depends mainly on the lifestyle, economic situation, waste management regulations and industrial structure.
- Sustainable and important sources for certain industrial chemicals can be obtained from the large amounts of the generated waste in the world. Food residues and waste such as kitchen refuse, garbage and swill are described; generally; the by-products and as solid wastes of food. Such wastes are produced from the processing, cooking, distribution, production, and consumption of food.
- Proper solid-waste collection is important for the protection of public health, safety, and environmental quality. It is a labor-intensive activity, accounting for approximately three-quarters of the total cost of solid-waste management.
- Once collected, municipal solid waste may be treated to reduce the total volume and weight of material that requires final disposal. Treatment changes the form of the waste and makes it easier to handle. It can also serve to recover certain materials, as well as heat energy, for recycling or reuse.
- Land disposal is the most common management strategy for municipal solid waste. Refuse can be safely deposited in a sanitary landfill, a disposal site that is carefully selected, designed, constructed, and operated to protect the environment and public health.
- The physical and chemical composition of solid wastes varies depending on the sources and types of solid wastes. The nature of the deposited waste in a landfill will affect gas and leachate production and composition by relative

proportions of degradable and non-degradable components, the moisture content and the specific nature of the biodegradable element.

- A landfill site, also known as a tip, dump, rubbish dump, garbage dump, trash dump, or dumping ground, is a site for the disposal of waste materials.
- Landfills are the primary method of waste disposal in many parts of the world. Bioreactor landfills are expected to reduce the amount of, and costs associated with the management of leachate, increase the rate of production of methane (natural gas) for commercial purposes and reduce the amount of land required for landfills. Bioreactor landfills are monitored and manipulate oxygen and moisture levels to increase the rate of decomposition by microbial activity.

## REVIEW QUESTIONS

1. What are the characteristics of municipal solid waste?
2. What are the three categories of municipal solid waste?
3. Give an example of municipal waste.
4. How is municipal solid waste disposed?
5. What are the environmental impacts of improper municipal solid waste management?
6. How does municipal solid waste affect human health?
7. What is the environmental law for solid waste management?
8. What are the methods of collection and disposal of solid waste?
9. What is a waste audit and why do we need one?

## REFERENCES


1. Bhalla, B., Saini, M., & Jha, M. K. (2013). Effect of age and seasonal variations on leachate characteristics of municipal solid waste. International Journal of Research in Engineering and Technology, 2, 223–232.
2. European Environmental Agency (EEA). (2013). Managing municipal solid waste—A review of achievements in 32 European countries. EEA Report No. 2/2013. Publications Office of the European Union.
3. Kenekar, A. (2021). Negative effects of improper solid waste disposal on human health. Organica Biotech. Retrieved from: https://organicabiotech.com/negative-effects-of-improper-solid-waste-disposal-on-human-health/#:~:text=Water%20and%20Air%20Pollution&text=During%20the%20rainy%20season%2C%20the,like%20cholera%2C%20diarrhoea%20and%20dysentery (accessed on 31 July 2024).
4. Letcher, T. M., & Vallero, D. A. (2019). Municipal landfill. In D. Vallero & G. Blight (Eds.), Waste: A Handbook for Management (pp. 235–249). Elsevier Academic Press.
5. Powell, J. T., Townsend, T. G., & Zimmerman, J. B. (2015). Estimates of solid waste disposal rates and reduction targets for landfill gas emissions. Nature Climate Change, 6(2), 162–165. https://doi.org/10.1038/nclimate2804.

6. Rawat, M., Ramanathan, A. L., & Kuriakose, T. (2013). Characterization of municipal solid waste compost from selected Indian cities: A case study for its sustainable utilization. Environmental Protection, 4, 163–171.

7. Sridevi, P., Modi, M., Lakshmi, M., & Kesavarao, L. (2012). A review on integrated solid waste management. International Journal of Engineering Sciences and Advanced Technology, 2, 1491–1499.

8. Sujauddin, M., Huda, M. S., & Rafiqul, A. T. M. (2008). Waste Management, 28(1688–1695). https://doi.org/10.1016/j.wasman.2007.06.013.

9. Tammemagi, H. (1999). The Waste Crisis: Landfills, Incinerators, and the Search for A Sustainable Future. Oxford: Oxford University Press.

10. Valkenburg, C., Walton, C. W., Thompson, B. L., Gerber, M. A., Jones, S., & Stevens, D. J. (2008). Municipal solid waste (MSW) to liquid fuels synthesis. Pacific Northwest National Laboratory, Richland, WA. https://doi.org/10.1016/j.ejpe.2018.07.003.

11. Wade, T. (2018). Effects of improper waste disposal on human health. Speedy Clearances. Retrieved from: https://www.speedyclearances.com/post/effects-of-improper-waste-disposal-on-human-health (accessed on 31 July 2024).

CHAPTER

# 5 Hazardous Waste

## LEARNING OBJECTIVES

- Illustrate the science of hazardous waste
- Discuss on hazardous waste management
- Find out the waste-to-energy conversion processes

## KEY TERMS FROM THIS CHAPTER

| | |
|---|---|
| Chemical neutralization | Explosive hazardous wastes |
| Flammable liquids | Household hazardous waste (HHW) |
| Ignitability | Industrial fuel |
| Living organisms | Nuclear power |
| Radioactive waste | Toxic fumes |
| Transportation | Violent chemical reactions |

# INTRODUCTION

Any waste material that presents a significant risk to the environment or public health is considered hazardous waste. The chemical, physical, or biological characteristics of this type of waste—which may include toxicity, corrosivity, reactivity, or ignitability—define its classification. Hazardous waste can come from a variety of places, such as homes, businesses, hospitals, and agricultural operations. Hazardous waste is distinguished by its capacity to cause serious harm if improperly managed, requiring strict laws and specialized disposal techniques. Hazardous chemicals found in toxic waste have the potential to cause cancer, neurological damage, or problems with reproduction when they are exposed to them. The drinking water sources, and food chain may become contaminated by these chemicals as they seep into the groundwater and soil. Strong acids and bases that can chemically react to destroy metals or living tissues are examples of corrosive hazardous waste. This kind of waste necessitates resistant containers and particular handling procedures because it can present risks during handling and storage.

Because of its extreme instability and propensity for violent chemical reactions, reactive hazardous waste can explode, catch fire, or release poisonous fumes. Some cyanides and peroxides are examples. Ignitable hazardous waste is made up of combustible liquids, solids, or gases that can ignite quickly in typical circumstances. In addition to being fire hazards, these materials can potentially cause harmful secondary reactions when ignited. Hazardous waste needs to be properly identified, categorized, stored, transported, treated, and disposed of. To supervise these procedures, regulatory frameworks are in place, such as the Resource Conservation and Recovery Act (RCRA) in the US. These laws mandate that organizations that produce hazardous waste adhere to certain standards to reduce the possibility of contaminating the environment and endangering people. This entails keeping thorough records, implementing suitable containment techniques, and making sure that disposal takes place at facilities that are approved and trained to handle hazardous materials safely.

The goal of treating hazardous waste is to lessen or neutralize its harmful characteristics so that it can be disposed of or reused safely. Typical methods of treatment include encapsulation, which isolates the waste from the environment; chemical neutralization, which changes the chemical composition to reduce toxicity; and incineration, which can destroy organic pollutants. Even with these precautions, the production of hazardous waste still presents major problems for the entire world, requiring continuous improvements in waste management techniques as well as more stringent enforcement of laws. Moreover, two essential elements of sustainable waste management are the reduction and recycling of hazardous waste. The strain on waste management systems can be greatly decreased by reducing the production of hazardous waste at the source, such as by switching to less hazardous materials and cleaner production technologies. Certain hazardous wastes, such as spent solvents and metals, can be recycled to reduce their negative effects on the environment and to save precious resources.

# 5.1. THE SCIENCE OF HAZARDOUS WASTE

Materials that can be potentially hazardous to the environment or human health are disposed of as hazardous waste. Hazardous wastes include discarded household items like paint thinners, cleaning fluids, and outdated batteries, as well as chemicals, heavy metals, and materials produced as byproducts during commercial manufacturing processes. Hazardous wastes can be solids, liquids, sludges, or contained gases. A large portion of this hazardous material is kept in containment areas like landfills. The public's health may be at risk if the contents of these hazardous waste sites are released into the environment due to improper management or design.

Waste classified as hazardous needs to be managed carefully to protect the environment and public health. Waste can be dangerous due to a variety of characteristics, such as toxicity, strong chemical reactions, or corrosiveness. Humanity will generate 300–500 million metric tons of hazardous waste yearly by 2022. Paints, batteries, and electronics are a few typical examples. Safe disposal is a key component of managing hazardous waste. Hazardous waste can be burned, recycled into new materials, or kept in hazardous waste landfills. Handling dangerous waste is crucial to achieving global sustainability. Both nationally and internationally, hazardous waste is governed by national governments, international treaties, and the United Nations (UN).

## 5.1.1. Types

### *5.1.1.1. Universal Wastes*

A unique class of hazardous waste known as "universal wastes" exists in the U.S. and are widely produced, generally less dangerous than other hazardous wastes, and produced in huge quantities by numerous generators. Fluorescent light bulbs, certain types of specialized batteries e.g., batteries made of lithium or lead, cathode ray tubes, and equipment made of mercury. Regulations pertaining to universal wastes are a little less strict. Conditionally exempt small quantity generators (CESQGs) are a subset of small quantity generators of universal waste that are exempt from some regulatory requirements regarding the handling and storage of hazardous wastes. Even so, universal wastes need to be disposed of correctly.

### *5.1.1.2. Household Hazardous Waste*

Waste produced by residential households is known as household hazardous waste (HHW), sometimes known as domestic hazardous waste or home-generated special materials. Only waste resulting from the use of products that are marketed and sold for residential use is covered by HHW. HHW does not include waste produced in an industrial or business setting (Figure 5.1).

Figure 5.1. Household hazardous waste is separated for proper disposal.

Source: https://upload.wikimedia.org/wikipedia/commons/thumb/b/b6/December_3%2C_2012_%E2%80%93_Household_Hazardous_Waste_separated_for_proper_disposal_%288268772840%29.jpg/1024px-December_3%2C_2012_%E2%80%93_Household_Hazardous_Waste_separated_for_proper_disposal_%288268772840%29.jpg.

The following list includes categories often applied to HHW. It is important to note that many of these categories overlap and that many household wastes can fall into multiple categories:

- Paints and solvents
- Automotive wastes (used motor oil, antifreeze, etc.)
- Pesticides (insecticides, herbicides, fungicides, etc.)
- Mercury-containing wastes (thermometers, switches, fluorescent lighting, etc.)
- Electronics (computers, televisions, mobile phones)
- Aerosols/Propane cylinders
- Caustics/Cleaning agents
- Refrigerant-containing appliances
- Some speciality batteries (e.g., lithium, nickel cadmium, or button cell batteries)
- Ammunition
- Asbestos
- Car batteries

- Radioactive wastes (some home smoke detectors are classified as radioactive waste because they contain very small amounts of radioactive isotope americium, e.g., americium-241).
- Smoke from chimneys

## 5.1.2. Characteristics of Hazardous Waste

- Ignitability. If a waste has a flash point of less than 60°C, ignites easily, and burns violently when exposed to air, or if it's an oxidizer or compressed gas that can ignite, it's classified as an ignitable hazardous waste. Examples include naphtha, oil-based paints, epoxy resins, lacquer thinner, and adhesives.
- Corrosivity. Corrosive hazardous waste is defined as any kind of liquid waste with a pH of less than or equal to 2 or more than or equal to 12 point 5. Many industries frequently use hydrochloric acid (low pH) and sodium hydroxide (high pH) to clean or degrease metal parts. Solvents that are disposed of before painting without being treated add to the hazardous waste that is corrosive.
- Reactivity. Reactive hazardous waste is defined as any unstable material, reacts violently with water, releases toxic gases when it encounters water or other corrosive materials, or blows up when it comes into contact with heat or flame. Gunpowder waste, sodium metal waste, and wastes containing cyanides or sulfides are a few examples of reactive wastes.
- Toxicity. Toxicity of a hazardous waste can be determined by taking a representative sample of the material and subjected to a test conducted in a certified laboratory and toxic characteristics can be determined.

## 5.1.3. Categories of Hazardous Wastes

### 5.1.3.1. Radioactive Substance

Hazardous waste that contains radioactive material is known as radioactive waste. Nuclear medicine, nuclear research, nuclear power, manufacturing, construction, coal and rare-earth mining, and the reprocessing of nuclear weapons are some of the industries that produce radioactive waste. Any material that has the potential to release ionizing radiation is referred to as radioactive, and it is dangerous since extended exposure frequently causes harm to living things. Because they last a long time and their disposal is dependent on their half-life, radioactive substances are of particular concern. The half-lives of various uranium compounds, for instance, vary from 72 years for U232 to 23,420,000 years for U236.

### 5.1.3.2. Chemicals

The five groups of hazardous chemical wastes are salts, acids and bases, flammables and explosives, synthetic organics, and inorganic metals. Certain chemicals pose a risk to human health and safety.

### 5.1.3.3. Bio-Medical Wastes

Biological research facilities and hospitals are the primary sources of hazardous biological waste. In addition to producing toxins, biological waste is capable of infecting other living things. Biomedical waste primarily consists of contaminated items like bandages, hypodermic needles, and out-of-date medications, as well as malignant tissues that are discarded during surgical procedures.

#### *5.1.3.4. Flammable Wastes*

Flammable waste is included in the category of hazardous waste. The risk associated with the collection, storage, and disposal of flammable wastes makes this grouping necessary. The combustible waste can be solid, liquid, or gaseous. Organic sludges, oils, plasticizers, and solvents are a few examples of flammable waste.

#### *5.1.3.5. Explosives*

Most explosive hazardous waste is made up of artillery or ordnance. Additionally, there is a significant risk associated with the storage, gathering, and disposal of explosives. These wastes can be in the form of solid, liquid, or gas.

### 5.1.4. Disposal

In the past, regular landfills were used to dispose of some hazardous waste. For hazardous wastes to be disposed of and stabilized, they frequently need to be treated differently and stabilized before being placed in a landfill. Most combustible substances can be recycled into industrial fuel. Lead acid batteries are one example of a material that can be recycled despite having hazardous ingredients. Countermeasures against groundwater contamination are necessary for many landfills. For instance, to contain any hazardous materials that might still be present in the disposal of waste, a barrier needs to be built along the landfill's foundation. (Fazzo et al., 2017)

#### *5.1.4.1. Recycling*

It is possible to recycle some hazardous waste into new goods. Lead-acid batteries and electronic circuit boards are two possible examples. These kinds of ashes contain heavy metals that, with the right processing, may bind to other pollutants and turn them into solids that are easier to dispose of or be used as pavement filler. These processes recycle the safe product while lowering the risk of dangerous substances like fly and bottom ash.

#### *5.1.4.2. Incineration*

Hazardous waste is burned in incinerators at temperatures between 1600° and 2500°F (870° and 1400°C), which significantly reduces the amount of waste by turning it into ash and gases. Hazardous waste can take many forms, including contaminated soil, sludge, liquids, and gases, which can be burned. Hazardous waste sites can be constructed with an incinerator on site, or, more frequently, the waste can be moved from the site to a permanent incinerator facility. Additionally dangerous are the gases and ash left over from burning. Metals are not destroyed; instead, they can either continue to burn in the furnace or turn into gas and release emissions. Despite taking up less space than the original waste, the ash must be stored in a hazardous waste landfill. Ammonia, carbon dioxide, nitrogen oxides, and volatile organic compounds are among the gases released during incineration. Sulfur dioxide and hydrochloric acid gas can also be produced in the furnace through reactions. These days, incinerators are built with systems to catch emissions of solid waste suspended in hazardous gases, so they can't release them into the air.

### 5.1.4.3. Landfill

Hazardous waste may be sequestered in a hazardous waste landfill or permanent disposal facility. "In terms of hazardous waste, a landfill is defined as a disposal facility or part of a facility where hazardous waste is placed or on land and which is not a pile, a land treatment facility, a surface impoundment, an underground injection well, a salt dome formation, a salt bed formation, an underground mine, a cave, or a corrective action management unit (40 CFR 260.10)."

### 5.1.4.4. Pyrolysis

Certain hazardous waste types can be removed by high-temperature pyrolysis, which doesn't always require an electrical arc but instead requires oxygen starvation to prevent combustion. However, all materials (waste) added to the process will melt into a molten slag when an electrical arc is used to generate the necessary ultra-heat (above 3000 degrees C temperature). This technology is called plasma, not pyrolysis. When cooled, inert materials created by plasma technology solidify into a material resembling rock. These treatment techniques are highly costly, but in some situations, such as the destruction of concentrated organic waste types, such as PCBs, pesticides, and other persistent organic pollutants, they might be better than high-temperature incineration.

## 5.1.5. In Society

### 5.1.5.1. Management and Health Effects

Improper handling and disposal of hazardous waste can have serious repercussions. Hazardous gaseous materials can be released into the air if they are disposed of incorrectly, increasing morbidity and mortality. These gaseous materials may contain carbon monoxide, hydrogen chloride, sulfur dioxide, nitrogen oxides, and heavy metals in some cases. Many organizations (RCRA, TSCA, HSWA, and CERCLA) developed an identification scale in response to the possibility of gaseous material being released into the atmosphere. This scale categorizes hazardous materials and wastes to enable prompt detection and mitigation of potential leaks. Materials on the F-List were classified as waste from non-specific industrial practices; materials on the K-List were wastes from specific industrial processes, such as the petroleum, explosive, and pesticide industries; and materials on the P and U lists were wastes produced for commercial use and shelf-stable pesticides. Hazardous waste mishandling can contaminate soil and groundwater in addition to having a negative direct health impact through air pollution. According to an Austrian study, residents close to industrial sites have lower levels of education, are twice as likely to be immigrants, and are unemployed more frequently. For those who rely significantly on the land for harvests and streams for drinking water—such as Native American populations—this causes disproportionately more problems. The facts put Native Americans at a multiplied risk of toxic exposure, even though all lower-class and/or social minorities are at a higher risk (Brook, 1998).

In certain tribes, improper handling of hazardous waste has led to numerous severe health issues. The Mohawk Nation at Akwesasne has seen an increase in cancer rates due to elevated blood levels of PCBs (polychlorinated biphenyls).

### 5.1.5.2. Global Goals

The UN is tasked with regulating hazardous materials and wastes, and it provides advice to nations on how to handle hazardous waste. To halt the transfer of hazardous waste from industrialized to developing nations with laxer environmental laws, 199 nations signed the Basel Convention in 1992. By incorporating it into Sustainable Development Goal 12, the global community has established responsible handling of chemicals and hazardous waste as a crucial component of sustainable development. Achieving the environmentally sound management of chemicals and all wastes throughout their life cycle is target 12.4 of this goal. The amount of hazardous waste generated per person and the percentage of treated hazardous waste broken down by type of treatment are two indicators for this goal.

## 5.1.6. Regulatory History

### 5.1.6.1. In the United States

- Resource Conservation and Recovery Act (RCRA)

Wastes classified as hazardous have characteristics that make them potentially dangerous to the environment or human health. Hazardous wastes can take the form of sludges, liquids, solids, or contained gases. They may be leftovers from production processes or just commercial items that were thrown away, such as insecticides or cleaning solutions. According to regulations, wastes that show at least one of the following four qualities—ignitability, corrosivity, reactivity, or toxicity—or that are listed on one of the four hazardous waste lists (F-list, K-list, P-list, or U-list) are considered RCRA hazardous wastes. Subtitle C of the Resource Conservation and Recovery Act (RCRA) governs hazardous waste regulations in the United States. The EPA concluded that waste that is hazardous by definition and is included in the lists that the Agency releases. These lists fall into three categories: the F-list, which is found in the regulations at 40 CFR 261.31; the K-list, which is found in the regulations at 40 CFR 261.32; and the P-list and U-list, which are found in the regulations at 40 CFR 261.33, which are lists of discarded commercial chemical products. Hazardous waste life cycle tracking is made easier by the RCRA record-keeping system, which also lowers the quantity of hazardous waste that is illegally disposed of.

- Comprehensive Environmental Response, Compensation, and Liability Act

The Comprehensive Environmental Response, Compensation, and Liability Act (CERCLA) was enacted in 1980. The primary contribution of CERCLA was to create a "Superfund" and provide for the clean-up and remediation of closed and abandoned hazardous waste sites. CERCLA addresses historic releases of hazardous materials but does not specifically manage hazardous wastes.

## 5.1.6.2. In India

- Environmental Act and Hazardous Waste RulesIndia's environmental consciousness was heightened in 1984 by the Bhopal disaster, a deadly methyl isocyanate gas leak. The Indian government responded by creating the Hazardous Waste Regulations in 1989 and the Environmental Act in 1986. Under these regulations, businesses are only allowed to produce hazardous waste by the state if they can properly dispose of it. State governments, however, failed to make these regulations effective. The period between the request and construction of hazardous waste landfills was approximately ten years. Businesses at the time disposed of hazardous waste in a variety of makeshift locations, like canal pits and the sides of roads, without having any immediate plans to transfer it to appropriate facilities.

- Supreme Court action

The Supreme Court intervened to safeguard the right to life by preventing harm caused by hazardous waste. The Supreme Court established the High-Powered Committee (HPC) of Hazardous Waste in response to a 1995 petition by the Research Foundation for Science, Technology, and Natural Resource Policy, which claimed that data from earlier government boards was useless. This committee discovered research that connected increased levels of lead, hexavalent chromium, and other heavy metals to pollution and inadequate waste treatment. These studies were essentially being disregarded by industries and regulators. Furthermore, the state's actions were not in compliance with the Basel Convention, an international agreement pertaining to the transportation of hazardous waste. To monitor its rulings, the Supreme Court started the Supreme Court Monitoring Committee and changed the Hazardous Waste Rules. The Court can now compel the closure of hazardous waste-polluting companies thanks to this committee.

## 5.1.7. Country Examples

### *5.1.7.1. United States*

The Resource Conservation and Recovery Act (RCRA) governs how hazardous waste is handled, stored, and disposed of in the US. Hazardous wastes are classified into two main categories under 40 CFR 261 (RCRA): characteristic wastes and listed wastes. Every business that produces hazardous waste and every business that stores or disposes of hazardous waste in the US is subject to the RCRA's regulations. Hazardous waste is generated by a variety of business types. Hazardous waste may be produced by dry cleaners, auto repair shops, hospitals, exterminators, and photo processing facilities. Some larger businesses, like chemical manufacturers, electroplating companies, and oil refineries, are also producers of hazardous waste. A U. S. Under the RCRA, any facility that handles, holds, or gets rid of hazardous waste needs a permit. Hazardous waste generators and carriers are subject to tight regulations regarding waste handling, management, and tracking. Congress gave the US Environmental Protection Agency (EPA) instructions to develop hazardous waste management regulations through the RCRA. The EPA created stringent guidelines for managing hazardous waste in all its forms, including handling, storing, and disposing of it, because of this mandate. States may create stricter regulations with a wider scope than federal regulations in addition to these federal requirements. Moreover, RCRA permits states to create regulatory programs that are at least as strict as RCRA, and the states may assume control of enforcing the RCRA's requirements once the EPA has reviewed them. Using this authority, most states create their hazardous waste programs that are just as strict, if not stricter than, federal programs.

The U.S. government provides several tools for mapping hazardous wastes to particular locations. These tools also allow the user to view additional information.

- Data from the US Environmental Protection Agency's (EPA) Toxics Release Inventory and Superfund Basic Research Program could be visually explored by users with the aid of maps of the country provided by TOXMAP, a Geographic Information System (GIS) service from the Division of Specialized Information Services Archived 2019-03-21 at the Wayback Machine of the United States National Library of Medicine (NLM). US Federal Government funding was provided for this resource. TOXMAP gathered its chemical and environmental health data from reliable sources such as PubMed and the Toxicology Data Network (TOXNET) of the National Library of Medicine.
- The US Environmental Protection Agency (EPA) "Where You Live" allows users to select a region from a map to find information about Superfund sites in that region.

# 5.2. HAZARDOUS WASTE MANAGEMENT

Hazardous waste management is the gathering, handling, and disposal of waste that, if not done correctly, can seriously endanger the environment or public health. Hazardous wastes are produced mostly by manufacturing, processing, and other industrial processes. They can be solids, liquids, sludges, or contained gases. When improperly stored, transported, treated, or disposed of, they could cause harm. Hazardous waste that is not properly stored or disposed of can lead to dangerous land pollution as well as harmful water pollution that contaminates surface water and groundwater supplies. Those who live in houses built close to outdated and abandoned landfills may be especially at risk. Governments strictly control the practice of managing hazardous waste to address current issues and stop harm from hazardous wastes in the future.

## 5.2.1. Hazardous Waste Features

Hazardous wastes are categorized according to their physical, chemical, and biological characteristics. Materials with these characteristics can be radioactive, corrosive, toxic, reactive, ignitable, or infectious. Even in trace or extremely small quantities, toxic wastes are poisons (Brook, Daniel July 3, 2006). They could cause violent sickness or death in the case of acute effects, or they could cause irreversible harm over time in the case of chronic effects. Some are carcinogenic, meaning that long-term exposure can result in cancer. Some are mutagenic, meaning that exposed humans and wildlife will have significant biological changes in their progeny. Chemically unstable, reactive wastes react violently with water or air. They release poisonous vapors or explode. Ignitable waste has the potential to start a fire right away because they burn at a low temperature. Corrosive waste includes highly acidic or alkaline materials. They react chemically to demolish both solid objects and living tissue when they come into contact. Used bandages, hypodermic needles, and other items from medical facilities or biological research labs are examples of infectious wastes. Ionizing radiation from radioactive waste can be harmful to living things. Controlling radioactive waste is a major concern because some of these materials can remain in the environment for thousands of years before completely decomposing. However, the local municipal government is not in charge of handling or disposing of radioactive material. The management of radioactive waste, especially nuclear fission waste, is usually regarded as an engineering task distinct from other forms of Hazardous Waste management due to the problem's complexity and scope. This topic is covered in the article nuclear reactor.

## 5.2.2. Principles and Methods of Hazardous Waste Management

The term "hazardous waste management" refers to practices and regulations related to hazardous waste management that do not pose a risk to human health or the environment. Hazardous waste is typically disposed of by burning and dumping in an open area. Open burning and incineration release particulates, nitrogen oxides, offensive odors, and other pollutants into the air, while open dumping pollutes the land, water, and atmosphere. Water pollution results from the disposal of solid waste residues. At the turn of the century, large cities in Europe and America were using sophisticated energy recovery systems for municipal incineration, but these systems quickly went out of use because of the high operating costs. Incineration of hazardous solid waste has declined in popularity in recent years due to the risk involved with stricter regulations on air pollution. The management of hazardous waste is becoming a bigger concern due to fast industrialization. The waste produced by different home and industrial processes can have a detrimental effect on the environment as well as pose serious health risks. The process that follows demonstrates the typical approach to waste management in a developed society. Figure 5.2 illustrates several phases in the management of hazardous waste. (V. Misra, S.D. Pandey 2005)

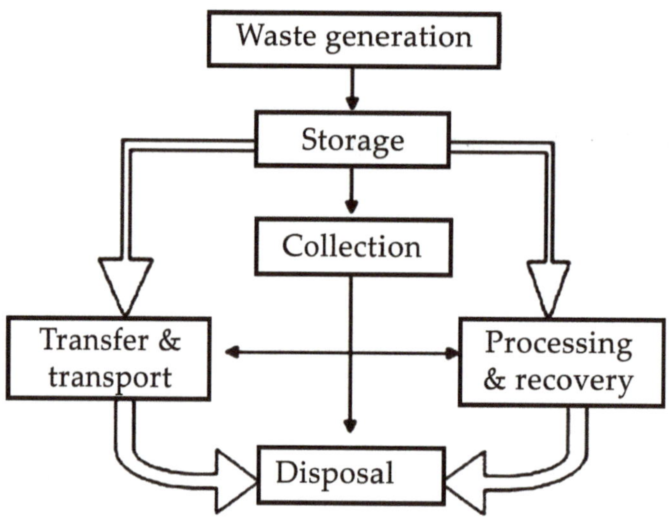


Figure 5.2. Steps involved in hazardous waste disposal.

Source: Kanagamani, K., Geethamani, P., & Narmatha, M. (2021). Hazardous Waste Management. IntechOpen. doi: 10.5772/intechopen.94080.

### *5.2.2.1. Handling of Hazardous Wastes*

It is recommended that those who handle hazardous waste take preventative measures to shield their health from potential harm. Hazardous waste exposure causes skin dermatitis, long-term asthma attacks, eye irritation, and tightness in the chest.

### 5.2.2.2. Transport of Hazardous Waste

Hazardous waste generated often requires transport to a particular site for an approved treatment, storage, or disposal facility (TSDF). Because of potential threats to public safety and the environment, transport is given special attention by governmental agencies to avoid any occasional accidental spill.

### 5.2.2.3. Disposal

Disposal of hazardous waste is the final stage of a hazardous waste management system. The different waste disposal methods include secure landfills, deep wells and bedrock disposal. (Amadi et al., 2017)

- Secure landfill

Unfavorable levels of hazardous materials seeped into the ground because of some hazardous wastes being disposed of in conventional landfills. These substances finally find their way into hydrologic systems in nature. Thus, a barrier for gathering hazardous materials that might be left in the disposal of waste is needed at landfills to keep chemicals from getting into the soil. Now, hazardous waste is stabilized and made into solids and placed in landfills and this process depends upon the type of hazardous waste. Hazardous wastes are disposed of in landfills, which are locations where the waste is buried and stored. A suggested design is exemplified in Figure 5.3. Before being disposed of, the waste is emptied into sealed drums.

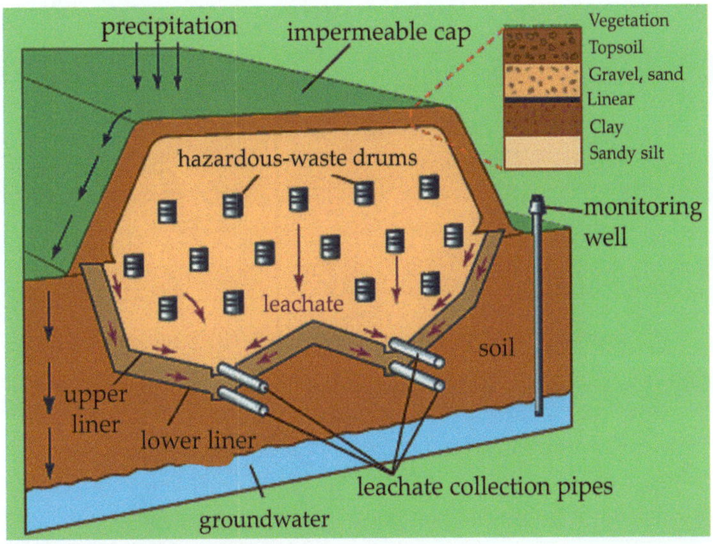


Figure 5.3. Secure landfill method.

Source: Kanagamani, K., Geethamani, P., & Narmatha, M. (2021). Hazardous Waste Management. IntechOpen. doi: 10.5772/intechopen.94080.

Leachate collection systems and two impermeable liners make up the hazardous waste landfill configuration. The network of pipes that make up the double leachate collection system is positioned above each liner. The lower layer serves as a backup while the upper layer lessens the buildup of leachate trapped in the fill. The collected leachate is sent to a treatment facility for additional processing. An impermeable cap or cover is placed over a finished landfill to reduce the amount of leachate in the fill and minimize the potential for environmental degradation.

The main components in the leachate from landfill sites are grouped as follows

- Major elements and ions such as calcium, magnesium, iron, sodium, ammonia, carbonate, sulfate and chloride.
- Trace metals such as manganese, chromium, nickel, lead and calcium
- A wide variety of organic compounds
- Biological agents

Hazardous waste mainly from industries will give rise to leachate. Heavy metals concentration in the leachate is of greater concern compared to other components of leachate.

- Deep well disposal

As seen in Figure 5.4, injecting liquid industrial waste into a deep well is an additional alternative method of disposal. One method of disposing of liquid waste is deep well injection.

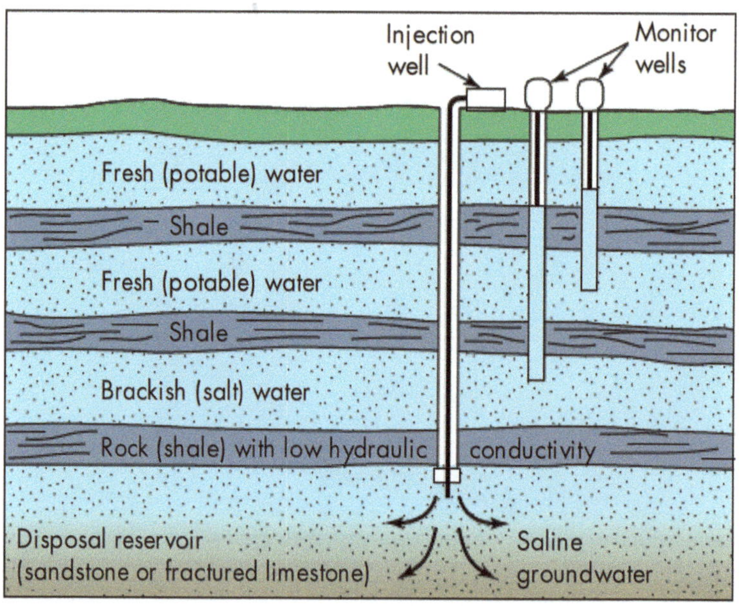


Figure 5.4. Deep well disposal method.

Source: Kanagamani, K., Geethamani, P., & Narmatha, M. (2021). Hazardous Waste Management. IntechOpen. doi: 10.5772/intechopen.94080.

This alternative involves injecting treated or untreated liquid waste using injection wells into geologic formations that are unlikely to permit contaminant migration into prospective aquifers of drinkable water. High pressures are used to push the liquid into the cracks and pores in the rock. The chosen rock unit must be separated above and below by low permeability layers (such as shale) since it is porous and permeable (typically sandstone or fractured limestone). Although deep-well injection is less expensive and doesn't require much or any pretreatment of the waste, there is a chance that hazardous waste will leak out and eventually contaminate subterranean water supplies.

- Bedrock disposal

Several different types of bedrocks are being researched as potential host rocks, and bedrock disposal is primarily intended for solid hazardous waste. Figure 5.5 illustrates the layout of a hazardous waste repository or bedrock disposal site. Its foundation is the multiple barrier (also known as multi barrier) concept, which entails enclosing solid hazardous waste and sealing it with a variety of materials to stop waste leakage or groundwater invasion. The nature of the host rock and a few possible disadvantages are major concerns. The technique is popular for handling high-level radioactive waste sealed inside canisters made of stainless steel or spent fuel rods that have been buried deep beneath stable rock formations and covered in metals that resist corrosion, like copper or stainless steel. Numerous geological formations will work, including thick basalts like the Columbia River plateau basalt, shale, salt, volcanic tuff, and granite.

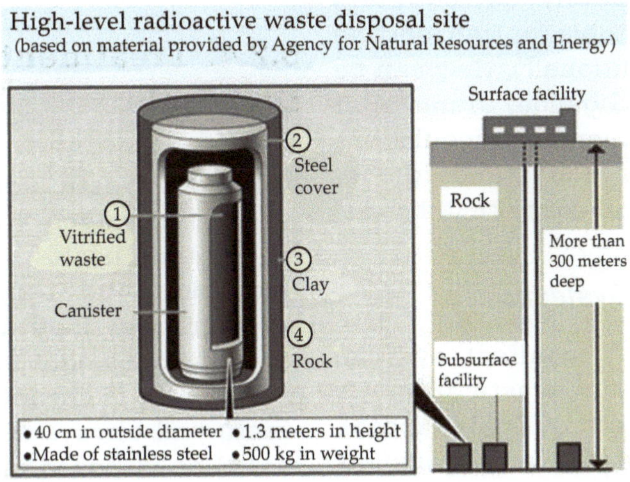


Figure 5.5. Bedrock disposal method.

Source: Kanagamani, K., Geethamani, P., & Narmatha, M. (2021). Hazardous Waste Management. IntechOpen. doi: 10.5772/intechopen.94080.

## 5.2.3. Transport of Hazardous Waste

Transporting hazardous waste to a designated treatment, storage, or disposal facility (TSDF) is frequently necessary when it is generated at a specific location. Governmental organizations pay particular attention to transportation because of possible risks to the environment and public safety. Hazardous waste has historically been purposefully spilt or abandoned at random places in a practice known as "midnight dumping," in addition to the rare unintentional spill. The passing of laws requiring the correct labeling, transportation, and tracking of all hazardous wastes has significantly reduced this practice.

#### 5.2.3.1. Transport Vehicles

Usually, trucks are used to transport hazardous waste across public roads. Very little is moved by air or by inland waterways, and very little is moved by rail. Highway shipment is the most popular since most industrial sites and authorized TSDFs are accessible to road vehicles. Trains made of railroad parts are only appropriate for very large shipments of waste and require costly siding facilities. Tank trucks composed of steel or aluminum alloy, with a maximum capacity of approximately 34,000 liters (9,000 gallons), are suitable for transporting hazardous waste. Additionally, they can be shipped in 200-liter (55-gallon) drums after being containerized. Government regulations include specifications and standards for shipping containers and cargo tank trucks.

#### 5.2.3.2. The Manifest System

The "cradle-to-grave" manifest system, which tracks hazardous waste from its point of origin to its site of final disposal, is a crucial component of waste transport regulations in the US and other nations. The issue of midnight dumping is lessened with the use of the manifest system. It also offers a way to figure out what kind and how much hazardous waste is being produced, along with the best course of action in the event of an unintentional spill. The person responsible for creating the hazardous waste, such as a chemical manufacturer, must create a manifest, which is a record-keeping document. The generator bears the primary responsibility for the final disposal of the waste and is required to provide the manifest to a licensed waste transporter along with the waste. The transporter is required to provide a copy of the manifest to the waste recipient at a TSDF that has been authorized. A copy of the manifest needs to be signed each time the waste is transferred. Each party involved keeps a copy of the manifest, and extra copies are forwarded to the relevant environmental agencies. Hazardous waste transporters are required to take prompt and appropriate action in the event of a leak or accidental spill during the waste's transportation, which includes reporting the discharge to local authorities. To contain the waste, an area may need to be diked. The waste also needs to be removed to lessen the risks to the environment and public health.

### 5.2.4. Treatment, Storage, and Disposal

There are numerous choices for managing hazardous waste. Reducing the amount of waste at its source or recycling the materials for another useful purpose is preferable. Reduction and recycling are desirable solutions, but they are not thought to be the only ones that can address the issue of disposing of hazardous waste. Hazardous waste will always need to be treated and stored or disposed of in some way.

#### 5.2.4.1. Treatment

Treatment options for hazardous waste include chemical, thermal, biological, and physical processes. Ion exchange, precipitation, oxidation, reduction, and neutralization are examples of chemical processes. High-temperature incineration is one of the thermal methods that can both destroy and detoxify some organic wastes. Waste can be burned in liquid, sludge, or solid form using specialized thermal equipment. These consist of the liquid-injection incinerator, rotary kiln, multiple-hearth furnace, and fluidized-bed incinerator. The possibility of air pollution is one issue associated with the incineration of hazardous waste. It is also possible to treat some organic wastes biologically, such as those from the petroleum industry. Landfarming is one technique for biologically treating hazardous waste. Using a suitable piece of land, the waste is carefully mixed with the surface soil in this technique. Nutrients and microbes that can break down the waste could be added. Sometimes a bacterial species that has undergone genetic engineering is employed. Forage and food crops are not planted on the same plot of land. On previously contaminated sites, microbes can also be utilized to stabilize hazardous wastes; in this case, the procedure is known as bioremediation. The waste material's molecular form is altered by the above-described chemical, thermal, and biological treatment techniques. Conversely, physical treatment condenses, solidifies, or lessens the waste's volume. Filtration, flotation, sedimentation, and evaporation are examples of physical processes. Solidification is a further procedure that is accomplished by encasing the waste in plastic, asphalt, or concrete. A solid mass of material that is resistant to leaching is produced by encapsulation. Waste can also be combined with fly ash, lime, and water to create a cement-like substance that is solid.

##### *5.2.4.2. Surface Storage and Land Disposal*

It is necessary to properly dispose of hazardous wastes that are not destroyed by burning or other chemical processes. Land disposal is the destination for many of these wastes, even though it is not a practice that is particularly appealing due to the inherent environmental risks involved. There are two primary techniques for disposing of land: subterranean injection and landfilling. Surface storage or containment systems are frequently used as a stopgap measure before disposing of land. Ponds and lagoons, as well as open waste piles, are examples of temporary on-site waste storage facilities. New waste piles need to adhere to regulations like those governing landfills and be carefully built over an impermeable base. The piles need to be shielded from erosion and wind dispersal. Systems for monitoring and control must be installed if leachate is produced. A new waste pile can only hold noncontainerized solid, nonflowing waste, and when the pile gets too big to handle, the waste must be landfilled. Lagoons are a common form of holding pond or open pit used for temporary storage of hazardous liquid waste. Groundwater protection requires the lining of new lagoons with flexible membrane liners and impervious clay soils. It is necessary to install leachate collection systems in between the liners and to drill wells for groundwater monitoring. Open lagoons do not treat the waste except for some surface aeration, some sedimentation, and the evaporation of volatile organics. Sludge that has accumulated needs to be periodically removed and handled further as hazardous waste. Numerous older, unlined waste piles and lagoons pose serious risks to the environment and public health because they are situated above aquifers that supply public water. Around the world, a significant number of these abandoned locations have been located and are slated for cleanup or remediation.

##### *5.2.4.3. Secure Landfills*

Compared to municipal solid waste, the disposal of hazardous solid or containerized waste is subject to stricter regulations. Hazardous wastes must be placed in so-called secure landfills, which have a minimum of three meters (10 feet) separating the landfill's bottom from the bedrock or groundwater table beneath it. Leachate collection systems and two impermeable liners are essential components of a secure hazardous waste landfill. Above each liner is a network of perforated pipes that make up the double leachate collection system. While the lower system acts as a backup, the upper system keeps leachate trapped in the fill from building up. Leachate collected is pumped to a facility for treatment. An impermeable cap or cover is placed over a finished landfill to minimize the amount of leachate in the fill and reduce the possibility of environmental damage.

It's also necessary to have a groundwater monitoring system that consists of several deeply drilled wells inside and outside the property. The wells enable a regular testing and sampling program to find any leaks or contaminated groundwater. In the event of a leak, the contaminated water can be captured by pumping the wells to the surface where it can be treated (Figure 5.6).

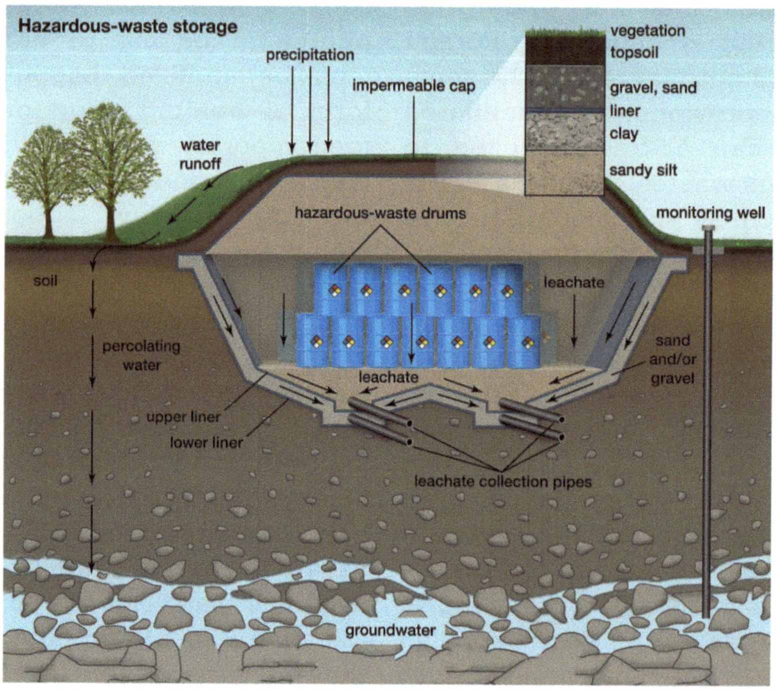


Figure 5.6. Schematic diagram of a secure hazardous waste landfill with a double leachate collection system.

Source: https://cdn.britannica.com/58/23958-050-8B583150/diagram-landfill-leachate-collection-system.jpg.

Deep-well injection is one method of disposing of hazardous liquid waste; it entails pumping the waste liquid through a steel casing and into a porous layer of sandstone or limestone. To permanently store the liquid, high pressure is used to force it into the rock's pores and fissures. The injection zone may extend more than 0.8 km (0.55 mi) below the surface, but it must be located beneath a layer of impermeable rock or clay. Although deep-well injection is comparatively cheap and doesn't require much, if any, pretreatment of the waste, there is a chance that hazardous waste will leak out and eventually contaminate subsurface water supplies.

### 5.2.5. Remedial Action

Hazardous waste disposal in open pits, ponds, or lagoons is dangerous for the environment and public health. Numerous such uncontrolled disposal locations have been used in the past and are now vacant. Remedial action at those sites might be required based on the assessment of risk. There are situations where the risk calls for immediate action. In other cases, before taking any corrective action, engineering studies might be necessary to fully evaluate the situation. Remediation may involve hauling all the waste away from the site and treating it somewhere else before disposing of it properly. Typically, the costliest choice is the so-called off-site solution. On-site remediation is an option that lowers the likelihood of groundwater contamination and decreases leachate production. The construction of a secure landfill on the same property, the temporary removal of hazardous waste, and the appropriate replacement of the waste are all examples of on-site remediation. Treatment of any contaminated groundwater or soil may also be a part of it. It is possible to refill treated soil on the spot and inject treated groundwater back into the aquifer using a deep well. Complete waste containment is a less expensive option. This is accomplished by covering the hazardous waste site with an impermeable cover and using subsurface cutoff walls to stop groundwater from flowing laterally. Cutoff walls can be used for this purpose if the site is naturally covered in a layer of impermeable rock or soil. The site's walls are built deep enough to reach the impermeable layer around the perimeter. Trenches can be dug out of them to surround the site without displacing or upsetting the waste. To prevent them from collapsing during construction, the trenches are filled with a slurry made of bentonite clay. This mixture is then filled with soil and cement, which solidifies to create an impermeable barrier. As a result, impermeable layers act as barriers at the bottom and cutoff walls act as vertical barriers to the flow of water.

### 5.2.6. Environmental Impact of Waste Management

Waste management practices can have significant environmental impacts at various stages, from waste generation to disposal. Here are some key environmental considerations related to waste management:

- Resource Depletion: The production of waste frequently entails the use of resources like raw materials, energy, and water. By encouraging recycling, reuse, and the recovery of valuable materials, efficient waste management can help reduce the depletion of natural resources.
- Landfill Emissions: Although landfills are a common place to dispose of waste, they may have negative environmental effects. Methane, a strong greenhouse gas that contributes to climate change, is produced when organic waste breaks down in landfills. Installing suitable landfill gas capture and utilization systems can reduce methane emissions.
- Air Pollution: Harmful pollutants can be released into the atmosphere by improper waste management practices, such as burning waste

outdoors. Pollutants from burning waste include sulfur dioxide, nitrogen oxides, particulate matter, and volatile organic compounds, all of which can be harmful to human health and the quality of the air.

- Water Pollution: Groundwater and surface water contamination can result from poor waste disposal techniques, such as poor landfill construction or inadequate wastewater treatment. Landfill leachate, heavy metals, and hazardous chemicals can seep into water sources and endanger aquatic ecosystems as well as public health.
- Soil Degradation: Loss of fertile land and deterioration of the soil can be caused by improper waste management. When a landfill is not built and maintained correctly, harmful materials can seep into the ground. The quality of the soil can also be negatively impacted by incorrect composting techniques or the improper disposal of hazardous waste.
- Energy Consumption and Greenhouse Gas Emissions: The transportation, treatment, and disposal of waste require energy, leading to greenhouse gas emissions. By adopting waste management practices that focus on waste reduction, recycling, and energy recovery, the overall environmental impact can be minimized.
- Ecological Impacts: Improper waste management practices can harm wildlife and natural habitats. Dumping waste in sensitive ecosystems, such as rivers, forests, or coastal areas, can disrupt ecological balance, endanger species, and degrade biodiversity.

Adopting sustainable waste management techniques, such as waste reduction, recycling, composting, and energy recovery using technologies like waste-to-energy systems, is essential to reducing these negative environmental effects. To reduce the environmental impact of waste management, it is also essential to implement efficient policies and regulations and to increase public awareness of the issue.

# 5.3. WASTE-TO-ENERGY CONVERSION PROCESSES

A range of technologies are used in waste-to-energy conversion processes to convert waste materials into useful forms of energy. These methods produce fuel, electricity, or heat concurrently and provide a sustainable and effective way to manage waste. Here is a summary of the three main methods for converting waste into energy: gasification, anaerobic digestion, and incineration.

- Incineration: Incineration is the process of burning waste materials under controlled conditions at high temperatures. Waste is burned in specially constructed incinerators, sometimes referred to as waste-to-energy plants, and the heat produced there is used to create steam. After that, the steam is used to turn turbines and produce energy. Incineration aids in the management of hazardous material disposal in addition to reducing the amount of waste produced. To reduce the number of pollutants released into the atmosphere, it is crucial to use the right emissions control technologies.
- Anaerobic Digestion: An oxygen-free biological process, anaerobic digestion takes place. Anaerobic digesters are fed organic waste materials like food scraps, animal manure, and agricultural residues. Waste is broken down by microorganisms, which produce digestate and biogas. Methane and carbon dioxide make up the majority of biogas, which is a renewable energy source. It can be converted to biomethane for injection into natural gas networks or burned to produce heat or electricity. Because of its high nutrient content, the digestate can be applied as an agricultural fertilizer. The benefit of anaerobic digestion is that it produces energy and lowers greenhouse gas emissions while keeping organic waste out of landfills.
- Gasification: Gasification is a thermochemical process that breaks down waste materials at high temperatures while adding steam or oxygen in moderation to create syngas, or synthesis gas. The waste is exposed to extreme heat in gasifiers, which causes it to disintegrate into its constituent parts. A combination of hydrogen, carbon monoxide, and other flammable gases make up the syngas that is created. It can be used as a fuel source for several purposes, including heating, producing electricity, and serving as a feedstock for the synthesis of chemicals and fuel for vehicles. Gasification is a versatile method for managing various waste materials such as industrial waste, biomass, and municipal solid waste.

Every waste-to-energy conversion method has pros and cons of its own. An established technique for disposing of waste and recovering energy is incinerating it. Anaerobic digestion addresses landfill methane emissions and is especially useful for organic waste. It also produces a renewable energy source. In addition to reducing waste volume and producing valuable energy resources, gasification is flexible and capable of handling a variety of waste streams. It is crucial to remember that to reduce possible emissions and guarantee overall sustainability, waste-to-energy conversion processes should be put into place with the proper environmental controls and monitoring systems. Waste reduction, recycling, and appropriate segregation should also be given top priority in waste management strategies to optimize the performance of waste-to-energy technologies.

### 5.3.1. Process Principles

Process principles refer to the fundamental concepts and mechanisms underlying waste-to-energy conversion processes. Understanding these principles is essential for designing, operating, and optimizing waste-to-energy systems. Here are key aspects of process principles in waste-to-energy conversion:

- Thermodynamics: Thermodynamic principles, which define the movement and transformation of energy, control waste-to-energy processes. The system's energy transfer, heat production, and energy conversion efficiency can all be understood using the first and second laws of thermodynamics. These guidelines aid in assessing waste-to-energy processes' viability and potential for energy recovery.
- Combustion Chemistry: One popular waste-to-energy method is incineration, which depends on burning waste. Understanding the chemistry of combustion is essential to ensuring controlled and effective waste burning. It entails comprehending the role of oxygen, the stoichiometry of combustion reactions, and the creation of combustion byproducts like carbon dioxide and water vapor as well as pollutants like sulfur dioxide ($SO2$) and nitrogen oxides ($NOx$).
- Microbiology and Biochemistry: In the absence of oxygen, microorganisms are necessary for the breakdown of organic waste in anaerobic digestion, a biological waste-to-energy process. Optimizing process performance requires a thorough understanding of the microbial ecology, enzymatic reactions, and metabolic pathways involved in anaerobic digestion. Microbial activity and the production of biogas are influenced by variables like pH, temperature, retention time, and substrate composition.
- Thermochemical Reactions: There are two thermochemical waste-to-energy processes: pyrolysis and gasification. Waste materials are partially oxidized during gasification, but thermal breakdown without oxygen occurs during pyrolysis. Reaction kinetics, operating conditions, and specific chemical reactions must all be understood for waste to be converted into syngas, biochar, or other valuable products efficiently.
- Heat Transfer and Energy Recovery: Mechanisms for heat transfer are important in waste-

to-energy conversions. Conduction, convection, and radiation are examples of heat transfer principles that are useful in the design and optimization of heat exchangers, boilers, and other heat recovery systems. Energy recovery processes like district heating or the production of steam for electricity are made easier by efficient heat transfer.

- Process Integration: Processes that convert waste into energy frequently have several steps and unit operations. To maximize energy efficiency and reduce waste generation, process integration principles concentrate on optimizing the overall system design and integrating various process components. This covers factors like feedstock handling, waste pre-treatment, gas cleaning, energy use, and waste residue management.

Understanding the process principles of waste-to-energy conversion enables engineers, researchers, and operators to develop efficient and sustainable waste management solutions. By applying these principles, it becomes possible to improve energy recovery, minimize environmental impacts, and enhance the overall efficiency of waste-to-energy systems.

## 5.3.2. Efficiency and Energy Recovery

Efficiency and energy recovery are crucial considerations in waste-to-energy conversion processes. These processes aim to extract maximum value from waste materials by converting them into usable forms of energy. Here are key aspects related to efficiency and energy recovery in waste-to-energy conversion:

- Conversion Efficiency: The percentage of waste energy that is successfully transformed into usable energy is known as conversion efficiency. A more efficient process is indicated by a higher conversion efficiency. Conversion efficiency is affected by several variables, such as the kind and makeup of waste, process technology, operational parameters, and heat transfer effectiveness. To maximize energy recovery, constant efforts are made to increase the efficiency of waste-to-energy processes.
- Calorific Value of Waste: Waste's energy content is determined by its calorific value. It displays the maximum quantity of thermal energy that can be released during combustion or other types of conversion. Higher calorific values of waste materials result in higher energy recovery rates and greater energy potential. To identify high-calorific-value waste streams suitable for energy conversion, proper waste characterization and sorting are crucial.
- Energy Generation: Processes that convert waste into energy produce energy in a variety of forms, including fuel, electricity, and heat. Heat energy can be used for industrial or district heating, and it is typically produced through combustion processes. Using steam turbines or other devices, heat energy is first converted to mechanical energy and subsequently to electrical energy in the process of producing electricity. The synthesis of syngas or biofuels for use in industrial or transportation processes is a form of fuel production.

- Combined Heat and Power (CHP): Cogeneration, or combined heat and power, is a technique that produces electricity and useful heat at the same time to maximize energy recovery. By using waste heat generated during the production of electricity to heat adjacent buildings or industrial processes, combined heat pump (CHP) systems can achieve a high level of overall energy efficiency. This raises waste-to-energy projects' overall sustainability and financial viability.
- Waste Heat Recovery: Using the extra heat produced during waste-to-energy processes is known as waste heat recovery. Heat energy is extracted and transferred from cooling systems, exhaust streams, and flue gases using heat exchangers, boilers, and other heat recovery systems. By using the recovered heat for space heating, feedwater preheating, and other industrial uses, energy efficiency can be increased even further.
- System Integration and Optimization: A waste-to-energy system's various components must be integrated and optimized for effective energy recovery. This entails appropriate equipment sizing and design, optimal operating conditions, and efficient waste feedstock management. To increase overall system efficiency and energy recovery rates, control systems, process improvements, and ongoing monitoring are put in place.

For waste-to-energy conversion processes to be successful overall and to be economically viable, efficiency and energy recovery are essential. These procedures help to minimize greenhouse gas emissions, lessen dependency on fossil fuels, and facilitate the shift to more environmentally friendly and circular waste management techniques by optimizing energy recovery.

## 5.3.3. Environmental Impacts

Waste-to-energy conversion processes have the potential to significantly impact the environment. While these processes offer benefits in terms of waste management and energy generation, it is essential to carefully manage and mitigate their environmental impacts. Here are some key environmental considerations associated with waste-to-energy conversion:

- Air Emissions: Combustion-based waste-to-energy processes, such as incineration and gasification, release emissions into the air. These emissions can include greenhouse gases (e.g., carbon dioxide and methane), particulate matter, heavy metals, nitrogen oxides (NOx), sulfur dioxide (SO2), and volatile organic compounds (VOCs). Proper emission control technologies, such as scrubbers, bag filters, and catalytic converters, are employed to minimize air pollution and meet regulatory standards.
- Ash Disposal: Waste incineration and gasification generate ash residues, which require proper handling and disposal. Depending on the composition of the waste, ash can contain heavy metals, dioxins, and other potentially hazardous substances. Careful management, including proper containment, treatment, and disposal of ash, is necessary to prevent environmental contamination and

ensure compliance with waste management regulations.

- Water Management: Waste-to-energy processes may generate wastewater and leachate that require appropriate treatment and disposal. Wastewater from scrubbers and other pollution control systems may contain pollutants such as heavy metals, organic compounds, and dissolved solids. Proper treatment methods, such as sedimentation, filtration, and chemical precipitation, are implemented to remove contaminants before discharge or recycling.
- Land Use and Resource Consumption: Waste-to-energy facilities require land for construction and operation. The choice of site should consider potential impacts on local ecosystems, biodiversity, and surrounding communities. Additionally, waste-to-energy processes consume resources, such as water, fuel, and chemicals. Strategies for resource conservation and optimization, including recycling and reuse, should be implemented to minimize the overall environmental footprint.
- Waste Management Hierarchy: Environmental impacts should be considered within the broader waste management hierarchy. Waste-to-energy processes should complement waste reduction, recycling, and reuse efforts. Prioritizing waste prevention and recycling can reduce the amount of waste requiring energy recovery and minimize the overall environmental burden.
- Life Cycle Assessment (LCA): Conducting life cycle assessments can help evaluate the environmental impacts of waste-to-energy processes holistically. LCA considers the entire life cycle of the system, including raw material extraction, energy consumption, waste treatment, and disposal. It enables the identification of potential environmental hotspots and supports decision-making for process optimization and sustainability improvements.

Environmental impacts associated with waste-to-energy conversion processes can be effectively managed through robust regulatory frameworks, technological advancements, and best management practices. Continual research and development focus on improving emission control technologies, optimizing resource utilization, and enhancing overall sustainability. By implementing appropriate measures, waste-to-energy processes can contribute to sustainable waste management practices and the transition to a low-carbon, circular economy.

### 5.3.4. Waste Suitability and Feedstock Selection

Waste suitability and feedstock selection are crucial aspects of waste-to-energy conversion processes. The proper identification and selection of waste materials for energy recovery play a significant role in process efficiency, environmental sustainability, and the overall success of waste-to-energy projects. Here are key considerations related to waste suitability and feedstock selection:

- Waste Composition: Understanding the composition of waste is essential for determining its suitability for energy conversion. Waste materials vary in terms of organic content, moisture content, calorific value, and presence of contaminants. Analyzing waste composition helps identify suitable feedstocks for specific conversion processes. High organic content and calorific value are desirable for efficient energy recovery.

- Moisture Content: The moisture content of waste can significantly impact the efficiency of energy conversion processes. High moisture content requires additional energy for drying or pre-treatment, reducing overall process efficiency. Therefore, waste with lower moisture content is generally preferred, as it facilitates better combustion, digestion, or gasification.
- Calorific Value: Calorific value is a measure of the energy content of waste material. Waste with higher calorific value has greater energy potential and contributes to higher energy recovery rates. Waste streams with low calorific value may require blending with other materials to improve their energy content. Proper waste characterization and sorting help identify high-calorific-value waste streams suitable for energy conversion.
- Contaminants and Hazards: Waste materials may contain contaminants such as heavy metals, pollutants, hazardous substances, or non-combustible materials. These contaminants can impact the efficiency of conversion processes and pose environmental and health risks. Waste sorting and pre-treatment processes are employed to remove or mitigate the presence of contaminants, ensuring safe and efficient energy recovery.
- Waste Segregation: Proper waste segregation is essential to optimize feedstock selection. Segregation at the source helps separate recyclable materials, organic waste, and other non-combustible or hazardous waste streams. By segregating waste, it becomes possible to channel appropriate materials into specific waste-to-energy conversion processes and maximize energy recovery while minimizing environmental risks.
- Regulatory Compliance: Waste suitability and feedstock selection should align with local regulations and waste management policies. Regulatory frameworks may specify waste classifications, acceptable feedstocks, and limits for contaminants or hazardous substances. Compliance with these regulations ensures the safe and sustainable operation of waste-to-energy facilities.

Optimizing waste suitability and feedstock selection involves a comprehensive understanding of waste characteristics, waste management practices, and the requirements of different conversion technologies. It requires collaboration between waste management authorities, waste generators, and waste-to-energy operators to ensure proper waste characterization, segregation, and feedstock optimization. By selecting suitable waste materials, waste-to-energy conversion processes can maximize energy recovery, minimize environmental impacts, and contribute to a more sustainable approach to waste management.

### 5.3.5. Technology Advances and Innovations

Technology advances and innovations play a vital role in improving waste-to-energy conversion processes, enhancing their efficiency, and addressing environmental concerns. Here are some key technology advancements and innovations in the field of waste-to-energy:

- Advanced Combustion Systems: Modern waste incineration facilities employ advanced combustion systems, such as fluidized bed combustion and grate combustion, that offer better control over combustion processes and reduce emissions. These systems enhance energy conversion efficiency, minimize air pollution, and allow for the co-combustion of different types of waste.
- Gasification and Pyrolysis Technologies: Gasification and pyrolysis processes have advanced significantly, enabling the conversion of various waste materials into syngas, biochar, and other valuable products. Improved reactor designs, better control over operating parameters, and integration with downstream processes have increased the efficiency and versatility of these technologies.
- Anaerobic Digestion Enhancements: Anaerobic digestion, a biological waste-to-energy process, has seen advancements in digester designs, mixing systems, and process control. Enhanced biogas production and higher organic waste degradation rates have been achieved through the introduction of advanced feedstock pretreatment methods, improved reactor configurations, and optimization of microbial consortia.
- Waste Sorting and Recycling Technologies: Efficient waste sorting and recycling technologies have emerged, allowing for the separation of valuable materials from the waste stream. Advanced sorting techniques, including optical sorting, automated separation systems, and advanced sensor technologies, enable the recovery of recyclable materials, reducing the amount of waste destined for energy recovery.
- Thermal and Energy Recovery Systems: Innovative heat exchange systems, such as heat pumps, economizers, and heat exchangers, optimize the recovery of heat from waste-to-energy processes. These systems maximize energy efficiency by utilizing waste heat for various applications, such as district heating, water heating, or industrial processes, thereby reducing the reliance on additional energy sources.
- Integration of Renewable Energy Technologies: Waste-to-energy facilities are increasingly incorporating renewable energy technologies into their systems. Integration with solar panels, wind turbines, or other renewable energy sources helps reduce reliance on conventional energy sources and further enhances the environmental sustainability of waste-to-energy projects.
- Advanced Monitoring and Control Systems: The use of advanced monitoring and control systems, including remote sensing, real-time data analysis, and predictive modeling, allows for efficient process monitoring, optimization, and troubleshooting. These systems enable operators to make informed decisions,

detect anomalies, and improve overall process performance.

- Circular Economy Approaches: Innovations in waste management focus on adopting a circular economy approach, where waste is viewed as a resource. Technologies such as waste-to-biofuels, chemical recycling, and advanced waste conversion methods aim to recover valuable resources from waste and minimize environmental impacts.

These technological advances and innovations contribute to the continuous improvement of waste-to-energy conversion processes. They enhance energy recovery efficiency, reduce emissions and environmental impacts, and support the transition to a more sustainable and circular approach to waste management. Continued research, development, and implementation of these technologies are crucial for achieving a greener and more efficient waste-to-energy sector.

### 5.3.6. Integration and Sustainability

Integration and sustainability are key considerations in the development and operation of waste-to-energy systems. By integrating waste management practices with energy generation, these systems aim to create a sustainable and circular approach to waste management. Here are important aspects related to integration and sustainability in waste-to-energy:

- Waste Management Integration: Waste-to-energy systems should be integrated into a broader waste management framework. This involves coordinating waste collection, sorting, and treatment processes to ensure a consistent and efficient supply of feedstock for energy conversion. Collaboration with waste management authorities, recycling facilities, and other stakeholders is essential for effective integration.
- Energy Grid Integration: Waste-to-energy facilities should be connected to the energy grid to maximize the utilization of generated energy. Integration with the grid allows for the distribution and supply of electricity, heat, or other forms of energy to consumers, including nearby communities, industries, or district heating networks. Proper planning and coordination with energy infrastructure providers are necessary for seamless grid integration.
- Environmental Impact Mitigation: Sustainable waste-to-energy systems strive to minimize their environmental footprint. This includes implementing emission control technologies to reduce air pollutants, managing waste ash and residues properly, and treating wastewater and leachate to prevent contamination. Regular monitoring and compliance with environmental regulations are essential for minimizing negative impacts on air, water, and soil quality.
- Resource Recovery and Recycling: Integration with resource recovery and recycling processes is crucial for maximizing the sustainability of waste-to-energy systems. Efforts should be made to recover valuable materials from waste streams before energy conversion, promoting recycling and reuse. Integrated systems can facilitate the extraction of recyclable materials, such as

metals and plastics, leading to the reduction of waste sent for energy recovery.

- Circular Economy Approach: Waste-to-energy systems contribute to a circular economy by valuing waste as a resource rather than a burden. This approach involves recovering energy from non-recyclable waste while simultaneously promoting waste prevention, recycling, and resource recovery. It aims to close the loop in the waste management cycle by minimizing waste generation, maximizing material reuse, and achieving energy recovery from the remaining waste.
- Economic Viability: Integration and sustainability in waste-to-energy systems require economic viability. These systems should be financially feasible, considering factors such as capital investment, operational costs, revenue generation, and market conditions for energy products. Proper financial planning and feasibility assessments are necessary to ensure the long-term sustainability and economic viability of waste-to-energy projects.
- Social Acceptance and Stakeholder Engagement: Successful integration and sustainability of waste-to-energy systems depend on social acceptance and engagement with stakeholders. It is crucial to involve local communities, waste management authorities, environmental organizations, and other stakeholders in the planning, decision-making, and operation of waste-to-energy facilities. Transparent communication, public awareness programs, and stakeholder consultations are essential for addressing concerns, ensuring community participation, and fostering support.

Integration and sustainability considerations in waste-to-energy systems are vital for creating a balanced approach to waste management and energy generation. By optimizing resource recovery, minimizing environmental impacts, and engaging stakeholders, these systems can contribute to a more sustainable and circular economy, reducing reliance on fossil fuels and promoting efficient waste management practices. (Martínez et al., 2022)

# CASE STUDY

## HAZARDOUS WASTE FLOWS: THE CASE OF CHINA

China plays a significant role in addressing global hazardous waste issues. China's economic miracle made it the world's largest waste producer. The annual growth rate of hazardous waste generation in China is even 4 higher than that of its economy (measured by gross domestic product, GDP). Meanwhile, China was the leading importer of global waste as well. A decade ago, up to 70% of the world's e-waste ended up in China, exacerbating the challenges of hazardous waste management in China. In this context, in 2017, the Chinese government enacted an unprecedented ban (i.e., Prohibiting the Imports of Foreign Garbage: The Reform Plan on Solid Waste Import Management, herein referred to as "the Chinese import ban"), restricting the import of many types of waste from other countries. Furthermore, in response to interprovincial physical transfers, China has implemented stricter policies to monitor the cradle-to-grave process of hazardous waste. Under the policy, hazardous waste producers are expected to minimize waste at the source. However, expanding and increasingly complex trade networks offer another option. That is, developed regions may comply with transboundary waste regulations by avoiding transporting their local hazardous waste to other regions; instead, they may directly import waste-intensive products and raw materials from less-developed areas. Such interregional trade does not only impact waste disposal processes but also transfers activities generating hazardous waste from the consuming region to the producing region. Since the disposal duty of hazardous waste belongs to the generator, the virtual transfer through trade implies a more radical transfer of waste management responsibilities than that caused by the physical waste transfer. In this context, it is necessary to shift the focus of waste management from end-of-life disposal to the entire supply chain.

### *Unequal Consumption- and Production-Based Hazardous Waste*

China's industrial production generated 39.76 million tons (mt) of hazardous waste in 2015, approximately 10% of the world's total. Manufacture of chemical products (CHEMI), smelting & processing of metals (SM_ME), mining and processing of nonmetal and other ores (MN_NM), and manufacture of paper and printing (PAPER) are the main sectors that generate hazardous wastes, accounting for 71% of China's total hazardous waste generation in that year.

From a consumption perspective, there is a huge disparity in hazardous waste generation across provinces (shown in Figure 1A and the length of bars on the left in Figure 2). The top 10% provinces are responsible for 23% of the national total, while the bottom half contributes only 28%. The greatest consumption-based contributors are Shandong (3.15 mt, 10% of the total), Xinjiang (2.24 mt, 7%), and Guangdong (1.79 mt, 6%).

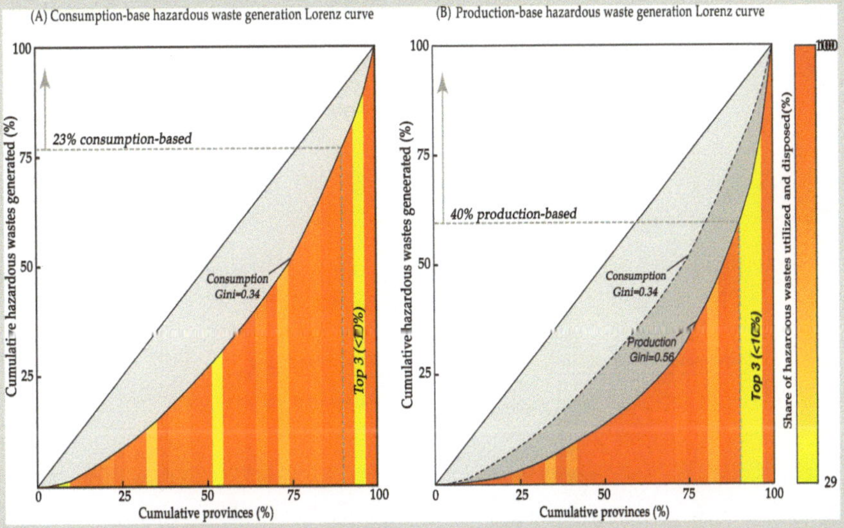


**Figure 1.** *Lorenz curves for consumption-based (A) and production-based (B) hazardous waste generation of 31 Chinese provinces. for consumption- and production-based lorenz curves, provinces are sorted by consumption- and production-based hazardous waste generation from the lowest on the left to the highest on the right, respectively.*

*Source: Li R, Liu M, Shan Y, Shi Y, Zheng H, Zhang W, Yang J, Fang W, Ma Z, Wang J, Bi J, Hubacek K. Large Virtual Transboundary Hazardous Waste Flows: The Case of China. Environ Sci Technol. 2023 May 30;57(21):8161-8173. doi: 10.1021/acs.est.2c07962. Epub 2023 May 16. PMID: 37192406; PMCID: PMC10233792.*

From the production perspective, the disparity of hazardous waste among provinces is even starker. Figure 1 shows Lorenz curves for consumption- and production-based hazardous waste generation in 31 Chinese provinces. The horizontal axis represents the cumulative share of provinces, and the vertical axis shows the cumulative share of consumption- and production-based hazardous waste generation. As shown in Figure 1B, the top 10% provinces generated 40% of China's industrial hazardous waste in 2015 (19% in Shandong, 13% in Qinghai, and 8% in Xinjiang). In contrast, provinces ranked in the bottom 50% produce only 14% of China's hazardous waste, with each accounting for less than 2%. Furthermore, we calculate the Gini coefficient to measure the inequalities of hazardous waste generation across provinces. The results show that the Gini coefficient from the production perspective is as high as 0.56, 1.65 times that on the consumption side. Such exacerbated inequalities in provinces' production-based hazardous waste are also captured when looking at per capita hazardous waste generation and waste generation intensities (measured by hazardous waste generation per unit of industrial output, shown in Figure 2). Taking Qinghai, the province with the highest per capita hazardous waste generation and intensities from both perspectives as examples, consumption-based per capita hazardous waste generation and hazardous waste intensity in Qinghai are 5 and 8 times the national average, while the numbers on the production side soar to an astonishing 28 and 54, respectively.

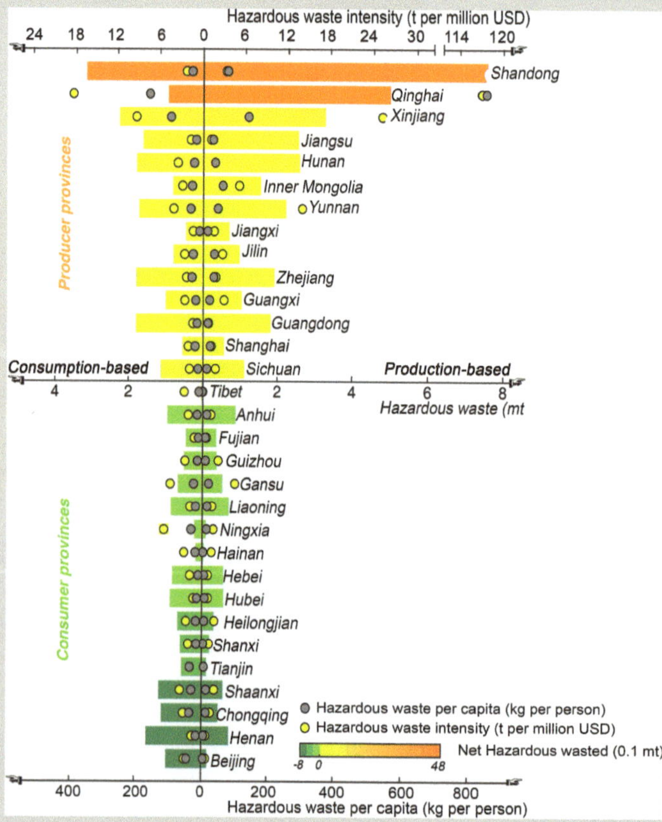


**Figure 2.** *Consumption- and production-based hazardous waste generation. The length of the bars illustrates the consumption- and production-based hazardous waste generated in each province. The bars are ranked and color-coded according to the amount of net hazardous waste generation from green (consumer provinces, consumption-based waste > production-based waste) to orange (producer provinces, production-based waste > consumption-based waste).*

Source: Li R, Liu M, Shan Y, Shi Y, Zheng H, Zhang W, Yang J, Fang W, Ma Z, Wang J, Bi J, Hubacek K. Large Virtual Transboundary Hazardous Waste Flows: The Case of China. Environ Sci Technol. 2023 May 30;57(21):8161-8173. doi: 10.1021/acs.est.2c07962. Epub 2023 May 16. PMID: 37192406; PMCID: PMC10233792.

Notably, the extreme inequality on the production side poses great challenges to waste management in hotspot provinces, especially in their key industries (e.g., manufacture of paper, printing (PAPER) in Shandong and mining and processing of nonmetal and other ores (MN_NM) in Qinghai and Xinjiang). More details on sectoral contributions to each province's hazardous waste. In recent years, hazardous wastes from all sectors in China must be reused or disposed of at specialized facilities with hazardous waste operation permits (mainly including incineration, landfilling, etc.). However, in general, these facilities do not target specific types of hazardous waste but are able to reduce or eliminate the hazards of a relatively wide range of wastes. The colors below the Lorenz curves in Figure 1 indicate the share of hazardous waste reused and disposed for the corresponding province from low (yellow) to high (orange). We find that in nearly 2/3 of Chinese provinces, more than 95% of the total generated hazardous waste can be reused or disposed of. However, shockingly, for two of China's top 3 hazardous waste producers, only around 1/3 of hazardous waste generated in 2015 was reused or disposed (29% in Qinghai, 38% in Xinjiang) due to their insufficient treatment capacities. Moreover, given the current production of such waste and limited capacities for reusing or processing, stockpiles of hazardous waste will continue to rise over time, threatening the surrounding environment and the well-being of residents.

### *Impact of Trade on Hazardous Waste in China*

Figure 2 also indicates a significant mismatch between consumption and production based hazardous waste generation. Fourteen provinces are defined as producer provinces where production-based hazardous waste is higher than consumption-based waste.

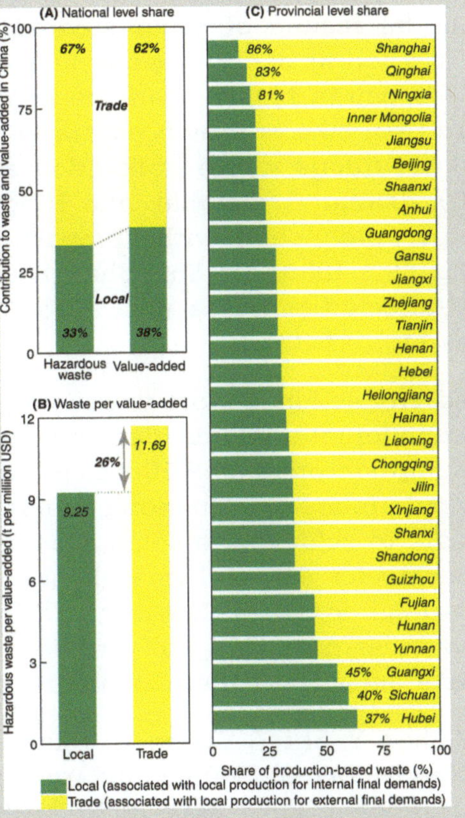


**Figure 3.** *Impact of trade on China's hazardous waste generation. (A) Impact of trade on national hazardous waste generation and value-added. (B) Hazardous waste per value-added associated with local production for internal (local) and external final demands (trade). (C) Impact of trade on provincial hazardous waste generation.*

Source: Li R, Liu M, Shan Y, Shi Y, Zheng H, Zhang W, Yang J, Fang W, Ma Z, Wang J, Bi J, Hubacek K. Large Virtual Transboundary Hazardous Waste Flows: The Case of China. Environ Sci Technol. 2023 May 30;57(21):8161-8173. doi: 10.1021/acs.est.2c07962. Epub 2023 May 16. PMID: 37192406; PMCID: PMC10233792.

In contrast, the remaining 17 are consumer provinces that induce much more hazardous waste than they generate within their geographical boundaries. It is noteworthy that consumption-based hazardous waste generation is 4–7 times higher than the corresponding production-based generation in Beijing, Hainan, and Tianjin, which are either developed megacities or islands and are dominated by tertiary industry. These consumer provinces are often overlooked in hazardous waste management on the production side, despite their roles as vital drivers.

The separation of waste-intensive commodities production from their consumption is induced by expanding interregional trade. Figure 3A shows a dramatic spatial shift in China's industrial hazardous waste generation facilitated by trade. In 2015, 67% of the industrial hazardous waste in China was generated during the production of goods that were ultimately consumed in another province or even abroad. The proportion is significantly higher than the results of other environmental consequences found by previous studies (e.g., 50% for carbon emissions in 2012, 50–60% for air pollutant emissions in 2007) and is 5% points higher than the share of industrial value-added embodied in trade. For more than 90% of provinces, trade-related (i.e., local production for external final demands) hazardous waste generation accounts for over half of the local generation (Figure 3C). Shanghai, Qinghai, and Ningxia are the three provinces most severely affected by export production, with a proportion of trade-related hazardous waste generation exceeding 80% (86% for Shanghai, 83% for Qinghai, and 81% for Ningxia). Even in the least-affected province (i.e., Hubei), the contribution of trade-related production is still as high as 37%.

In addition, trade-related production is more waste-intensive than locally consumed production (i.e., local production for internal final demand, shown in Figure 3B). Industrial activities serving local consumption produce 9.25 t of hazardous waste per one million USD value-added, whereas trade-related production generates an average of 11.69 t (i.e., 26% more) of hazardous waste to achieve the same economic outcome. For around 2/3 of Chinese provinces, producing for external final demands is generally more waste-intensive than for their own final demands. Notably, in Qinghai, driven by mining and processing of nonmetal and other ores (MN_NM), trade-related production generates 145% higher hazardous waste per value-added than locally consumed production.

Reference: Li R, Liu M, Shan Y, Shi Y, Zheng H, Zhang W, Yang J, Fang W, Ma Z, Wang J, Bi J, Hubacek K. Large Virtual Transboundary Hazardous Waste Flows: The Case of China. Environ Sci Technol. 2023 May 30;57(21):8161-8173. doi: 10.1021/acs.est.2c07962. Epub 2023 May 16. PMID: 37192406; PMCID: PMC10233792.

# CLASS ACTIVITY

## HAZARDOUS WASTE IDENTIFICATION AND MANAGEMENT SIMULATION

### *Objective*

Students will learn to identify different types of hazardous waste, understand their characteristics, and develop strategies for proper management and disposal.

### *Materials Needed*

Printed images, safety data sheets (SDS), protective gloves and goggles, whiteboard, markers, role cards, etc.

### *Assessment*

Evaluate students based on their participation, accuracy in identifying and classifying hazardous waste, and the effectiveness of their management plans. Provide feedback on their presentations and teamwork.

## SUMMARY

- Hazardous waste refers to any waste material that poses substantial or potential threats to public health or the environment. This category of waste is characterized by its chemical, physical, or biological properties, which can include toxicity, corrosivity, reactivity, or ignitability.
- Reactive hazardous waste is highly unstable and can undergo violent chemical reactions, potentially causing explosions, fires, or the release of toxic fumes.
- Proper management of hazardous waste involves its identification, classification, storage, transportation, treatment, and disposal. Regulatory frameworks, such as the Resource Conservation and Recovery Act (RCRA) in the United States, are in place to oversee these processes.
- Hazardous wastes are discarded materials with properties that make them potentially harmful to human health or the environment. Hazardous wastes can include things such as chemicals, heavy metals, or substances generated as byproducts during commercial manufacturing processes, as well as discarded household products like paint thinners, cleaning fluids, and old batteries.
- Universal wastes are a special category of hazardous wastes that (in the U.S.) generally pose a lower threat relative to other hazardous wastes, are ubiquitous and produced in very large quantities by many generators.
- Household Hazardous Waste (HHW), also referred to as domestic hazardous waste or home-generated special materials, is waste that is generated from residential households.
- Explosive hazardous wastes are mainly ordnance (artillery) materials. Explosives also involve a high potential for hazards in the case of storage, collection and disposal. These types of waste may exist in solid, liquid or gaseous form.
- Hazardous wastes are wastes with properties that make them dangerous or potentially harmful to human health or the environment. Hazardous wastes can be liquids, solids, contained gases, or sludges. They can be by-products of manufacturing processes or simply discarded commercial products, like cleaning fluids or pesticides.
- Hazardous waste management is the collection, treatment, and disposal of waste material that, when improperly handled, can cause substantial harm to human health and safety or to the environment.

## REVIEW QUESTIONS

1. What are the defining characteristics of hazardous waste?
2. How are the different categories of hazardous waste classified, and what distinguishes them from each other?
3. What are the key principles and methods for managing hazardous waste effectively?

4. What processes and regulations are involved in the safe transport of hazardous waste?
5. How does hazardous waste management impact the environment, and what measures can mitigate negative effects?

## REFERENCES


1. Amadi, C. C., Okeke, O. C., & Amadi, D. C. (2017). Hazardous waste management: A review of principles and methods. International Journal of Advanced Academic Research Sciences, Technology & Engineering, 3(8), 1–20.
2. Bharati, V., Singh, J., & Singh, A. P. (2017). A review on solid waste management methods and practices in India. International Journal of Engineering Research & Technology (IJERT), 10(21), 4065–4067.
3. Brook, D. (2006). Environmental genocide: Native Americans and toxic waste. The American Journal of Economics and Sociology, 57(1), 105–113. https://doi.org/10.1111/j.1536-7150.1998.
4. Enger, E. D., & Smith, B. F. (2004). A study of interrelationships. In Environmental Science. Edward E. Bartell.
5. Fazzo, L., Minichilli, F., Santoro, M., Ceccarini, A., Della Seta, M., Bianchi, F., Comba, P., & Martuzzi, M. (2017). Hazardous waste and health impact: A systematic review of the scientific literature. Environmental Health, 16. https://doi.org/10.1186/s12940-017-0311-8.
6. Iyyanki, V. M., & Vali, M. (2017). Science and engineering for industry. In Environmental Management (1st ed.). Elsevier Inc.
7. Malviya, R., & Chaudhary, R. (2006). Factors affecting hazardous waste solidification/stabilization: A review. Journal of Hazardous Materials, 137(1), 267–276. https://doi.org/10.1016/j.jhazmat.2006.01.065.
8. Martínez, J. H., Romero, S., Ramasco, J. J., & Estrada, E. (2022). The world-wide waste web. Nature Communications, 13(1), 1615. https://doi.org/10.1038/s41467-022-28810-x.
9. Misra, V., & Pandey, S. D. (2005). Hazardous waste, impact on health and environment for development of better waste management strategies in future in India. Environment International, 31(3), 417–435. https://doi.org/10.1016/j.envint.2004.08.005.
10. Roger, B., James, E. S., & David, D. (1989). The safe disposal of hazardous wastes: The special needs and problems of developing countries. A World Bank Technical Paper, 1(93), 154.

CHAPTER

# 6 Medical Waste

## LEARNING OBJECTIVES

- Understand healthcare waste
- Discuss the environmental impacts of biomedical waste
- Explain the impact of Covid-19 on waste infrastructure
- Describe bio-medical waste management

## KEY TERMS FROM THIS CHAPTER

| | |
|---|---|
| Biomedical waste | Biomedical waste treatment |
| Chemical waste | Genotoxic waste |
| Health consequences | Health risk |
| Healthcare facilities (HCFs) | Healthcare units |
| HIV infections | Infectious waste |
| Personal protective equipment (PPE) | Post incineration process |

# INTRODUCTION

Medical waste is a subset of the waste produced by healthcare institutions, including hospitals, clinics for doctors, dentists, blood banks, veterinary hospitals, and research and development centers. Medical waste, also known as regulated medical waste, is generally defined as healthcare waste that may have been tainted by bodily fluids, blood, or other potentially infectious materials.

Waste workers and the public may both be at risk for health problems if needles and other sharps are not disposed of properly. For instance, when containers inside garbage trucks break open or needles are inadvertently sent to recycling facilities, discarded needles may expose waste workers to potential needle stick injuries and infections. If stray sharps pierce plastic garbage bags, janitors and housekeepers run the risk of getting hurt as well. Serious illnesses like hepatitis and the human immunodeficiency virus (HIV) can be spread through used needles.

# 6.1. HEALTHCARE WASTE

The primary producers of healthcare waste are healthcare facilities (HCFs). Healthcare waste (HCW) is the term most frequently used to characterize waste produced by HCFs. Other terms that can be used are clinical waste, biomedical waste, medical waste, and waste from health facilities. All waste produced from HCFs, whether it is a chemical or not, hazardous or non-hazardous, or both, is referred to as hazardous construction waste (HCW). HCWs are thought to make up approximately. 1-2 of all urban waste is generated. 85 of the waste produced because of healthcare operations is classified as non-hazardous.

The remaining fifteen percent comprises of radioactive, toxic, or infectious hazardous materials. Hospitals, clinics, labs, veterinary clinics, research facilities, mortuaries, blood banks, and assisted living facilities make up the bulk of HCW generators. The amount of hazardous waste produced per hospital bed per day (kg/bed/day) varies from 6 kg in low-income countries to nearly 11 kg in high-income countries. However, HCW is frequently not separated into hazardous and non-hazardous waste in low-income nations, which results in a far higher actual amount of hazardous waste produced. (Yazie ., Tebeje, Chufa., 2019) (Figure 6.1).

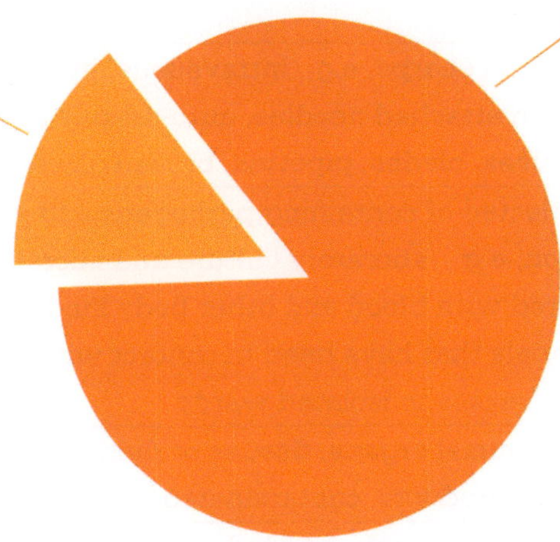


Figure 6.1. Standard waste composition in health facilities.

Source: Janik-Karpinska E, Brancaleoni R, Niemcewicz M, Wojtas W, Foco M, Podogrocki M, Bijak M. Healthcare Waste-A Serious Problem for Global Health. Healthcare (Basel). 2023 Jan 13;11(2):242. doi: 10.3390/healthcare11020242. PMID: 36673610; PMCID: PMC9858835.

An important consideration in HCW management is the state of the economy. Institutions that produce medical waste are required by law to manage this kind of waste in many developed nations. As a result, structures are in place to handle different kinds of waste, and the quantity of hazardous waste produced is continuously tracked. HCW management issues are more common in developing nations that generate hundreds of tons of waste every day. Research conducted in Ethiopia showed that 35 medical facilities gather and discard needles, syringes, and other sharp objects in a way that continuously increases the risk of exposure and injury to medical staff and the public. HCW management techniques including landfilling, recycling, incineration, and storage are commonly used in these nations. Despite being illegal, the most popular method of disposing of hazardous waste is to landfill it without any pre-treatment because it is inexpensive and simple. HCW is disposed of in open pits, combined with municipal waste, and burned after storage. HCW can have a long-term impact on public health and the health of those who handle, collect, and recycle waste. Additionally, the burning of surface waste or untreated medical waste pollution is contaminating fresh water supplies and soil, harming the ecosystem. (Hasan & Rahman, 2018)

## 6.1.1. Categories and Sources of HCW

HCW and its byproducts are produced during medical diagnosis, treatment, medical intervention, or animal or human immunization. Table 6.1 provides an overview of the various categories and materials they cover.

Table 6.1. Categories of Healthcare Waste

| Waste Categories | Descriptions and Examples |
|---|---|
| Hazardous HCW | |
| Sharps waste | Used or unused sharps, e.g., intravenous, hypodermic or other needles, syringes with attached needles, auto-disable syringes, infusion sets, scalpels, knives, blades, pipettes, broken glass and vials |
| Infectious waste | Waste suspected of containing pathogens and presenting a risk of disease transmission, e.g., laboratory culture and microbiological stocks, waste contaminated with blood and other body fluids, waste including excreta and other materials that have been in contact with infected patients with highly infectious diseases in isolated wards |
| Pathological waste | Human and animal organs, tissues or fluids, body parts, fetuses, placentas, unused blood products, contaminated animal carcasses |
| Pharmaceutical waste | Expired pharmaceuticals, unused, contaminated vaccines and drugs, items contaminated by or containing pharmaceuticals |
| Cytotoxic waste | Waste containing substances with genotoxic properties (mutagenic, carcinogenic or teratogenic substances), e.g., cytotoxic drugs used in cancer therapy and their metabolites |
| Chemical waste | Waste containing chemicals, e.g., expired or unused disinfectants or laboratory reagents, unused, solvents, waste with high heavy metal content, such as batteries, blood pressure gauges, mercury-containing equipment and devices (e.g., old thermometers) |

| | |
|---|---|
| Radioactive waste | Waste containing radioactive material, e.g., unused radiotherapy liquids, radioactive diagnostic material, contaminated packages, absorbent paper or glassware, urine and excreta from patients tested or treated with radionuclides, radioactive sealed sources |
| Non-hazardous or general HCW | Waste, which does not pose any biological, chemical, physical or radioactive hazard |

Source: Janik-Karpinska E, Brancaleoni R, Niemcewicz M, Wojtas W, Foco M, Podogrocki M, Bijak M. Healthcare Waste-A Serious Problem for Global Health. Healthcare (Basel). 2023 Jan 13;11(2):242. doi: 10.3390/healthcare11020242. PMID: 36673610; PMCID: PMC9858835.

Healthcare workers are produced in a variety of settings, including clinics for animals, hospitals, medical facilities, private medical practices, clinical laboratories, and pharmacies. Table 6.2 summarizes the various types of HCW that are generated, depending on the source.

Table 6.2. Sources and Examples of HCW

| Source | Sharp Medical Instruments | Infectious Waste | Medications, Chemicals and Potentially Toxic Waste | Non-Hazardous Waste |
|---|---|---|---|---|
| Medical department | Intravenous needles, hypodermic needles, broken ampules and vials | Bandages, dressings, gauze, gloves and masks contaminated with blood or body fluids | Broken thermometers and blood pressure gauges; split medications and disinfectants | Empty bottles, non-blood intravenous tubing, non-blood diapers and bags, packaging, flowers, paper, food scraps |
| Operating room | Needles, blades, scalpels, surgical power tools | Masks, gowns, gloves, gauze and medical equipment contaminated with blood and body fluids; organs and tissues | Anesthetic gases, disinfectant products | Package, uncontaminated medical clothing and medical personal protective equipment |

| | | | | |
|---|---|---|---|---|
| Veterinary clinics | Needles, broken glass and vials, pipettes, Petri dishes, slides and cover slips, needles set blades, scalpels, syringes | Dressings, bandages, masks, gloves, sample tubes and containers contaminated with blood and body fluids; infected animal carcasses | Anesthetic gases, disinfectants, broken blood pressure gauges and thermometers, split medicaments and vaccines, dental amalgam fillings, vaccines; contaminated urine and excreta | Package, containers, paper, uncontaminated masks, gloves, hats, shoe covers and gowns, empty bottles, empty bottles, non-blood intravenous tubing, non-blood diapers and bags, paper, animal food scraps |
| Laboratory | Broken glass, vials, pipettes and slides | Blood and body fluids, microbiological cultures, sample tubes and containers contaminated with blood and body fluids; infected animal carcasses | Broken thermometers, chemicals, such as methanol, fixatives, formalin, toluene, xylene, medications, disinfectants | Package, containers, paper, uncontaminated masks, gloves, hats, shoe covers and gowns, empty bottles |
| Oncology department | Syringes and needles | | Bulk chemotherapy drugs; gloves and materials contaminated with cytotoxic drugs; contaminated urine and excreta | Package, paper |
| Pharmacy store | | | Expired and split medicaments and their components; gloves | Package, paper, empty containers |
| Vaccination process | Syringes and needles | | | |

Source: Janik-Karpinska E, Brancaleoni R, Niemcewicz M, Wojtas W, Foco M, Podogrocki M, Bijak M. Healthcare Waste-A Serious Problem for Global Health. Healthcare (Basel). 2023 Jan 13;11(2):242. doi: 10.3390/healthcare11020242. PMID: 36673610; PMCID: PMC9858835.

### 6.1.2. HCW Production Rate

Globally, the rate of HCW production varies and is influenced by numerous factors. These variables include ways of waste management, the kinds of healthcare facilities and specializations, the quantity of reusable equipment the facility has on hand, and the daily patient volume. Nonetheless, compared to developed nations, registered HCW production is lower in developing nations. Table 6.3 provides comprehensive data on the rate of HCW production in various continents and chosen nations.

Table 6.3. Example of HCW Production Rate in Various Countries Worldwide

| Continent | Countries | HCW Generation (kg/bed/day) |
|---|---|---|
| North America | USA | 8.4–10.7 |
| | Canada | 8.2 |
| Europe | Spain | 3.5–4.4 |
| | Norway | 3.9 |
| | Greece | 0.3–3.6 |
| | France | 2.7–3.3 |
| Asia | Kazakhstan | 5.34–5.4 |
| | China | 0.6–4.03 |
| | Jordan | 2.5–6.10 |
| | India | 0.8–2.31 |
| South America | Argentina | 2.7–3.0 |
| | Brazil | 2.94–3.3 |
| | Ecuador | 2.09–2.1 |
| Africa | Ethiopia | 1.1–1.8 |
| | Egypt | 0.7–1.7 |
| | Sudan | 0.38–0.9 |
| | Morocco | 0.4–0.7 |

Source: Janik-Karpinska E, Brancaleoni R, Niemcewicz M, Wojtas W, Foco M, Podogrocki M, Bijak M. Healthcare Waste-A Serious Problem for Global Health. Healthcare (Basel). 2023 Jan 13;11(2):242. doi: 10.3390/healthcare11020242. PMID: 36673610; PMCID: PMC9858835.

### 6.1.3. Risk Related to HCW

HCW can be harmful and polluting. Individuals near this waste, whether they work in healthcare facilities, handle medical waste, or are exposed through careless actions, may be at risk. The primary at-risk groups are doctors, nurses, healthcare support staff, patients, visitors in healthcare facilities, and workers in support services such as laundry, waste management, transportation, and waste disposal. Over two million medical professionals globally are exposed to pathogens due to their daily work routines. In summary, HCW constitutes a significant danger to human health and lives, particularly in low- and middle-income countries. It is estimated that at least 5.2 million people worldwide die annually, including 4 million children, due to diseases caused by improperly managed medical waste.

#### *6.1.3.1. Infectious Waste and Sharps*

Hazardous waste can include infectious waste, which is dangerous because it is pathogenic and can harm people's health. Always assume that various pathogenic microorganisms may be present in infectious waste. HCWs have the potential to spread over thirty harmful blood-borne infections. When infectious waste is not properly managed, pathogens can enter the human body by ingestion, mucous membranes, rubbing, puncturing, or cutting the skin. Research has shown that medical waste contains a variety of pathogens and that there may be pathways for those pathogens to spread. Thus, it can be said that there is a significant risk to human health associated with this kind of waste.

Needlestick and sharp injury (NSSI) is the leading cause of blood-borne pathogen transmission. An estimated 600,000 to 800,000 percutaneous injuries, including needle stick injuries, are reported in the United States each year. Furthermore, there are about 100,000 NSSIs in the UK annually. According to estimates, improper handling by healthcare workers (HCWs) may have contributed to up to 30% of cases of hepatitis B, 1-3 of cases of hepatitis C, and 0.3 of HIV cases. HBV is about 100 times more contagious than HIV and other blood-borne viral pathogens. Therefore, HBV presents the biggest risk to non-immune healthcare workers in their line of work. Furthermore, compared to other healthcare workers, non-medical waste handlers, or the public, medical waste handlers are most susceptible to HBV infection. The study revealed that compared to non-clinical waste handlers, the prevalence of HBV and HCV was significantly higher in medical waste. We made it very evident why this is the case. Inadequate waste management systems have been linked to an increase in the number of acute injuries and bodily fluid spills. A. Although Sub-Saharan Africa accounts for 70% of the global HIV-positive population, the region only reports 4% of all occupational HIV infections worldwide. Up to 5% of HIV infections in Africa are thought to result from unsafe injection practices, including handling medical waste carelessly and encountering sharp objects. A study carried out in China revealed that nurses had low-risk awareness regarding the possibility of contracting HIV and did not follow standard precautions when working daily. NSSIs can also spread more than 20 other infections, such as malaria, herpes, and syphilis. Despite preventive measures like standard operating procedures and real-time injury-monitoring systems, non-surgical surgical infections (NSSIs) are still reported in developed countries even though most occur in developing nations. In addition to raising the risk of detrimental health effects, these injuries cause mental stress, anxiety, tension, and fear in medical staff members. Implementing safety procedures and requiring healthcare personnel to complete mandatory training programs can lower the incidence of NSSIs and related infections.

#### 6.1.3.2. Chemical and Pharmaceutical Waste

Numerous drugs and chemicals used in healthcare systems can be dangerous. When unwanted or expired chemicals and pharmaceuticals are disposed of, they are typically found in larger quantities than in medical waste. Contact with chemical waste can cause intoxication and has a detrimental effect on human health. The secondary outcome is poisoning from chemical or pharmaceutical substances absorbed through the skin, mucous membranes, inhalation, or ingestion. Thirdly, contact with flammable, corrosive, or reactive chemicals (formaldehyde and other volatile substances) can result in injuries to the skin, eyes, or respiratory tract mucous membranes. When pharmaceuticals are handled carelessly and are primarily disposed of theto sewage systems, they leak into the environment. There have been reports of pharmaceuticals in soil, surface water, and groundwater, among other locations. Primary classes of pharmaceuticals found in environmental specimens include antibiotics, hormones, non-steroidal anti-inflammatory drugs, beta-blockers, cholesterol regulators, and antidepressant drugs. Over a prolonged period, these pharmaceuticals in the environment lead to an acute and chronic damage, alterations in behavior, reproductive issues, and impediments in cell division in animals. The adverse effects of pharmaceuticals on the environment are further demonstrated by the emergence of antibiotic resistance in certain bacterial strains, leading to an accumulation of antibiotics in the environment. Consequently, it is crucial to decontaminate chemical and pharmaceutical waste before disposal in landfills, as inappropriate disposal enables interaction between environmental bacteria and antibiotics, potentially fostering antibiotic-resistant mechanisms among them.

#### 6.1.3.3. Genotoxic Waste

Primary means of encountering genotoxic waste include inhaling and absorbing through the skin. Additionally, ingestion, accidental injections, or other sharp injuries are potential sources of exposure. Contact with a patient's bodily fluids and secretions during chemotherapy is another route of exposure. Cytotoxic drugs, or anticancer medications, are categorized as hazardous substances. Short-term exposure typically results in temporary symptoms like dizziness, headache, nausea, and fatigue. Furthermore, cytotoxic drugs have strong irritating properties, and direct contact can lead to local symptoms such as rashes, dermatitis, skin irritation, mucous membrane ulceration, and throat or eye irritation. Prolonged or repeated exposure to these drugs can have severe consequences, including an increased risk of spontaneous abortions during pregnancy and birth defects in children of women with occupational exposure to anticancer medications. Cytotoxic drugs also pose a threat to the environment, particularly aquatic environments, as some are not fully metabolized and are poorly biodegradable. They can resist conventional biological and chemical processes used in wastewater treatment and challenge water-decontamination technology. Although aquatic cytotoxic drug levels may be below detection limits, they can accumulate and magnify in fauna and flora through bioaccumulation and biomagnification processes, potentially causing harm. Therefore, their impact should be thoroughly evaluated, as delayed effects may appear in offspring. Kovacs et al., found that long-term exposure of zebrafish to anticancer drugs affected their DNA integrity and induced extensive whole-transcriptome changes, which could impact entire zebrafish populations.

#### 6.1.3.4. Radioactive Waste

The kind and degree of exposure determines the type and severity of illness caused by radioactive waste. This can involve far more serious issues along with headache, nausea, and dizziness. In addition to being genotoxic, radioactive waste can also alter genetic material at sufficiently high radiation doses. When radiation diagnostic equipment is handled carelessly, it can result in far more severe harm, such as tissue destruction and, in extreme circumstances, the need to amputate body parts. Severe situations may even be lethal. (Janik-Karpinska et al., 2023)

# 6.2. ENVIRONMENTAL IMPACTS OF BIOMEDICAL WASTE

If biomedical waste is not properly disposed of or burned, airborne particles of infectious diseases can become airborne hazards. Radioactive material diagnostic labs have the potential to contaminate landfills and the atmosphere.

## 6.2.1. The Syringe Tide Environmental Disaster

Medical waste became more widely known during the 1987–1988 syringe tide environmental disaster in Connecticut, New Jersey, and New York, because of medical syringes washing ashore. The syringes threatened people who visited the beach and endangered marine life. The crises compelled scientists and legislators to devise procedures, guidelines, and legislation to ensure that healthcare providers handled their biowaste in an ecologically responsible manner.

## 6.2.2. Effects of Medical Waste on the Environment

Health professionals, the public, and the environment may all suffer direct or indirect health effects from improper handling of medical waste. Indirect effects include the production of millions of used syringes within a period of three to four weeks due to an inadequately planned mass immunization campaign, or toxic emissions from improper burning of medical waste.

Biomedical waste encompasses more than just medical equipment; it also includes medications, waste placed in red biohazard bags, and supplies like bandages and cotton used for patient care. The release of toxins into the waters, which ocean life may subsequently consume, is the most dangerous impact that biomedical waste has on our seas. When humans eat marine life, toxins will eventually end up in the food chain and into our bodies. Exposure to these toxins can lead to birth defects and stunted human growth and development.

The environment is also seriously threatened by the medical industry's heavy reliance on plastic. All medical equipment is composed of 85% disposable plastic materials, claim North and Halden. Because plastic materials are so unique in that they can be strong, lightweight, and economical without sacrificing the sterility of medical equipment, we currently rely heavily on them. This quantity of single-use plastics can accelerate the cumulative negative health effects of macro- and microplastics, in addition to the grave health risks associated with the discharge of toxic chemicals into the environment from medical waste disposal sites.

## 6.2.3. Incineration of Biomedical Waste

### 6.2.3.1. Methods of Biomedical Waste Incineration

There are three different types of medical waste incinerators: rotary kilns, excess air, and controlled air. Modular combustion, two-stage incineration, and starved-air incineration are other names for controlled air. In this process, waste is fed into a combustion chamber, where the combustion air starts to dry out and helps the waste to volatilize. Carbon dioxide and other surplus gases are consequently released into the atmosphere.

The excess air process is the second kind of incineration. The way the waste is dried, ignited, and burned by the heat from the main chamber burner is comparable to the controlled air process. The primary distinction is that the waste's volatile ingredients and moisture are evaporated.

The procedure in a rotary kiln is comparable to the two mentioned above, but it is more adaptable because it can combine dry and wet waste components, and many waste engineers believe it to be the most environmentally friendly.

### 6.2.3.2. Impact on the Environment

Hazardous ash residue is created during the incineration process and is frequently dumped in landfills. Since there is no barrier protecting these landfills, the residue may end up in subterranean water that is frequently used by people. Toxic gases released when plastic material is burnt, escape and mix with the surrounding air. Exposure to these gases in humans and animals can result in long-term respiratory and health problems. The health of plants and animals is also impacted by the rotation of toxic air, in addition to humans. Incinerator-related air pollution damages crops and forests, weakens the ozone layer, and exacerbates the effects of climate change. Continuous exposure to these pollutants and chemicals in the air may be harmful to trees and other plants, eventually resulting in the extinction of some species in particular regions. Tree fruits can become poisonous and thus unfit for consumption due to pollution and chemical leaks. (Abdo et al., 2016)

## 6.2.4. Environmentally Friendly Alternatives

Reusable RMW or sharps containers cut down on $CO_2$ emissions and the quantity of plastic dumped in landfills.

The four main methods used in non-incineration treatment are thermal, chemical, irradiative, and biological. The treatment technology's primary goal is to eliminate pathogens from waste to decontaminate it. Medical professionals and hospitals can now dispose of medical waste in an environmentally friendly manner thanks to the mechanics invented by modern technology. These methods include gasification, autoclaving, plasma pyrolysis, microwave irradiation, chemical methods, and gasification. These substitutes are quite adaptable and applicable to any kind of waste.

Like a pressure cooker, an autoclave kills microorganisms by penetrating waste materials with high-temperature steam. It is advised to autoclave waste related to microbiology and biotechnology, sharps waste, and soiled and solid waste. The idea behind producing high frequency waves is the foundation of microwave irradiation. The pathogens are killed from within by the heat produced by the vibrations these waves create in the waste material's particles. Chemical disinfection is a straightforward but efficient technique; one percent hypochlorite can eradicate living microorganisms. An eco-friendly process called plasma pyrolysis turns organic waste into byproducts that can be sold. It can safely and reliably dispose of all waste types, including hazardous waste, biomedical waste, and municipal solid waste, thanks to the intense heat produced by the plasma. Gasification can reduce the carbon footprint of biomedical waste treatment by providing energy generation and carbon sequestration.

# 6.3. IMPACT OF COVID-19 ON WASTE INFRASTRUCTURE

Globally, the rates of solid waste generation in urban areas are rising quickly due to factors such as population growth, economic expansion, shifting consumption habits, and urbanization. With a per capita waste footprint of 0.79 kg per person per day, more than 2.24 billion tons of solid waste were generated globally in 2020. In a scenario where business as usual continues, annual waste generation is projected to rise by 73 from 2020 levels to 3 point88 billion tons in 2050. This generated waste presents many risks to humans and the ecosystem at different scales if it is not appropriately collected, disposed of, or treated.

The sustainability of urban systems depends on the environmentally sound management of solid waste. Countries, income levels, consumption and lifestyle patterns, seasonal variations, and different sectors and infrastructures all have different rates and compositions of waste generation. Furthermore, the infrastructure of municipal waste management faces several challenges due to political and financial factors. Special situations, like natural disasters, create a significant amount of unique waste, usually in a short amount of time, adding to the already known difficulties. The COVID-19 pandemic was one such instance that presented significant difficulties for infrastructure systems at all levels, including the waste management infrastructure. In contrast to natural calamities like earthquakes and tropical storms, the COVID-19 pandemic disaster persisted for more than two years, upending every system on the planet.

The COVID-19 pandemic caused an unusual spike in waste, mostly infectious medical waste, which had a significant effect on municipalities' waste infrastructure. The pre-COVID-19 waste management systems were not equipped to handle the significant variations in quantity and quality, as they were intended to handle only mild variations. Although appropriate waste management is essential for both health and safety, it becomes even more crucial during pandemics to make sure that potentially infectious materials are disposed of properly to stop the virus from spreading. As a result, all nations were forced to promptly modify their waste management protocols to meet the increased demand and safeguard vital personnel and the public from the risks posed by the massive amounts of infectious waste generated. The inherently uncertain nature, dynamic nature, and dearth of recent historical precedents to inform informed decision-making further impeded the waste management sectors' capacity to effectively address these novel challenges. (Das et al., 2021)

The United Nations Environment Program report highlights two well-known sayings in the emergency management infrastructure: "The onset of a critical incident is the wrong time to exchange business cards" and "In the pandemic is usually the wrong time to try to install new healthcare waste management systems and practices from scratch." It is, therefore, crucial to examine waste management systems from a resilience standpoint. Although it may not be as commonly used in the context of waste management, resilience is not a novel concept. Waste management systems that are resilient should be able to continue operating at a satisfactory level even in the face of ongoing strains and unexpected shocks. To do this, the system needs to be able to respond and adjust to changes in the infrastructure that is available as well as, most importantly, changes in the kinds and quantity of waste that is generated.

Although the COVID-19 pandemic is no longer as serious of a threat, the lessons learned from it could still present opportunities for building resilient urban waste management systems that will allow them to handle similar situations in future. Some of the literature that has already been published has covered the effects and difficulties of COVID-19 on the short-term waste management procedures for infectious medical waste; however, the research community has not given much thought to the lessons waste management specialists have learned from the pandemic or to the strategies that have the best chance of being implemented in the long run to create resilient waste management systems. Therefore, by examining the dynamics of waste generation during COVID-19 and identifying long-term opportunities in developing resilient and sustainable waste management strategies, this study aimed to close knowledge gaps.

To observe the patterns and trends in waste generation, a thorough literature review of the case studies that have already been completed was carried out. Two major waste streams were considered: the infectious medical waste produced by healthcare facilities and the non-medical waste streams produced by residential and other sectors. The review was expanded to examine the short-term waste management plans implemented during COVID-19 and the difficulties that went along with them. After identifying a few significant long-term opportunities, this study offered suggestions for possible upgrades to the waste management infrastructure. From a long-term operational standpoint, the lessons learned from COVID-19 and the opportunities that may arise are significant because they may help policymakers and urban planners better support innovative and systematic approaches to waste management.

## 6.3.1. Waste Generation Dynamics and Associated Waste Management Challenges during the COVID-19 Pandemic

Numerous studies already conducted on COVID-19's effects on waste have demonstrated that the pandemic significantly affected the waste infrastructure. Although the overall patterns are visible on a global scale, depending on the severity of the impact, each municipality faced different difficulties and experiences. Some researchers have tried to tabulate the estimated generation of pandemic-related medical waste across countries, even though the data regarding the percentage of the change in waste generation before and during the pandemic are limited. We will only discuss a few key global statistics and region-specific trends related to the two main waste streams generated during the COVID-19 pandemic: infectious medical waste generated from healthcare-related facilities and non-medical waste generated from residential and other sectors to gain a comprehensive understanding of the variations in the waste generation trends across those regions. Table 6.4 provides a summary of some of the pertinent published data on the dynamics of waste generation; the subsection that follows goes into more detail about the patterns.

There are eight main categories into which medical waste can be divided: pathogenic, infectious, chemical, cytotoxic, radioactive, pharmaceutical, sharps, and non-hazardous waste. Pre-COVID healthcare waste was generally reported to be non-hazardous at a rate of 85%. This certainly changed during the COVID-19 pandemic; all the contaminated medical waste was deemed dangerous, even though the exact percentage of the increase was not disclosed. (Mondal et al., 2022)

All frontline and healthcare workers were required to wear a full PPE set, which included gloves, face masks, head coverings, goggles, surgical gowns, and shoe covers, to reduce the associated risk. The World Health Organization estimated that 89 million medical masks, 76 million examination gloves, and 11.6 million goggles would be needed globally each month for healthcare professionals during the early stages of the COVID-19 pandemic. A different study calculated that 65 billion gloves and 129 billion facemasks would be needed each month, globally. During the pandemic, the average density of medical waste was changed while the quantity kept increasing due to the use of personal protective equipment (PPE). Disposable personal protective equipment (PPE) such as surgical gowns and goggles, as well as N95 masks and single-use masks are predominantly composed of plastics (>80%), with polypropylene, polyurethane, or polyacrylonitrile frequently found in them. Thus, incorrect COVID-19 medical waste disposal also added to the plastic pollution problem.

The Asian Development Bank estimated that every infected patient in China produced 3.4 kg of biomedical waste per day, based on initial experience managing COVID-19 patients. Given the total number of COVID-19 cases in that region, the hospitalization rate, and the days of isolation, this estimation served as the foundation for calculating the overall waste generation rates in numerous case studies.

The changes in the waste generation rates for the medical and non-medical waste streams during COVID-19 for a subset of case studies are summarized in Table 6.4.

Table 6.4. Dynamics of Waste Generation During COVID-19

| Region | Waste Type | Change in Waste Generation During COVID-19 |
|---|---|---|
| Wuhan City, China | Medical Waste | 6 times more waste |
| Healthcare Facility, Jordan | Medical Waste | 10-fold increase |
| Malaysia | Medical Waste | 27% increase |
| India | Medical Waste | 10% more waste |

| | | |
|---|---|---|
| Bangladesh | Medical Waste | 80% increase |
| Tehran, Iran | Medical Waste | 17.6–61.8% increase |
| Bangkok, Thailand | Medical Waste | 5 times more waste |
| Jakarta, Indonesia | Medical Waste | 5 times more waste |
| Manila, Philippines | Medical Waste | Additional 280 tons/day |
| Hanoi, Vietnam | Medical Waste | Additional 160 tons/day |
| Kuala Lumpur, Malaysia | Medical Waste | Additional 154 tons/day |
| Survey from 23 Countries | Food Waste | 43% increase |
| Survey from 23 Countries | Plastic Waste | 53% increase |
| Ontario, Canada | Residential Organic Waste | 20% increase |
| Ontario, Canada | Residential Garbage | 15% increase |
| Ontario, Canada | Residential Recyclables | Unchanged |
| Ontario, Canada | ICI Waste | Declined, data not reported |
| City of New York | Organic Waste | 13.3% more waste |
| Brazil | Residential Recyclables | 25% increase |
| Tokyo, Japan | Residential Waste | 110% increase |
| Tokyo, Japan | ICI Waste | 57% decline |

Source: Jayasinghe PA, Jalilzadeh H, Hettiaratchi P.The Impact of COVID-19 on Waste Infrastructure: Lessons Learned and Opportunities for a Sustainable Future. Int J Environ Res Public Health. 2023 Feb 28;20(5):4310. doi: 10.3390/ijerph20054310. PMID: 36901318; PMCID: PMC10001637.

Hazardous medical waste reached a peak of 240 tons per day during the outbreak in Wuhan City, China, which is almost six times more than what it was before the pandemic. Nevertheless, their daily capacity for incinerating medical waste was limited to 49 tons. According to a study done using data from a Jordanian healthcare facility, the rate at which waste was generated during the pandemic increased ten times in comparison to pre-pandemic levels. Malaysia reported a 27% rise in healthcare waste generation linked to COVID-19-related activities. India produced 600 tons of biomedical waste per day because of the pandemic, an increase of roughly 10%. Within a year, medical waste generation increased by 80%, according to a study done in Bangladesh. The COVID-19 pandemic caused a 17.6–61.8 increase in medical waste generation in Tehran. Similarly, additional medical waste in the cities like Manila, Jakarta, Kuala Lumpur, Bangkok, and Hanoi reached 150–280 tons per day. The waste infrastructure of major cities faced several challenges due to the disparity between the amount of medical waste produced and the treatment capacity that was available.

Most cities saw a rise in the production of non-medical waste due to the lockdowns, orders to stay at home, travel restrictions, and employee layoffs. The other general waste streams, such as residential, commercial, and institutional (ICI), and construction, renovation, and demolition (CRD) waste, are referred to as non-medical waste streams. The higher level of consumption, which included buying food in bulk, sanitizers and cleaning supplies on a whim, placing internet orders, and the associated packaging, was linked to the increases in household waste. Both the residential and ICI waste generated during the COVID-19 pandemic included potentially infectious materials, such as masks, gloves, and tissues; however, quantitative information on the infectious vs. non-infectious categories in the residential and ICI waste streams is not available.

Investigated by Filho et al., revealed that, in comparison to pre-pandemic levels of waste generation, there had been increases in food waste of 43% and plastic packaging waste of 53%. However, there appeared to be some variance in the amount of waste generated by various subsectors, and there was a possible shift in the potential tonnage burden between the residential and ICI sectors. The patterns in ICI and CRD waste showed that decreased activities were usually the cause of the tonnage of ICI and CRD waste that was collected declining. Considering Roy et al., in Ontario, Canada, the rate at which residential waste is generated increased dramatically by 15% for garbage and 20% for organic waste; recyclables did not change, but there was a decrease in ICI waste. New York produced 13.1 more organic waste during the COVID-19 pandemic. Brazil saw a nearly 25% increase in the amount of recyclable residential waste material in its cities. Similarly, Tokyo reported a 110% increase in residential waste over pre-COVID levels and a 57% decrease in ICI waste generation because of lockdown procedures and business closures. Some cities displayed contradictory behavior, with a decreasing trend during different waves of the pandemic, despite the majority showing a similar increasing trend of residential waste. This behavior may have been caused by variations in consumption patterns and the associated socioeconomic stresses.

When compared to the residential and other sectors, the infectious medical waste produced by healthcare facilities had the biggest influx of changes in waste generation dynamics, according to statistics from numerous case studies conducted globally. More possibly infectious materials were found in waste streams from the residential and other sectors than in the past; however, because gathering this data is difficult, there isn't much of it in the literature. During the COVID-19 pandemic, there were several associated challenges with the waste infrastructure. These included the absence of appropriate medical waste management systems to handle the sudden surge in infectious biomedical waste, the rise in demand for single-use plastic due to the need for personal protective equipment (PPE) and other healthcare items, a decrease in recycling efforts and their diversion to illegal dump sites and landfills, an increase in mixed waste that included infectious waste with minimal source segregation, residential waste volumes that exceeded the capacity of the municipal solid waste management systems, and a decrease in ICI waste as a result of the slowdown in the economy. Waste management systems and procedures needed to be quickly adjusted to account for these variations and safeguard the public and workers from infection.

To comprehend the intricacy of COVID-19's effects on waste infrastructure and to condense the lessons learned from the case studies, we created a causal loop diagram (CLD), as illustrated in Figure 6.2. A few fundamental components constitute the

CLD: the variables, the links that connect them, the signs that indicate the links, and the loops that illustrate the relationships between the variables. One of the main objectives for handling waste during the pandemic, according to the CLD in Figure 6.2, is to reduce the risk of transmission. As was previously mentioned, the general trends of the waste shifts showed that residential waste was increasing while ICI waste was decreasing. On the other hand, in terms of the overall quantity, the increases and decreases across various subsectors would have largely cancelled each other out. The case studies demonstrate that once these cities got past the initial obstacles related to the capacity of the existing waste treatment facilities, they were able to continue with their regular operations. As a result, reducing the risk of transmission also requires effective treatment and disposal procedures.

Reducing the risk of transmission can also be achieved by designing automated waste collection systems, encouraging reuse at higher levels, and manipulating the system to produce as little waste as possible. These strategies may also prove useful in the future for handling situations that may be similar.

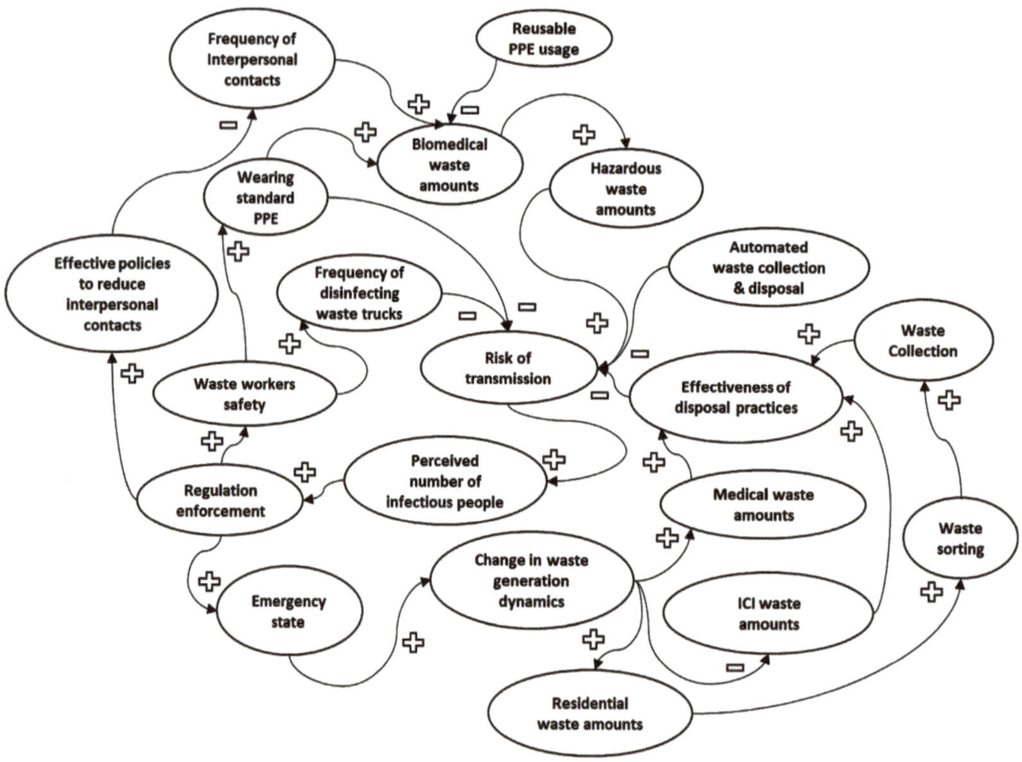


Figure 6.2. Causal loop diagram for COVID-19-related waste generation.

Source: Jayasinghe PA, Jalilzadeh H, Hettiaratchi P. The Impact of COVID-19 on Waste Infrastructure: Lessons Learned and Opportunities for a Sustainable Future. Int J Environ Res Public Health. 2023 Feb 28;20(5):4310. doi: 10.3390/ijerph20054310. PMID: 36901318; PMCID: PMC10001637.

## 6.3.2. Key Lessons Learnt and Opportunities

The COVID-19 pandemic has had a significant impact on environmental and public health, posing numerous challenges at various scales to every aspect of municipal infrastructure. During the pandemic, authorities worldwide needed to prioritize the public health sector; however, other equally important sectors should not be disregarded. The waste infrastructure was one of these little-known but crucial sectors that eventually required attention to avoid severe disruptions to the environment and society. The main obstacles were a lack of suitable components in the waste infrastructure system, such as safe disposal methods that could handle the unexpected increase in plastic- and medical-based waste streams, thorough and organized instruments, flexible and alternative technologies, and regulations. If these difficulties are approached thoughtfully and constructively, they also present chances for advancements and innovations.

### *6.3.2.1. Healthcare Waste Management Systems*

The volume of waste generated by healthcare facilities worldwide increased faster than waste management systems could handle the increase in hospitalizations. Every nation or urban system's ability to handle the growing amounts of medical waste is dependent on several factors, including the creation and implementation of policies; the facilities and methods currently in place for waste collection, transportation, disposal, and treatment; the processes and capabilities currently in place for handling waste; and the funds allotted for medical waste management. Effective medical waste management must include the proper identification, collection, separation, storage, transportation, disposal, and treatment. Consequently, another area that requires special attention is productive medical waste administration, which includes knowledge of the required standards, guidelines for disinfection, personnel protection from related risks, training, and monitoring-related aspects.

Managing waste in healthcare has never been easy. Many nations, particularly developing ones, do not have appropriate protocols in place for the handling and disposal of medical waste, which can have detrimental effects on the environment, including air and groundwater contamination and the possibility of viral outbreaks. The World Health Organization reports that 30% of healthcare facilities—and 60% of them in the least developed nations—are ill-equipped to manage COVID-19 waste loads on top of the ones they already have. When seen positively, the pandemic has compelled these nations to address these overlooked issues and recognize healthcare waste management as a vital field of concern.

Some of the major obstacles to implementing safe waste management practices were the financial limitations to applying appropriate collection, transportation, disposal, and treatment processes as well as a general lack of education and guidelines on the proper handling of medical waste. Many organizations, such as the World Health Organization (WHO), the European Union (EU), the United Nations Environment Programme (UNEP), and the European Centre for Disease Prevention and Control (ECDC), released updated management guidelines for infectious waste in response to the COVID-19 outbreak. These guidelines emphasised on the infectious medical and household waste streams associated with the virus. Additionally, since the research community is now more interested than ever in this field, several researchers have proposed a set of best practices for the safe handling of medical waste. If these guidelines and best practices have not yet been incorporated into the current systems, now is the right time to do so while keeping in mind that medical waste management is one of the most important components of the waste infrastructure. (Ragazzi et al., 2020)

During the COVID-19 pandemic, another area of concern was the incapacity of the current treatment facilities to handle variations in the excess waste generated by healthcare centres. Some regions' treatment facilities were so full that they were forced to burn uncontrolled waste and dump waste illegally in suburban areas. The incorrect waste disposal caused by the pandemic may result in environmental destruction in the post-pandemic world, given the long-term effects on the environment and human health. It's also important to note that different approaches were taken in various regions to manage the excess waste caused by the pandemic. Some common practices to deal with the excess waste included providing mobile incineration or autoclaving as an alternative system, using cement kilns and other industrial furnaces as alternative treatment facilities, and using alternative technologies.

In the medical waste industry, safe landfill disposal and incineration are two of the most popular management strategies. Ash residues and other hazardous compounds, as well as harmful air pollutants, can be released during uncontrolled incineration or the burning of inappropriate materials. Before being dumped in a landfill, the solid waste requires chemical or thermal sanitization on site to make it non-pathogenic. Air, soil, and water contamination can result from the improper disposal of untreated medical waste in landfills. In terms of treatment, it might have been wiser to combine different technologies rather than relying solely on one technique to reduce the additional demand. Autoclaving, microwaving, gas sterilization, irradiation, thermal inactivation, chemical disinfection, and hydrothermal carbonization are a few of the practical substitute technologies. Any of these technologies could be adopted, but doing so would depend on several factors, such as their affordability, capability, adaptability, and technological maturity, as well as concerns about the environment and the economy. (Jayasinghe et al., 2023)

#### *6.3.2.2. Integrated and Decentralized Waste Management*

The COVID-19 pandemic was much more than just a worldwide health emergency; it also added both acute and chronic shocks impacted every system of infrastructure, the economy, and society at large. It is difficult to get back to normal; therefore, a long-term comprehensive plan for the solid waste infrastructure must be developed.

Some of the major obstacles in addressing the COVID-19-related waste were identified as the limited resources and technology options, as well as the limited adaptability and capability of the current waste generation, collection, transportation, and disposal systems for managing the increased volume of healthcare waste.

It takes more than one strategy to successfully reduce the risks from waste. The optimal strategy might involve combining a few different activities as described in the integrated solid waste management (ISWM) model. The ISWM, which addresses all the sources, actions, and facets related to waste management, is a strategic approach to the sustainable management of solid waste. The five Rs of waste reduction at the source, product reuse, material recycling, energy recovery, and residual management are referred to as the ISWM hierarchy in this context, and it may be a useful tool. In contrast to the more complex reactive approach, which treats waste after it has been produced, this strategy encourages a less complex proactive approach to waste management, which involves anticipating problems and preventing them from occurring by producing less waste or zero waste.

Due to their function as a "system of systems" and their simple to complex interdependencies, which can make critical functions susceptible to cascading failures, the waste infrastructure of an urban system is intricately linked to other critical infrastructures. To put it another way, a waste management system is set up like a network because it consists of a network with multiple nodes and support networks made up of roads, buildings like incinerator facilities, storage centers, and utilities. This is a reciprocal relationship: every urban system's infrastructure system produces waste of some kind, and the waste infrastructure is dependent on all other infrastructures as well. Landfills compete with other infrastructures for space, waste collection depends on trucks and transportation, and sorting facilities require energy. Waste treatment facilities are physical buildings. While acknowledging the negative effects of these interdependencies, it is crucial to view them positively because they offer chances to improve sustainability and resilience. An integrated infrastructure system approach offers significant advantages over the traditional silo-based approach of analyzing infrastructures in isolation. It does this by improving resilience and system-wide efficiencies, minimizing service disruptions and rehabilitation work and costs, promoting sustainability goals such as resource minimization and proper allocation, and examining the lifecycle of individual components. It also involves identifying and managing short- and long-term impacts in addition to many other environmental and economic benefits.

There are several ways that infrastructures can be integrated. Combining decentralization and integration—where integration occurs in a decentralized manner rather than solely in a centralized location—is one of the appealing aspects. Nevertheless, the growth of healthcare facilities and the construction of new facilities, such as hospitals, isolation wards, temporary quarantine camps, and testing centers, posed additional challenges in terms of adhering to the protocols related to waste segregation, storage, transportation, and disposal. During COVID-19, the surge in healthcare waste largely choked the existing waste management systems. The lack of in-situ or adjacent waste treatment facilities increased the transportation of large amounts of infectious waste over long distances, which increased contamination and jeopardized worker safety in addition to the capacity limitations of central waste management facilities. This emphasizes how crucial it is to incorporate onsite and decentralized waste management facilities into the waste infrastructure. Simple methods include composting in the backyard, in-situ biodigesters, incineration units, and domestic waste segregation. Urban areas may be able to become less reliant on central waste management facilities if decentralized systems are implemented with active citizen participation and adhere to the 5R principles in ISWM.

#### *6.3.2.3. Quantity and Quality of Waste, System Thinking, and Novel Tools*

Daily updates and worldwide statistics about public health and environmental issues, including the number of cases and deaths and real-time data from air pollution monitoring, were available during the COVID-19 pandemic. Nevertheless, there was no comparable format for widely accessible waste generation, composition, and recycling data. Even though it can be difficult, it is crucial to collect accurate, up-to-date data regarding the quantity and quality of waste produced, including its type, amount, and amount recycled thus far, as well as the remaining amounts that need to be disposed of. The waste management sector would benefit from having this information to quickly identify the main problems and to quickly identify and prioritize mitigating actions, particularly in emergency scenarios. However, through alterations in consumption and behavior patterns, public awareness of these issues may have had some beneficial effects on the waste infrastructure. (Naughton C.C., 2020)

Sector-specific waste generation trends were impacted differently by pandemic-related measures implemented in different sectors, as demonstrated by the case studies previously discussed. When working from home, waste from businesses and schools decreased but waste from homes increased. A comprehensive set of models, such as system dynamic models and input–output models, are required to comprehend and accurately evaluate the quantity (i.e., waste generation volumes or rates by a sector, and waste composition) and quality (i.e., infectious vs. non-infectious) of the waste. For a better understanding of these trends and the interdependencies among the waste sectors, for instance, a systems thinking approach combined with causal loop diagrams would be helpful; one such diagram is shown in Figure 6.2.

The urban metabolism approach is another tool that can be used to quantify the mass and energy balances of the inputs, outputs, production, consumption, and waste of an urban system. Both quantitative and qualitative models and tools should incorporate all the intricate and interdependent components to more accurately depict the system, while also being easy to use and validate in a variety of scenarios.

However, without trustworthy data, these models and tools are useless. Consistent and trustworthy data are also necessary for integrating new systems, enhancing efficiency, and modifying current waste management systems. Comprehensive models and innovative techniques could be used to predict and estimate the data, such as the generation rates and waste composition, even though data scarcity is still a crucial issue to investigate. For this reason, the absence of these models and tools creates an additional challenge in circumstances like the COVID-19 pandemic.

Instead of using traditional tools, there are chances to implement innovative and reliable methods, like utilizing tools and techniques enhanced by artificial intelligence (AI) in waste infrastructure, particularly for precise waste trend forecasting. Artificial neural networks (ANN), support vector machines (SVM), decision trees (DT), and genetic algorithms (GA) are a few examples of AI-coupled models that have been shown to have a significant impact on waste treatment modeling and optimization as well as waste prediction, classification, collection, and transportation.

Precise predictions of waste patterns through AI-powered instruments would be crucial for managing waste demand and optimizing the frequency and routes of waste collection, offering numerous advantages for the environment and the economy. Waste collection robots, autonomous waste collection trucks, automated sorting systems, and smart recycling bins are a few instances of how AI-based techniques are being used in waste management. These methods can be very helpful in the fight against infectious diseases like COVID-19 because they are immune to infection and prevent contamination from humans.

In addition, improved healthcare waste prediction models and tools—such as a non-linear, multi-level regression model—along with the related social and economic indicators would be helpful in the planning, designing, budgeting, and optimization processes, as well as in boosting sustainability in the healthcare waste management industry.

### *6.3.2.4. Product Design and the Circular Economy Framework*

The COVID-19 pandemic led to a notable surge in emand for single-use plastic due to the necessity of personal protective equipment (PPE), sanitary napkins, and other healthcare supplies, as well as packaging. Supply chain problems also arose in this scenario, particularly since these systems depended on disposables. Even though plastics were crucial to our survival during the pandemic, the amount of unmanaged plastic waste piling up is causing a global ecological disaster because it could result in microplastic pollution, which could have both short- and long-term effects on the environment and human health. Managing plastic waste has never been easy. A small portion of the plastic waste generated during the COVID-19 pandemic was recycled, mostly because recycling facilities prohibited the use of potentially contaminated plastics. Most of the waste was landfilled or burned. Plastic waste shouldn't be disposed of in landfills because it might not break down there and burning it could cause air pollution.

Therefore, even though it would be difficult, quick action is needed to move toward sustainable consumption and production as well as to investigate recycling and alternative treatment options for plastic waste management. In this context, moving away from the conventional and unsustainable linear economy approach and toward the circular economy approach has garnered a lot of attention in recent years. The circular economy encourages cyclical resource flows in the production-consumption system within a closed loop, focusing on minimizing energy and natural resource consumption, achieving zero waste goals, and mitigating associated environmental impacts throughout the system's lifecycle. It also offers opportunities to identify best practices and advance sustainability.

As the circular economy also uses the "cradle-to-cradle" approach as a framework, eco-designs and biodegradable products are essential. When designing a product, considering every stage of its lifecycle promotes best practices, such as using recycled and secondary resources as raw materials, creating eco-designs, using biodegradable materials in production, and boosting recycling and waste-to-energy recovery. To integrate products like bioplastics into the market, however, comprehensive techno-economic and environmental footprint assessments for industrial-scale applications are required. Initiatives during the COVID-19 pandemic, such as manufacturing reusable cloth masks and disinfectants from leftover products, provided evidence. These methods supported the supply chain by reducing reliance on imported goods and gave consumers financial savings in addition to using locally accessible resources to create sustainable products. Therefore, it is essential to integrate the circular economy into the healthcare sector to minimize import dependency in the crucial healthcare value chain, as well as to be resilient and sustainable over the long term.

Researchers in the past have also proposed that one of the most important ways to improve the sustainability of the healthcare industry is to adopt systematic and greenification strategies, such as reducing medical waste, adopting green energy sources, reusing, recycling, and refurbishing medical products, using resources efficiently, and sustainable procurement.

Furthermore, even though end-of-pipe plastic treatment is the least preferred option in the waste management hierarchy, it still needs to be improved. This means that new technological approaches, like integrated mechanical and chemical recycling processes, must be developed in addition to waste prevention, reduction, and recycling.

#### *6.3.2.5. Implications for Policy, Protocols, and Guidelines*

Some researchers have pointed out that the lack of adequate policies, protocols, and guidelines in the waste infrastructure prevented authorities from acting quickly for waste collection, classification, transportation, and disposal—especially in the early stages of the crisis. Among the problems noted are the following: illegal waste pickers posed a risk to the community; waste operators struggled to maintain manual collection and sorting procedures; and illegal waste disposal in some locations created an additional risk to the community. It is therefore essential to incorporate appropriate waste disposal procedures and give the populace protective logistics.

In keeping with other previously discussed measures, it is appropriate to mandate international waste management and local authorities to thoroughly review and enhance current policies, procedures, guidelines, and best practices as necessary to enhance the efficiency of the waste infrastructure following the pandemic. When creating policies to increase waste management's resilience and efficiency, it is necessary to thoroughly analyze the short-, long-, and worst-case scenarios to avoid approaching emergency management from a business-as-usual perspective. Examples include reviewing international and national guidelines for the management of biomedical waste; updating emergency preparedness measures in the waste infrastructure; assigning temporary, cost-effective facilities that can be used to continue waste management operations; reviewing local collection, handling protocols, and sanitation worker protection; prohibiting the sorting of waste by hand; improving funding for current waste management initiatives; giving consumers financial incentives to reduce waste generation; expanding public-private and national-international partnerships; and encouraging extended producer responsibility (EPR) and collaborations.

A summary of all the major prospects mentioned above can be found in Figure 6.3. Apart from the technological, economic, and environmental aspects were examined, maintaining proper waste management practices also requires behavioral shifts and social cooperation. One of the main ideas is that waste management requires a team effort from all citizens, who must each exercise caution and accountability for their own actions and consumption habits. This goes beyond the authority of waste managers. (Jayasinghe et al., 2023

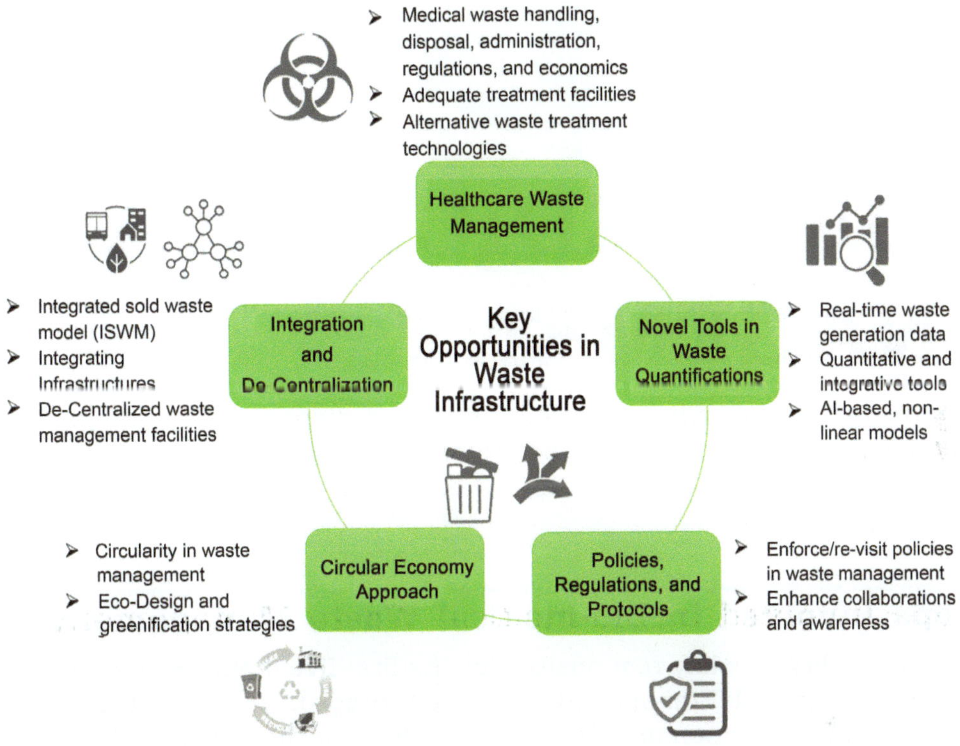


Figure 6.3. Opportunities in the post-pandemic waste infrastructure.

Source: Jayasinghe PA, Jalilzadeh H, Hettiaratchi P. The Impact of COVID-19 on Waste Infrastructure: Lessons Learned and Opportunities for a Sustainable Future. Int J Environ Res Public Health. 2023 Feb 28;20(5):4310. doi: 10.3390/ijerph20054310. PMID: 36901318; PMCID: PMC10001637.

# 6.4. BIO-MEDICAL WASTE MANAGEMENT

The appropriate handling, collection, treatment, and disposal of waste produced during healthcare operations is referred to as "biomedical waste management." Materials contaminated with blood, bodily fluids, or other potentially infectious materials fall under this category.

## 6.4.1. Steps Involved in Bio-medical Waste Management

The Healthcare Facility is solely responsible for the first five steps (Segregation, Collection, Pre-treatment, Intramural Transportation, and Storage). Treatment and disposal are primarily the operator's responsibility, except for lab and highly infectious waste, which it must pre-treat.

The following are the responsibilities of Healthcare Facilities (HCF) for the management and handling of bio-medical waste:

- Biomedical Waste should be segregated at the point of generation by the person who is generating the waste in a designated color-coded bin/ container
- Biomedical Waste & General Waste shall not be mixed. Biomedical Waste & General Waste shall not be mixed. Storage time of waste should be as little as possible so that waste storage, transportation and disposal are done within 48 hours.
- Phase out the use of chlorinated plastic bags (excluding blood bags) and gloves.
- No secondary handling or pilferage of waste shall be done at healthcare facilities. If Common Bio-medical Waste Treatment Facility (CBWTF facility) is available at 75 km from the HCF, bio-medical waste should be treated and disposed of only through such CBWTF operator.
- Only Laboratory and Highly infectious waste shall be pre-treated onsite before sending for final treatment or disposal through a CBWTF Operator.
- Provide bar-code labels on all color-coded bags or containers containing segregated bio-medical waste before such waste goes for final disposal through a CBWTF.
- The management of bio-medical waste can overall be summarized in the following steps;

- Waste Segregation in color-coded and barcode-labeled bags/ containers at the source of generation
- Pre-treat Laboratory and Highly infectious waste
- Intra-mural transportation of segregated waste to the central storage area
- Temporary storage of biomedical waste in the central storage area
- Treatment and Disposal of biomedical waste through CBWTF or Captive facility

## 6.4.2. Biomedical Waste Segregation

At the point of generation, biomedical waste produced by a healthcare facility must be separated according to the prescribed color coding. Following activities to be followed to ensure proper waste segregation:

- Waste needs to be separated at the source when it is generated, not afterwards. "Point of Generation," refers to the area where wastes originate, gather, and are under the supervision of medical personnel, nurses, etc. who are treating the patient and producing biomedical waste in the process.
- Posters/placards for bio-medical waste segregation should be provided in all the wards as well as in the waste storage area.
- An adequate number of color-coded bins/containers and bags should be available at the point of generation of bio-medical waste.
- Color-coded plastic bags should be in line with the Plastic Waste Management Rules
- Provide Personnel Protective Equipment to the bio-medical waste handling staff.

## 6.4.3. Biomedical Waste Collection

### *6.4.3.1. Time of Collection*

- Every hospital ward should have its biomedical waste collected every day at a predetermined time interval. Several ward collections may occur throughout the day.
- In addition to ensuring the collection, transportation, handling, and disposal of biomedical waste, HCF must also guarantee the 48-hour disposal of soiled, biotechnology, human, and animal anatomical waste.
- Collection times should be fixed and appropriate to the quantity of waste produced in each area of the healthcare facility.
- General waste should not be collected at the same time or in the same trolley in which bio-medical waste is collected.
- Collection should be daily for most wastes, with collection timed to match the pattern of waste generation during the day. For example, in an IPD ward where the morning routine begins with the changing of dressings, infectious waste could be collected midmorning to prevent soiled bandages from remaining in

the area for longer than necessary.

- General waste collection must be done immediately after the visiting hours of the HCFs, as visitors coming to facility generate a lot of general waste and to avoid accumulation of such general waste in the HCF. The collection timings must enable the HCF to minimize or nullify the use of interim storage of waste in the departments.
- Bio-medical waste collected by the staff should be provided with PPEs.

#### 6.4.3.2. Packaging

- Sharps containers and biomedical waste bags should be filled no more than three quarters of the way. When they get to this point, they ought to be sealed and prepared for pickup.
- Plastic bags can be sealed or tied with a plastic tag or tie instead of being stapled.
- Every waste collection site should have extra bags or containers on hand so that empty ones can be swapped out right away.
- To enable tracking until final disposal, color-coded waste bags and containers should be printed with the bio-hazard symbol and labeled with information such as the date, waste type, waste quantity, sender's name, recipient's name, and bar code.
- Making sure that each bag has a bar code sticker applied according to the instructions.

#### 6.4.3.3. Labeling

The Biohazard or Cytotoxic Hazard symbol, depending on the type of waste, must be labeled on all bags, containers, and bins used for the collection and storage of biomedical waste.

In compliance with the Guidelines for Barcode System for Effective Management of Biomedical Waste, biomedical waste bags and containers must be labeled with a bar code.

#### 6.4.3.4. Interim Storage

- Interim storage of Biomedical waste is discouraged in the wards/ different departments of HCF.
- If waste is needed to be stored on an interim basis in the departments, it must be stored in the dirty utility/ sections.
- No waste should be stored in patient care areas and procedures areas such as Operation Theatre. All infectious waste should be immediately removed from such areas.
- In the absence of dirty utilities/ sections must be stored in a designated place away from patient and visitor traffic or low-traffic areas.

### 6.4.4. In-House Transportation of Biomedical Waste

#### 6.4.4.1. Transportation Trolleys

Biomedical waste must be transported inside hospital premises using closed trolleys or containers, ideally with wheels for easy mobility, from the site of waste generation or interim storage to the central waste collection center. These trolleys or carts are exclusively intended for the collection of biomedical waste. BMW transportation is not permitted to use patient trolleys.

The size of these waste transport trolleys ought to correspond with the amount of waste produced by the HCFs.

### *6.4.4.2. Route of Intramural Transportation of Bio-Medical Waste*

Transporting Bio-Medical Waste from various wards or laboratories in healthcare facilities requires the use of covered trolleys or carts along a path with minimal patient and visitor traffic.

Route of transportation preferably be planned in such a way that:

- Transportation does not occur through high-risk areas
- Supplies and waste are transported through separate routes.
- Waste is not transported through areas having high traffic of patients and visitors
- Central Waste collection area can be easily accessed through this route
- Safe transportation of waste is undertaken to avoid spillage and scattering of waste

## 6.4.5. Central Waste Collection Room for Biomedical Waste

All healthcare facilities must designate a central waste collection room on their premises for the purpose of storing biomedical waste until it is collected and transported to the CBWTF for treatment and disposal. A designated person should oversee this room, which should be locked and secured.

The following points may be considered for the construction of the central waste collection room:

- The location of the central waste collection room must be away from public/ visitors access.
- The space allocation for this room must be as per the quantity of waste generated from the hospital.
- The planned space must be sufficient to store at least two days generation of waste.
- The central waste collection room must be roofed and manned and should be under lock and key under the responsibility of a designated person.
- The entrance of this centre must be accessible through a concrete ramp for easy transportation of waste collection trolleys.
- Flooring should be of tiles or any other glazed material with a slope to ease the cleaning of the area.
- Exhaust fans should be provided in the waste collection room for ventilation.
- It is to be ensured by the healthcare facility that such central storage room is safety inspected for potential fire hazards and based on such inspection preventive measures have to be taken by the healthcare facility like installation of fire extinguishers, smoke detectors etc.
- There should also be provision for a water supply adjacent to the central waste storage area for cleaning and washing of this station and the containers. The drainage from the storage and washing area should be routed to the Effluent Treatment Plant.
- Sign boards indicating relevant details such as contact person, and a telephone number should be provided.

- The entrance of this station must be labeled with "Entry for Authorized Personal Only" and the logo of Biomedical Waste Hazard.
- It is to be ensured that no general waste is stored in the central waste collection area.

### 6.4.5.1. Other Considerations for Central Waste Collection Area:

- To ensure there is no pilferage of recyclables, it is to be ensured that the central storage area is under lock & key, guarded by a designated person.
- Healthcare facilities should maintain a record of waste generated and handed over to the authorized recyclers.
- The healthcare facility is responsible for assuring that no stray animals are on the property and for installing cattle traps at the entrance to the facility to guarantee animal protection.
- The HCF assure that a pest control company is hired to regularly apply pest control measures in the central storage area to guarantee protection against pests.

### 6.4.5.2. Central Storage for HCFs Having Captive Treatment and Disposal System

The waste generated from the HCF must be kept in this central waste collection area until it is transported to the reception area of the captive waste treatment facility within the premises. This is especially important for healthcare facilities that have captive treatment facilities for the treatment and disposal of biomedical waste through incinerators, autoclaves/microwaves, and shredders within their premises. (Yazie et al., 2019)

# CASE STUDY

## MANAGEMENT OF MEDICAL WASTE IN PUBLIC HOSPITALS

Hosseinzadeh, A., Gitipour, S., Pazoki, M., Reshadi, M. A. M., Nazaripour, M., & Rezaei, M. (2022). Management of Medical Wastes in Public Hospitals: A Case Study. Journal of Advances in Environmental Health Research, 10(4), 319-324. doi: 10.32598/JAEHR.10.4.1269

In recent decades, due to an increase in health services and population growth, the generation of medical wastes in the world has received serious attention. Medical wastes are solid wastes generated in hospitals, clinics, medical laboratories, blood centers and so forth. Infectious waste is also any type of waste contaminated with pathogens that can cause a disease. Furthermore, hospitals are one of the most important sources of medical waste generation, which is called hospital waste.

The amount of medical waste produced in hospitals varies depending on factors such as the number of beds, the type of hospital, and the number of referrals. In Iran, the average generation of medical waste is 3.7 kg/bed/day. However, this amount can change in different hospitals, as 2 and 6 kg/bed/day have been reported in previous studies. In terms of quality, medical wastes can be classified into two general forms, common wastes and hazardous wastes. But, in better classification, medical wastes include general, infectious, sharp, and chemical and pharmacy wastes. The World Health Organization (WHO) defines medical waste as waste generated in the process of diagnosis, treatment and immunization of humans and animals. It is estimated that about 20% of this waste are due to properties such as infectivity, toxicity and radioactivity and can be classified as a hazardous waste.

The management of medical wastes is a regular process that includes the stages of generation control, separation, storage, collection, disinfection and disposal. Proper management of medical waste is effective in reducing environmental risk and health problems. One of the most important factors in proper management of medical wastes is having access to data showing the amount of waste generation and the ratio of infectious wastes. Hence, the purpose of this study was to investigate the quantity of medical waste in public hospitals in the city of Tehran and compare them with each other, as well as to investigate the ratio of their components.

### *Material and Methods*

This study investigated the quantity and composition of medical waste in 10 public hospitals in Tehran. The 10 studied hospitals were arranged alphabetically from A to J. The data were obtained by means of interviews with the waste management staff in hospitals as well as official statistics in the hospitals. The data on the quantity and composition of medical waste were obtained based on a physical analysis of them in one year. Also, only the weight of medical waste was considered. To investigate the composition of medical wastes, they were classified in three groups as follows and then the data were analyzed in statistical software:

- Common wastes such as nylon, plastic tumbler, food waste, and tea slag;
- Infectious and sharp wastes such as blood-contaminated gauze, latex gloves, syringes needles, and surgical blades;
- Chemical and pharmaceutical wastes such as used medicine ampoules, acrylic, and wax.

### *Results and Discussion*

The results showed that the average generation of medical waste in the hospitals studied was 3.53 kg/bed/ day. As can be seen in Table 1, the generation of medical waste in the hospitals ranged 107.5 to 2508 kg/day. Although the generation of medical waste in the hospitals was in the range of 1.22 to 7 kg/bed/day, it was in the range of medical waste generation in Iran. In previous studies, the rate of medical waste generation in Iran was reported to be less than 2 kg/bed/day to more than 6 kg/bed/day. However, in various studies, the average generation of medical waste in Iran was reported in the range between 3.1 and 3.7 kg/bed/day.

**Table 1.** *Quantity and composition of medical wastes in hospitals studied (kg/bed/day)*

| Hospitals | Number of Beds | Occupied Bed (%) | Common Waste | Infectious and Sharp Waste | Chemical and Pharmacy Waste | Total Waste |
|---|---|---|---|---|---|---|
| A | 452 | 95 | 1157 | 660 | 4 | 1821 |
| B | 143 | 84 | 748 | 252 | 2 | 1002 |
| C | 118 | 62 | 50 | 80 | 15 | 145 |
| D | 260 | 84 | 850 | 140 | 1.5 | 1005 |
| E | 186 | 83 | 250 | 200 | 1.5 | 451.5 |
| F | 290 | 91 | 945 | 402 | 7.5 | 1354.5 |
| G | 820 | 92 | 1750 | 723 | 35 | 2508 |
| H | 412 | 93 | 1100 | 703 | 10 | 18.13 |
| I | 227 | 89 | 460 | 260 | 2.5 | 722.5 |
| J | 72 | 79 | 62.5 | 40 | 5 | 107.5 |

*Source: Hosseinzadeh, A., Gitipour, S., Pazoki, M., Reshadi, M. A. M., Nazaripour, M., & Rezaei, M. (2022). Management of Medical Wastes in Public Hospitals: A Case Study.* Journal of Advances in Environmental Health Research, 10*(4), 319-324. doi: 10.32598/JAEHR.10.4.1269*

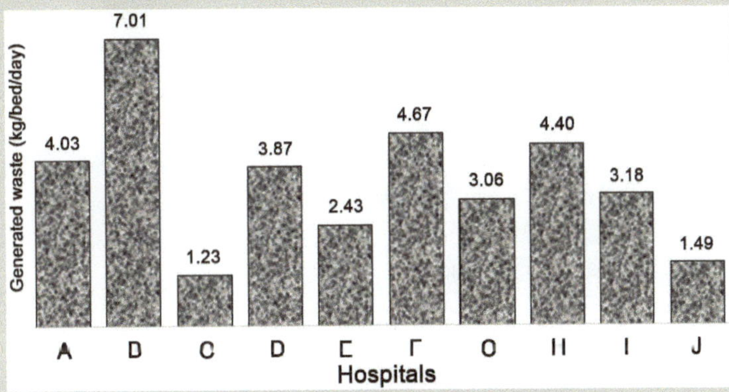


**Figure 1.** *Hosseinzadeh, A., Gitipour, S., Pazoki, M., Reshadi, M. A. M., Nazaripour, M., & Rezaei, M. (2022). Management of Medical Wastes in Public Hospitals: A Case Study.* Journal of Advances in Environmental Health Research, *10(4), 319-324. doi: 10.32598/JAEHR.10.4.1269.*

Considering the ratio of occupied beds in the hospitals studied the amount of medical waste generation in them can be expressed as 1.88 to 8.34 kg/patient/day. Although this criterion is not very common for the generation of medical waste, in similar studies the generation of medical waste in Iran has been reported as up to 10 kg/patient/day. Of course, this criterion has been used more to express the generation of medical wastes in medical laboratories; in previous studies in Iran, its rate was reported 0.5 to 0.6 kg/patient/day. Therefore, based on different criteria, the amount of medical waste generation in the hospitals studied was adopted with average of medical waste generation in Iranian hospitals. Of course, the reason for the difference between other studies and this study can be found in the type of medical centres, the difference between public and private hospitals, and the number of hospital beds.

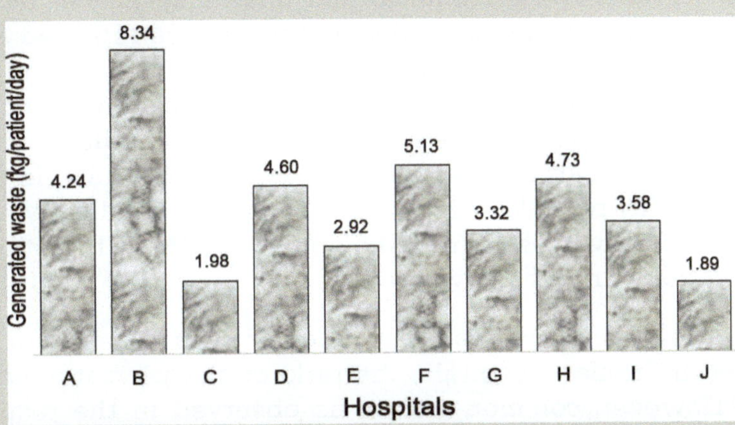


**Figure 2.** *Medical waste generation in hospitals studied (kg/patient/day).*

*Source: Hosseinzadeh, A., Gitipour, S., Pazoki, M., Reshadi, M. A. M., Nazaripour, M., & Rezaei, M. (2022). Management of Medical Wastes in Public Hospitals: A Case Study.* Journal of Advances in Environmental Health Research, *10(4), 319-324. doi: 10.32598/JAEHR.10.4.1269.*

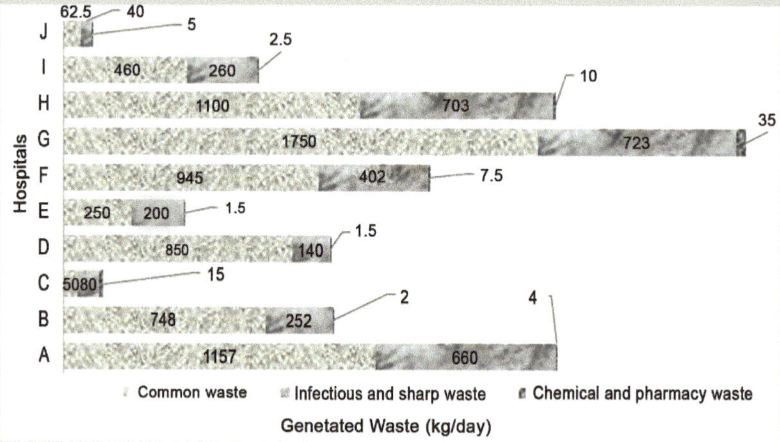


**Figure 3.** *Composition of medical waste in hospitals studied.*

*Source: Hosseinzadeh, A., Gitipour, S., Pazoki, M., Reshadi, M. A. M., Nazaripour, M., & Rezaei, M. (2022). Management of Medical Wastes in Public Hospitals: A Case Study.* Journal of Advances in Environmental Health Research, 10*(4), 319-324. doi: 10.32598/JAEHR.10.4.1269.*

As shown in Figure 3, most of the medical waste in the hospitals studied were common waste. This type of waste in different hospitals included 34.4 to 84.5% of medical wastes. Also, chemical and pharmacy waste was the smallest component of the medical waste in the hospitals ranging between 0.2 to 10.3%. However, infectious and sharp wastes as the most important medical wastes accounted for 13.9 to 55.1% of the medical waste in the hospitals. As Figure 4 shows, the average composition of medical wastes in the hospitals studied included general wastes (67.45%), infectious and sharp wastes (31.65%), and 0.76% chemical and pharmacy waste. In previous studies, the ratio of infectious waste in the medical waste mass was reported about 37% to 44% which shows that, in the studied hospitals in this study, the ratio of infectious waste was adopted with the reported ratio for other hospitals in Iran. Of course, this ratio is reported differently in different hospitals. For example, in two studies in Tehran hospitals, the ratio of infectious and sharp waste was 21% and 24 %, while, in a study in Shiraz, this ratio was reported to be 48%. Also, the results of a medical waste management study in hospitals of 13 provinces in Iran showed that 37% of the studied waste lay in the category of infectious and sharp wastes.

Of course, the ratio of sharp waste is much lower than the ratio of infectious waste; in studies conducted in Iranian hospitals, the ratio of sharp wastes was reported to be between 1 to 5%. However, common waste, as observed in the results of this study, contain the largest share of medical waste in Iran. In comparison with the findings of this study, in the reports of medical waste studies in Iranian hospitals, the share of common waste is more than 50% (49%-72%). The proportion of infectious waste in the medical waste mass depends on the amount of waste segregation in the hospital. If the segregation of this kind of waste in hospitals is done helpfully, the ratio of infectious waste will decline and most of the medical waste will be common waste. It is reported that, due to improved segregation and proper storage in a hospital, the share of infectious wastes was reduced by 13%.

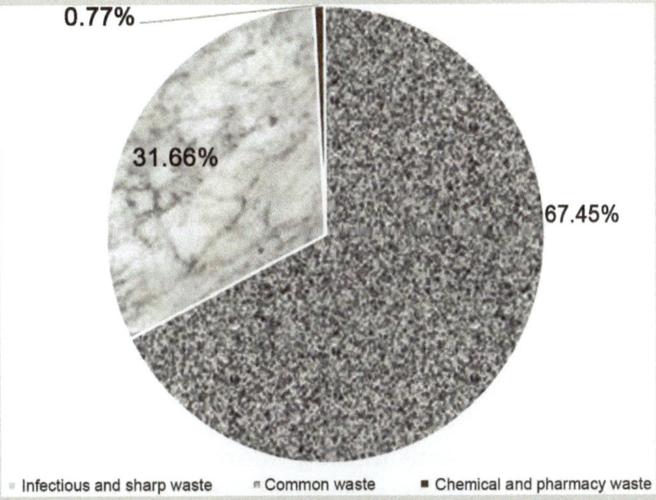


**Figure 4.** *Composition of medical waste in the hospitals studied.*

*Source: Hosseinzadeh, A., Gitipour, S., Pazoki, M., Reshadi, M. A. M., Nazaripour, M., & Rezaei, M. (2022). Management of Medical Wastes in Public Hospitals: A Case Study.* Journal of Advances in Environmental Health Research, 10*(4), 319-324. doi: 10.32598/JAEHR.10.4.1269.*

The amount of medical waste generation in the hospitals studied was in the common range of medical waste generation in Iran and the composition of this waste accorded with other reports. However, the amount of medical waste generated in the hospitals may be different from other countries. For example, it is reported that in the United States, 465,000 tons of biologically hazardous waste is generated annually by 377,000 healthcare centers. Hospital waste generation in some European countries and the United States is 0.25- 7.0 kg/bed/day. In comparison, this rate is 0.4-5.5 kg/ patient/ day for 12 developed and developing countries. Hospital waste generation in Japan, Turkey, Canada, India, Thailand and Bangladesh is also estimated at 0.11-3.9 kg/bed/ day. A study performed in Taiwan found that rates of infectious waste generation in four different facilities were, respectively, as follows: blood centers (3.14 kg/bed/day), private clinics (1.91 kg/bed/ day), medical laboratories (1.07 kg/bed/day) and public clinics (0.053 kg/bed/day).

One of the main concerns in the management of hospital waste is to choose a suitable disinfection method to reduce health and environmental threats. The results of this study indicated that a significant share of hospital waste was of infectious and sharp wastes; therefore, it is of great importance to select an appropriate disinfection measure. The observations of a study indicated that, in Iran, on-site disinfection methods are more desirable than out site disinfection methods. The results of our studies illustrated that the hospitals had an autoclave as an on-site disinfection system. The use of different disinfection methods for infectious waste has been reported in various parts of the world. In Spain, the methods utilized for disinfection included incinerator, sterilization, autoclave, chemical stabilization, and microwave. In the United States, 49%-60% of this type of waste has been disinfected by incineration, 20%-37% by autoclave, and 4%-5% by other technologies. In South Korea, the final disposal method for 52.7% of hospital waste is incinerators and for 46.4% of hospital waste is steam sterilization, and other methods such as microwave and chemical disinfection are used for less than 0.2% of hospital waste. In Japan, 218,000 tons of infectious waste generated per year, of which 82% is eventually incinerated. However, in Iran, according to experts, the application of microwaves and autoclaves for the disinfection of medical waste is a priority, and the use of incinerators is no longer desirable due to the emission of gases and the risk of contamination.

### Conclusion

In the current research, medical waste management was studied in 10 public hospitals in the city of Tehran. The observations showed that the average generation of hospital waste was 3.53 kg/bed/day. According to the ratio of occupied beds in the studied hospitals, the amount of medical waste was 4.07 kg/patient/day. Considering the average generation of medical wastes in Iran, which is about 3.7 kg/bed/day, it can be said that the amount of waste generation in the hospitals studied was in line with the general pattern of the country. 67.5% of medical waste in hospitals was common waste. And 36.5% was infectious and sharp wastes, and 0.8% was chemical and pharmacy wastes. The difference in the quantity of medical waste in the studied hospitals and the difference in the ratio of infectious wastes showed that waste management in them was different from each other. Due to the existence of disinfection equipment in the hospitals studied more attention to medical waste segregation to reduce the ratio of infectious wastes can be effective in reducing the risk and the cost of waste management.

R: Hosseinzadeh, A., Gitipour, S., Pazoki, M., Reshadi, M. A. M., Nazaripour, M., & Rezaei, M. (2022). Management of Medical Wastes in Public Hospitals: A Case Study. Journal of Advances in Environmental Health Research, 10(4), 319-324. doi: 10.32598/JAEHR.10.4.1269

# CLASS ACTIVITY

Instructors should ask participants how they would classify the following items and record responses by group.

- Sharps
- Infectious
- Pathological
- Pharmaceutical
- Chemical
- Radioactive
- General (non-hazardous)

| No | Item | Sharp | Infect | Patho | Pharm | Chem | Radio | General |
|---|---|---|---|---|---|---|---|---|
| 1 | Bandage saturated with blood | | | | | | | |
| 2 | Empty IV bag | | | | | | | |
| 3 | IV bag containing fluid blood | | | | | | | |
| 4 | Broken thermometer | | | | | | | |
| 5 | Unused (clean) hypodermic syringe | | | | | | | |
| 6 | Soiled disposable diaper | | | | | | | |
| 7 | Dirty glove | | | | | | | |
| 8 | Cotton swab with little dried blood | | | | | | | |
| 9 | Clean broken glassware | | | | | | | |
| 10 | Used disinfectant solution | | | | | | | |

| | | | | | | | | |
|---|---|---|---|---|---|---|---|---|
| 11 | Empty juice bottle used by patient | | | | | | | |
| 12 | Cotton swab with little alcohol (ethanol) | | | | | | | |
| 13 | Expired antibiotics | | | | | | | |
| 14 | Eye wash | | | | | | | |
| 15 | Old fluorescent lamp | | | | | | | |
| | | | | | | | | |
| | | | | | | | | |

## SUMMARY

- Medical waste is a subset of waste generated at healthcare facilities, such as hospitals, physicians' offices, dental practices, blood banks, and veterinary hospitals/clinics, as well as medical research facilities and laboratories.
- Healthcare facilities (HCFs) are the main healthcare waste producers. The most common term used to describe waste generated by HCFs is healthcare waste (HCW).
- HCW and by-products are generated because of the diagnosis, treatment, medical intervention or the immunization of humans or animals.
- Infectious waste is a variety of hazardous waste which, due to its pathogenic nature, poses a threat to human health. It should always be assumed that infectious waste may contain various pathogenic microorganisms.
- Many chemicals and pharmaceuticals used in healthcare systems can be hazardous. They are usually found in small amounts in medical waste, while larger amounts can be found when unwanted or expired chemicals and pharmaceuticals are directed for disposal.
- Biomedical waste contaminates air if not segregated or incinerated properly, resulting in highly hazardous airborne particles of contagious diseases.
- Improper management of healthcare waste can have both direct and indirect health consequences for health personnel, community members and the environment.
- Solid waste generation rates in urban areas around the world are rapidly increasing with the expansion of the population, economic development, changing consumption patterns, and urbanization.
- Waste generation rates and composition differ across countries, income levels, consumption and lifestyle patterns, seasonal variations, and various sectors and infrastructures.
- As the number of hospitalizations surged, the amount of waste from healthcare facilities around the world grew beyond the handling capabilities of waste management systems.
- The inability of the existing treatment facilities to meet the variations in the extra waste generated by healthcare centers was another area of concern during the COVID-19 pandemic.
- Biomedical waste management refers to the proper handling, collection, treatment, and disposal of waste generated during healthcare activities.
- The first five steps (Segregation, Collection, pre-treatment, Intramural Transportation and Storage) are the exclusive responsibility of the healthcare Facility.
- Bio-medical waste generated from a healthcare facility is required to be segregated at the point of generation as per the color coding stipulated.
- In house transportation of Biomedical Waste from the site of waste generation/ interim storage to the central waste collection center, within the premises of

the hospital must be done in closed trolleys/containers preferably fitted with wheels for easy maneuverability.

## REVIEW QUESTIONS

1. Describe standard waste composition in health facilities.
2. What are the categories and sources of HCW? Explain.
3. Discuss HCW production rate in various countries worldwide.
4. Explain the risk related to HCW.
5. Discuss the effects of medical waste on the environment.
6. Focus on waste generation dynamics and associated waste management challenges during the COVID-19 pandemic.
7. Discuss the opportunities in the post-pandemic waste infrastructure.
8. Explain the steps involved in bio-medical waste management.

## REFERENCES


1. Abdo, N., Khader, Y. S., Abdelrahman, M., Graboski-Bauer, A., Malkawi, M., Al-Sharif, M., & Elbetieha, A. M. (2016). Respiratory health outcomes and air pollution in the Eastern Mediterranean Region: A systematic review. Reviews on Environmental Health, 31(2), 259–280. https://doi.org/10.1515/reveh-2015-0076.
2. Aljabre, S. H. M. (2002). Hospital generated waste: A plan for its proper management. Journal of Family & Community Medicine, 9, 61–65.
3. Alvim-Ferraz, M. C., & Afonso, S. A. (2003). Incineration of different types of medical wastes: Emission factors for particulate matter and heavy metals. Environmental Science & Technology, 37(14), 3152–3157.
4. Bdour, A., Altrabsheh, B., Hadadin, N., & Al-Shareif, M. (2007). Assessment of medical wastes management practice: A case study of the northern part of Jordan. *Waste Management, 27*, 746–759. https://doi.org/10.1016/j.wasman.2006.03.004.
5. Bokhoree, C., Beeharry, Y., Makoondlall-Chadee, T., Doobah, T., & Soomary, N. (2014). Assessment of environmental and health risks associated with the management of medical waste in Mauritius. APCBEE Procedia, 9, 36–41. https://doi.org/10.1016/j.apcbee.2014.01.007.
6. Ciplak, N., & Kaskun, S. (2015). Healthcare waste management practice in the West Black Sea Region, Turkey: A comparative analysis with the developed and developing countries. Journal of the Air & Waste Management Association, 65, 1387–1394. https://doi.org/10.1080/10962247.2015.1076539.
7. Das, A. K., Islam, M. N., Billah, M. M., & Sarker, A. (2021). COVID-19 pandemic and healthcare solid waste management strategy—A mini-review. Science of the Total Environment, 778, 146220. https://doi.org/10.1016/j.scitotenv.2021.146220.
8. Dehghani, M. H., Ahrami, H. D., Nabizadeh, R., Heidarinejad, Z., & Zarei, A. (2019).

Medical waste generation and management in medical clinics in South of Iran. MethodsX, 6, 727–733. https://doi.org/10.1016/j.mex.2019.03.029.

9. Hasan, M. M., & Rahman, M. H. (2018). Assessment of healthcare waste management paradigms and its suitable treatment alternative: A case study. Journal of Environmental and Public Health, 2018, 6879751. https://doi.org/10.1155/2018/6879751.
10. Janik-Karpinska, E., Brancaleoni, R., Niemcewicz, M., Wojtas, W., Foco, M., Podogrocki, M., & Bijak, M. (2023). Healthcare waste – A serious problem for global health. Healthcare (Basel), 11(2), 242. https://doi.org/10.3390/healthcare11020242.
11. Jayasinghe, P. A., Jalilzadeh, H., & Hettiaratchi, P. (2023). The impact of COVID-19 on waste infrastructure: Lessons learned and opportunities for a sustainable future. International Journal of Environmental Research and Public Health, 20(5), 4310. https://doi.org/10.3390/ijerph20054310.
12. Kannangara, M., Dua, R., Ahmadi, L., & Bensebaa, F. (2018). Modeling and prediction of regional municipal solid waste generation and diversion in Canada using machine learning approaches. Waste Management, 74, 3–15. https://doi.org/10.1016/j.wasman.2017.11.057.
13. Kwikiriza, S., Stewart, A. G., Mutahunga, B., Dobson, A. E., & Wilkinson, E. (2019). A whole systems approach to hospital waste management in rural Uganda. Frontiers in Public Health, 7, 136. https://doi.org/10.3389/fpubh.2019.00136.
14. Lanphear, B. P., Linnemann, C. C., Jr., Cannon, C. G., DeRonde, M. M., Pendy, L., & Kerley, L. M. (1994). Hepatitis C virus infection in healthcare workers: Risk of exposure and infection. Infection Control & Hospital Epidemiology, 15, 745–750.
15. Meleko, A., Tesfaye, T., & Henok, A. (2018). Assessment of healthcare waste generation rate and its management system in health centers of Bench Maji Zone. Ethiopian Journal of Health Sciences, 28, 125–134. https://doi.org/10.4314/ejhs.v28i2.4.
16. Naughton, C. C. (2020). Perspective: Will the COVID-19 pandemic change waste generation and composition? The need for more real-time waste management data and systems thinking. Resources, Conservation and Recycling, 162, 05050. https://doi.org/10.1016/j.resconrec.2020.105050.
17. Ragazzi, M., Rada, E. C., & Schiavon, M. (2020). Municipal solid waste management during the SARS-CoV-2 outbreak and lockdown ease: Lessons from Italy. Science of the Total Environment, 745, 141159. https://doi.org/10.1016/j.scitotenv.2020.141159.
18. Sepetis, A., Zaza, P. N., Rizos, F., & Bagos, P. G. (2022). Identifying and predicting healthcare waste management costs for an optimal sustainable management system: Evidence from the Greek public sector. International Journal of Environmental Research and Public Health, 19, 9821. https://doi.org/10.3390/ijerph19169821.
19. Wu, F., Niu, D., Dai, S., & Wu, B. (2020). New insights into regional differences of the predictions of municipal solid waste generation rates using artificial neural networks. Waste Management, 107, 182–190. https://doi.org/10.1016/j.wasman.2020.04.015.
20. Xin, Y. (2015). Comparison of hospital medical waste generation rate based on diagnosis-related groups. Journal of Cleaner Production, 100, 202–207. https://doi.org/10.1016/j.jclepro.2015.03.056.

21. Yazie, T. D., Tebeje, M. G., & Chufa, K. A. (2019). Healthcare waste management current status and potential challenges in Ethiopia: A systematic review. BMC Research Notes, 12, 285. https://doi.org/10.1186/s13104-019-4316-y.

22. Zamparas, M., Kapsalis, V. C., Kyriakopoulos, G. L., Aravossis, K. G., Kanteraki, A. E., Vantarakis, A., & Kalavrouziotis, I. K. (2019). Medical waste management and environmental assessment in the Rio University Hospital, Western Greece. *Sustainable Chemistry and Pharmacy, 13*, 100163. https://doi.org/10.1016/j.scp.2019.100163.

CHAPTER

# 7 Nuclear Waste

## LEARNING OBJECTIVES

- **Understand the nuclear waste**
- **Explain the basics of nuclear science**
- **Define ramifications of nuclear accidents**
- **Identify sources of nuclear waste**
- **Discus nuclear waste disposal**
- **Explain the status of nuclear waste management**

## KEY TERMS FROM THIS CHAPTER

| | |
|---|---|
| Containment | Decommissioning |
| Deep geological repository | Disposal |
| Environmental impact | High-level waste |
| Long-term storage | Low-level waste |
| Nuclear reactor waste | Radioactivity |
| Repository | Reprocessing |

# INTRODUCTION

Nuclear waste, often referred to as radioactive waste, is a byproduct generated from nuclear reactors, nuclear weapons production, medical treatments, and various industrial processes utilizing radioactive materials. This waste is categorized based on its radioactivity, which determines its handling, storage, and disposal methods. The main types include low-level waste (LLW), intermediate-level waste (ILW), and high-level waste (HLW).

Low-level waste consists of materials with relatively low levels of radioactivity and includes items like protective clothing, tools, and reactor components. It typically requires minimal shielding and can often be disposed of in near-surface facilities designed to contain it safely for extended periods.

Intermediate-level waste comprises materials with higher levels of radioactivity, such as reactor components and resins used to filter radioactive liquids. This type of waste necessitates greater shielding and isolation from the environment due to its longer-lived radioactive isotopes.

High-level waste is the most hazardous category and primarily consists of spent nuclear fuel from reactors. It emits high levels of radiation and contains long-lived radioactive isotopes that remain hazardous for thousands of years. Due to its intense radioactivity, HLW requires robust containment and isolation strategies to prevent environmental contamination and human exposure.

Nuclear waste management involves several key processes: storage, treatment, transport, and disposal. Storage methods range from temporary on-site storage at nuclear facilities to long-term disposal in specially engineered repositories. Treatment techniques such as vitrification (encapsulation in glass) and ion exchange are employed to stabilize and reduce the volume of waste for safer handling and storage.

Transportation of nuclear waste requires stringent safety measures to prevent accidents and minimize radiation exposure to the public and the environment. Specialized containers and transportation routes are designed to ensure secure and monitored transit from generation sites to storage or disposal facilities.

The disposal of nuclear waste remains a complex and contentious issue globally. Deep geological repositories are considered the most promising long-term solution for HLW, where waste is isolated and contained in stable rock formations hundreds of meters below the Earth's surface. However, public concerns about safety, environmental impact, and long-term monitoring persist, necessitating transparent communication and regulatory oversight in waste management practices.

# 7.1. WHAT IS NUCLEAR WASTE?

In a reactor, the splitting of a nuclear fuel nucleus releases nuclear energy. The fission products, or leftover smaller nuclei, are the main constituent of nuclear waste (Figure 7.1).

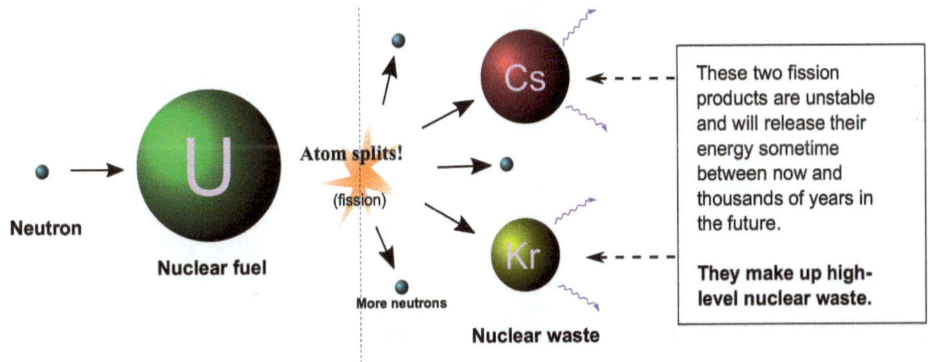


Figure 7.1. The fission process of a single atomic nucleus. Note that Krypton and Cesium are two possible fission products, but many other pairs of elements also occur.

Source: https://whatisnuclear.com/waste.html.

Nuclear waste appears exactly like the fuel that was put into the reactor from the outside; it usually consists of assemblies of metal rods that are cylindrical and contain fuel pellets. However, the contents have undergone nuclear reactions, so they aren't the same (Ojovan, 2019).

Nuclear waste is composed of elements from the periodic table, just like everything else. These elements include isotopes of iron, zinc, germanium, zirconium, silver, and iodine. As billions upon billions of atoms split, its hard to predict which two fission product elements will be produced from a particular fission event, but the average composition remains the same.

## 7.1.1. What Are the Hazards of Nuclear Waste?

Much of the energy released during a nucleus's split is immediately carried off by coolant to perform useful work. Nonetheless, thousands of years after an atom splits, energy is still released. The radioactive waste is dangerous because of this afterglow heat. The unique property of radioactive waste is that it loses toxicity over time, unlike most other toxic waste (Figure 7.2).

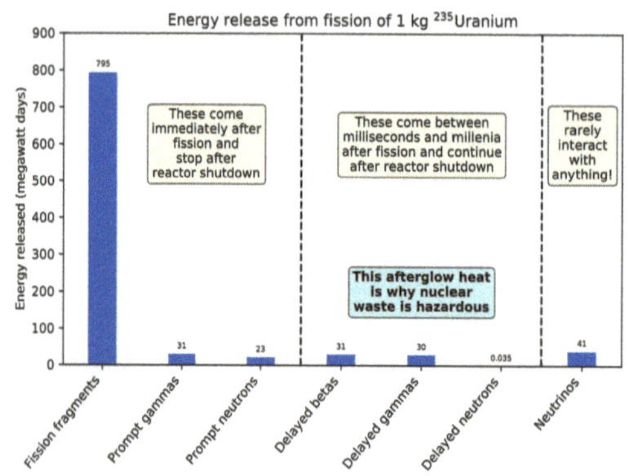


Figure 7.2. How much energy comes out via different mechanisms after you fission 1 kg of Uranium-235. This much fission is typical of one day of operation of a large gigawatt-scale power plant, powering a city of about 1 million people. Note that in the long term, energy from the alpha-decay of materials heavier than uranium produces much of the heat rather than fission (not shown).

Source: https://whatisnuclear.com/waste.html.

Due to these delayed energetic emissions, radioactivity from nuclear waste is very high. It is so dangerous when it first emerges from the reactor that standing near it unprotected would expose you to a fatal radiation dose in a matter of seconds and cause acute radiation sickness that would kill you in a matter of days. Every minute that passes as the energy manifests, the waste loses radioactivity and consequently becomes less dangerous. However, it takes thousands of years to change from dangerous to benign (Figure 7.3).

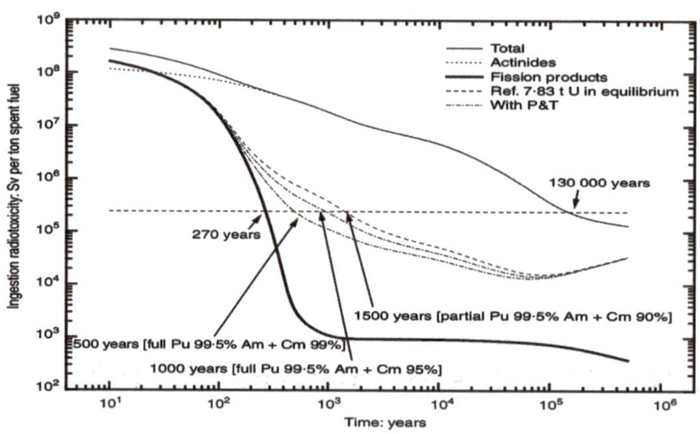


Figure 7.3. How dangerous nuclear waste is to eat as a function of time compared to eating natural uranium ore (dotted line). The various curves show that toxicity decreases faster if you recycle the waste and burn the long-lived radionuclides as fuel.

Source: https://whatisnuclear.com/waste.html.

Essentially, the question of nuclear waste revolves around humanity's ability to stop this radioactive material from endangering both humans and the environment.

# 7.2. BASICS OF NUCLEAR SCIENCE

The study of atomic nuclei, which are the basic building blocks of atoms, and their behavior is known as nuclear science.

## 7.2.1. Nuclear Structure

A cloud of negatively charged electrons surrounds the minuscule, positively charged nucleus of an atom. Positively charged protons and electrically neutral neutrons held together by the so-called strong or nuclear force make up an atom's nucleus, which is typically less than a ten-thousandth of the atom's size but contains more than 99.9 of the atom's mass. This force is restricted to distances of a few $x10^{-15}$ meters, but it is far stronger than the well-known electrostatic force that binds the electrons to the nucleus (Figure 7.4).

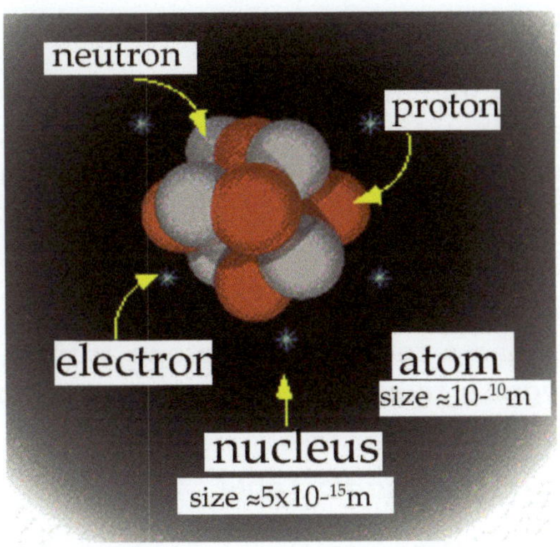


Figure 7.4. Nuclear structure.

Source: https://www2.lbl.gov/abc/Basic.html.

The term "atomic number" refers to Z, the number of protons in the nucleus. This establishes the atom's chemical element. N stands for the number of neutrons in the nucleus. Z + N equals the atomic mass of the nucleus or A. Numerous isotopes of a given element exist; these isotopes vary from one another based on how many neutrons are present in their nuclei. The number of protons in the nucleus and the number of electrons orbiting it are equal in a neutral atom. The atom's net charge is zero because the electric charges of the proton and electron are, respectively, +1 and -1 (in units of the proton charge). There are currently 112 known elements, ranging from hydrogen, the lightest element, to the recently found element 112, which has not yet been named. Every element heavier than uranium is created by humans. There are over 2000 unstable isotopes and about 270 stable isotopes among the elements (Figure 7.5).

Figure 7.5. Chemical element of the atom.

Source: https://www2.lbl.gov/abc/Basic.html.

## 7.2.2. Radioactive Elements

Nuclear fission occurs naturally and intentionally in nuclear reactors, particle accelerators, etc., producing radioactive elements as a byproduct. The geosphere contains naturally occurring radioactive elements as well as those created by human activity. Radium, radon, and other elements are a few instances of naturally occurring radioactive elements. Among the radioactive elements that humans have created are plutonium, neptunium, etc. There are radioactive isotopes in every element.

For example, hydrogen has a radioactive isotope tritium, carbon has a radioactive isotope carbon-14, and oxygen's radioactive isotope is oxygen-15 (Attix, Frank, 1986).

### *7.2.2.1. Radioactive Elements in Periodic Table*

Let us look at the radioactive elements that have been found and mentioned in the periodic table (Figure 7.6).

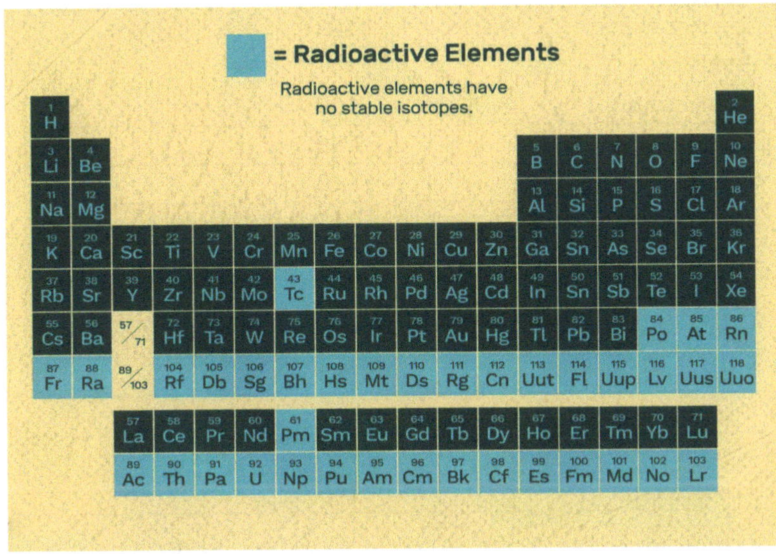


Figure 7.6. Radioactive elements.

Source: https://www.thoughtco.com/list-of-radioactive-elements-608644.

Radioactive elements are indicated in the table by elements that are highlighted. The periodic table contains references to 37 radioactive elements. The names of some of these elements are listed in Table 7.1.

Table 7.1. 37 Radioactive Elements

| Elements | Atomic Number |
|---|---|
| Technetium | 43 |
| Francium | 87 |
| Astatine | 85 |
| Neptunium | 93 |
| Plutonium | 84 |
| Americium | 89 |
| Curium | 96 |
| Promethium | 61 |
| Berkelium | 97 |
| Californium | 98 |
| Einsteinium | 99 |
| Fermium | 100 |
| Mendelevium | 101 |
| Nobelium | 102 |
| Rutherfordium | 104 |
| Dubnium | 105 |
| Seaborgium | 106 |

| | |
|---|---|
| Lawrencium | 103 |
| Bohrium | 107 |
| Meitnerium | 109 |
| Hassium | 108 |
| Darmstadtium | 110 |
| Roentgenium | 111 |
| Copernicium | 112 |
| Flerovium | 114 |
| Livermorium | 116 |
| Nihonium | 113 |
| Tennesine | 117 |
| Oganesson | 118 |
| Moscovium. | 115 |

Source: https://www.thoughtco.com/list-of-radioactive-elements-608644.

### *7.2.2.2. Radioactive Element Decay*

Radioactive elements are of three types:

- Alpha Decay
- Beta Decay
- Gamma Decay

#### *Alpha Decay*

When a nucleus with a mass number higher than 200 experiences alpha decay, the parent nuclei release the 4He4 nucleus, also called an alpha particle (⊠). Alpha decay is the process that radioactive elements go through to become stable. Let us see an example of this reaction:

$$^{238}U_{92} \rightarrow {}^{234}Th_{90} + {}^{4}He_{2}$$

#### *Beta Decay*

An electron with high energy is released as a beta particle when a neutron turns into a proton, also known as beta decay. Let us consider an example of the same:

$$^{14}C_{6} \rightarrow {}^{14}N_{7} + {}^{0}\beta_{-1}$$

#### *Gamma Emission*

The electromagnetic emission at the short-wave edge of the electromagnetic wave spectrum is known as gamma radiation. X-ray radiation comes from atoms' nuclei and is produced in their electron shells. Separate quanta can be released, absorbed, and transported in gamma emission. Let us consider an example-

$$^{238}U_{92} \rightarrow \ ^{234}Th_{90} + \ ^{4}He_{2} + \gamma_0^0$$

## 7.2.3. Reactions

The strong nuclear force allows nuclei to interact with one another and cause reactions if they are close enough to one another. Nuclear reactions, like chemical reactions, can be exothermic (i.e., energetic release) or endothermic (i.e., need to be supplied with energy). Fusion and fission are the two main classes of nuclear reactions that are significant.

### 7.2.3.1. Fusion

Two light nuclei fuse to form one heavier nucleus through fusion. An instance of a fusion reaction that holds significance in thermonuclear weapons and upcoming nuclear reactors is the conversion of two distinct hydrogen isotopes into an isotope of helium: (Figure 7.7).

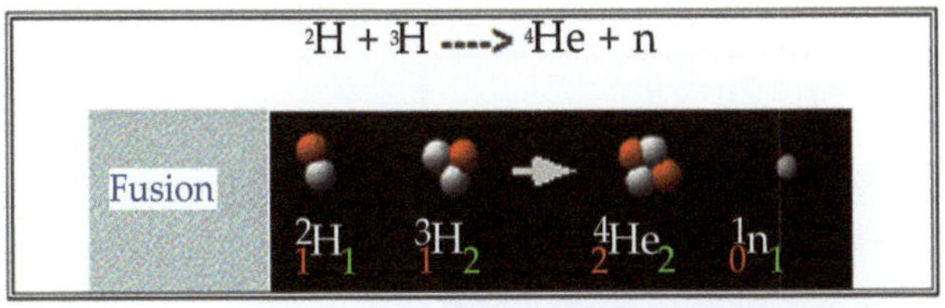


Figure 7.7. Nuclear process of fusion.

Source: https://www2.lbl.gov/abc/graphics/FISSION.GIF.

More than a million times more energy is released in this reaction than in a normal chemical reaction. Because the masses of the product nuclei in a fusion reaction are less than the masses of the original fusing nuclei, a significant amount of energy is released in fusion reactions when two light nuclei fuse. Once more, the mass that is lost is transformed into energy that is carried away by the fusion products, as explained by Einstein's equation $E=mc^2$.

Fusion is an energetically favorable reaction for light nuclei, but because it requires a significant energy investment, it does not happen on Earth under normal circumstances. As the reacting nuclei come closer together, there is a strong electrostatic repulsion between them because they are both positively charged. They don't sense the strong nuclear force that can overcome the electrostatic repulsion and lead to their fusion until they are squeezed extremely closely to one another.

In our universe, fusion reactions have been occurring for billions of years. Most stars, including our Sun, produce their energy through nuclear fusion reactions. It has only been approximately sixty years since scientists have been able to create fusion reactions on Earth. Small-scale studies were conducted at first, with very few fusion reactions occuring. But these early trials eventually resulted in the creation of hydrogen bombs, or thermonuclear fusion weapons.

The process that occurs in stars like our Sun is called fusion. We are witnessing the results of fusion every time we feel the warmth of the Sun and see its light. Because the Sun's light creates food and heats the Earth, we can know that life exists on Earth. Thus, we can state that the foundation of our existence is fusion.

The building blocks of a star are hydrogen and helium, which were produced during the Big Bang, the event that created the universe. In a star, hydrogen isotopes collide and fuse to form the nucleus of helium. Afterwards, heavier elements are created when the helium nuclei collide. Nuclei fuse to form a heavier nucleus in a nuclear reaction known as fusion. The fundamental response is what powers the Sun. Lighter elements combine to form heavier elements. Until the nuclei reach iron (roughly mass sixty), the nucleus with the greatest binding energy, these reactions persist. A star's nucleus stops fusing when it reaches mass sixty because producing higher masses is energetically unfavorable. A star is almost at the end of its life when a significant portion of its core mass is converted to iron.

There is not enough fuel left in the fusion chain to continue. Some stars continue to contract until they are reduced to the size of a cooling iron ember. On the other hand, a massive enough star may explode brilliantly and violently. A star can burst into a massive expansion that releases more energy in a brief period than our Sun will in a lifetime. We refer to a star as having turned into a supernova when this occurs.

During the supernova phase, a star goes through several significant reactions. The accelerated nuclei reach much higher velocities than that possible in a merging star. Nuclei can fuse and create elements with masses greater than iron thanks to the extra energy they receive from their speed. The higher mass element's energy barrier must be overcome by the extra energy in the explosion. Lead, gold, and silver, among other elements, were formerly the remnants of a supernova explosion. Supernovae and dead stars are the primary sources of the element iron that is found throughout the Earth and in its core.

Researchers are working to find more peaceful applications for fusion in the hopes of controlling fusion reactions soon to produce cheap, clean power.

### 7.2.3.2. Fission

A heavy nucleus can split into two smaller nuclei through a nuclear process called fission. Nuclear reactors continue to use fission reactions, one of which was utilized in the first atomic bomb (Figure 7.8).

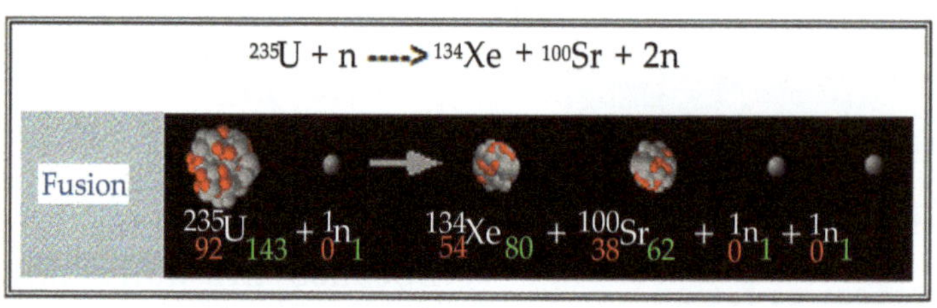


Figure 7.8. Nuclear process of fission.

Source: https://www2.lbl.gov/abc/graphics/FISSION.GIF.

There are many potential product nuclei; the products displayed in the equation above are just one such exapmle. If the total number of protons and neutrons in the products equals the number in the original fissioning nucleus, fission reactions can result in any combination of lighter nuclei. Like fusion, fission releases a significant amount of energy because, in the case of heavy nuclei, the mass of the fissioning nucleus is greater than the sum of the masses of the lighter product nuclei.

The high concentration of positively charged protons in a heavy nucleus causes electrostatic repulsion, which leads to fission. Internal electrostatic repulsion between two smaller nuclei is lower than between one larger nucleus. It can therefore fission once the larger nucleus overcomes the strong nuclear force holding it together. Fission can be compared to a tug-of-war between the powerful electrostatic force and the strong, attractive nuclear force. Electricity repels electrostatically in fission reactions.

For billions of years, fission has been taking place in the universe. As was previously mentioned, fission is used peacefully every day in nuclear power plants to produce energy. It has also been used to create nuclear bombs. It's interesting to note that, even though the first nuclear reactor was created by humans only around 50 years ago, the Earth had a natural fission reactor approximately two billion years ago in a uranium deposit in West Africa!

# 7.3. RAMIFICATIONS OF NUCLEAR ACCIDENTS

The explosion at Chernobyl and the improper handling of a radioactive source in Brazil are the two biggest nuclear accidents of the past 20 years. The situation in Brazil was the least well-known of these accidents, but it was possibly the most appropriate in terms of waste management.

The uncontrolled radiation source was eventually thrown away as scrap after going missing in an abandoned medical facility. Scavengers in a junkyard compromised the platinum capsule and the stainless-steel jacket containing the radioactive cesium. Due to the luminescence of the cesium, it was given out to the public as "carnival glitter" The substance was applied straight onto people's faces, skin, and clothes. For the most part, those who were exposed immediately showed signs of severe illness. By the spring of 1988, four people had died from exposure, and it was anticipated that more would perish in the coming months. Less than 1 g of radioactive cesium was released, contaminating over 40 tons of materials, including shoes, clothes, and building supplies.

## 7.3.1. Biological Effects of Radiation

The relationship between ionizing radiation and living things is still largely unknown, but the mechanism underlying radiation damage to molecules, cells, and organ systems is better understood than that of many other environmental hazards. Figure 7.9 displays the warning signs for radioactive materials. Years of research into the various uses of radionuclides have resulted in the accumulation of a massive amount of quantitative dose-response data. With risks no higher than those associated with any other technology, the nuclear technology industry has been able to continue thanks to this information. A brief explanation of the various forms of ionizing radiation and the potential consequences of excessive exposure to radioactive materials can be found in the upcoming subsections. (Anderson, Mary; Woessner, William, 1992)

Figure 7.9. Radioactive material warning sign.

Source: Mary K. Theodore Louis Theodore; environmental management; ISBN-13: 978-1-4200-8908-0.

## 7.3.2. Radioactive Transformations

Several methods, chief among them the emission of alpha, beta, and gamma particles, are used to carry out radioactive transformations. These mechanisms are all examples of spontaneous nuclear transformations. Different, more stable elements are formed due to these transformations. The type of nuclear instability and the mass/energy relationship determine the transformation that will occur for any given radioactive element. The ratio of neutrons to protons determines the nuclear instability; a distinct kind of decay will take place to enable a more stable daughter product. According to the mass/energy relationship, the principles of mass and energy conservation must be adhered to during any radioactive transformation.

A helium nucleus with energy is called an alpha particle. A radioactive element with a low neutron to proton ratio releases the alpha particle. There are two protons and two neutrons in the helium nucleus. The absence of electrons in the alpha particle makes it different from helium atoms in terms of emission. This kind of transformation results in a daughter product with an atomic mass number that is 4 less and an atomic number that is 2 less than its parent. The polonium (Po) used in the example of alpha decay below has an atomic mass number of 210 (protons and neutrons) and an atomic number of 84.

$$^{210}84\text{Po} \rightarrow {}^{4}2\text{He} + {}^{206}82\text{Pb}$$

Lead and helium are denoted by the terms "Pb" and "He," respectively. Because the lead daughter product is stable and won't degrade further, this is an instructive example. Just enough was altered from 1.5 to 1.51 in the neutron to proton ratio to produce a stable element. It is well known that alpha particles have a high linear energy transfer (LET). The alphas will only release energy over a brief distance. Alpha particles can be stopped by a piece of paper or the outermost layer of skin. Therefore, although alpha particles are not dangerous outside, they can be very dangerous if swallowed or inhaled. When an ordinary electron is released from an atom's nucleus, beta particle emission takes place. When a neutron (n) inside the nucleus changes into a proton, the electron (e) emerges.

$$^{1}0n \rightarrow {}^{1}1H + {}^{0}(-1)e$$

Take note that the hydrogen (H) nucleus represents the proton. All the resultant particles' total charges must be preserved during this transformation. In contrast to alpha emission, elements with an excess of neutrons experience beta emission. A beta emitter's daughter product is one atomic number higher than its parent but retains the same atomic mass number. A gamma ray is simultaneously released by many elements that decay via beta emission. We refer to these components as beta-gamma emitters. Because strong beta radiation can permeate bodily tissue, it poses a risk from the outside. Positron emission, in which the parent releases a positively charged electron, is comparable to beta decay. Beta-positive decay is the term used to describe positron emission. When the neutron-to-proton ratio is too low and alpha emission is not energetically feasible, this decay scheme takes place. The positron, or positively charged electron, travels quickly until it encounters another electron. Every particle will vanish upon contact, producing two gamma rays. This type of formation of two gamma rays is known as annihilation radiation.

Gamma radiation is an electromagnetic wave with a specific range of wavelengths, unlike beta and alpha radiation. Gamma rays can only be lessened in intensity with increased shielding; they cannot be entirely blocked. Usually, the photoelectric effect, Compton scattering, pair production, or direct interactions with the nucleus are how gamma rays interact with matter.

### 7.3.3. Dose–Response

There has been much research done on how people react to different radiation dosages. Depending on the dose received and the duration of that dose, the observed radiation effects can be classified as either stochastic or non-stochastic. Radiation effects are typically classified as stochastic effects, in contrast to most biological effects. Three characteristics distinguish non-stochastic effects: an effect's magnitude increases with dose size; a clear causal relationship can be established between the dose and the subsequent effects; and a minimum dose or threshold dose must be received before the specific effect is observed. Cember draws a comparison between consuming alcohol and encountering a toxic substance. For instance, before exhibiting signs of alcohol consumption, an individual must consume more than a specific amount of alcohol. After that, if the person drinks, the alcohol's effects will only get stronger. Lastly, their behavior is a direct result of their drinking if they show signs of inebriation. Conversely, stochastic effects are the result of chance. A portion of the exposed population and a portion of the unexposed population will both exhibit stochastic effects. Thus, contrary to what the example above suggests, stochastic effects are not inherently linked to a toxic substance. There is no cutoff point for stochastic effects; exposure increases the likelihood of an effect but does not guarantee that one will occur. The two most frequent side effects associated with radiation exposure are cancer and genetic consequences. Damage to a somatic cell can result in cancer, whereas damage to a germ cell that leads to a pregnancy can have genetic effects.

# 7.4. SOURCES OF NUCLEAR WASTE

## 7.4.1. Naturally Occurring Radioactive Materials

The concentrations of NORMs in the earth's crust vary. Radium, uranium, and radon are the main naturally occurring radionuclides to be concerned about. It has been discovered that phosphate fertilizers, petroleum scale, and sludge from water treatment plants are the main sources of these radionuclides.

Public water supplies in the United States are thought to distribute 40 billion gal of water every day. Since water originates from a variety of places, including lakes, reservoirs, aquifers, and streams, its naturally occurring radioactivity varies. When radioactivity encounters geologic materials that contain uranium and thorium, it leaches into the ground or surface water. Radium, uranium, and radon, along with the byproducts of their decay, are the main radionuclides which are found in water.

Water is typically treated to ensure its quality before being consumed by the public for reasons related to public health. Water treatment involves putting it through a series of filters and equipment that use chemicals to remove any pollutants and living things. There is a chance that waste sludges or brines containing high concentrations of radioactive materials will be produced if one or more of these systems are used to treat water with elevated radioactivity. Even when removal of radionuclides isn't the goal of the treatment process, such wastes could still be produced.

Phosphate rock (phosphorite) mining is the fifth-largest mining industry in the US. With Florida, North Carolina, and Tennessee holding more than 90% of the country's capacity for rock production, the southeastern United States is the hub of the domestic phosphate rock industry. Phosphoric acid and elemental phosphorus are produced by processing phosphate rock. Following their combination with other materials, these two products are used to create phosphate fertilizers, detergents, animal feed, various food products, and other materials. Eighty percent of the phosphorite in the United States is used in fertilizer, the most significant application of phosphate rock.The range of uranium in US phosphate ores is 20–300 parts per million (ppm), or roughly 7–100 pCi/g. Lower concentrations of thorium are found between 1 and 5 ppm or roughly 0.1–0.6 pCi/g. Based on each radionuclide's activity, the concentration of each radionuclide is expressed in picocuries per gram (pCi/g). A predetermined number of radioactive transformations in a second are represented by the units of curies. The main waste byproduct produced during the phosphoric acid production process is phosphogypsum. The main waste product left over after producing elemental phosphorus is phosphate slag. The wastes in question have been found to contain elevated concentrations of uranium and thorium, along with the products of their decay. Because phosphate industry wastes are produced in large quantities, there is concern that these materials could pose a radiological risk to people who are exposed to such waste dispersed throughout the environment.

Spread across vast tracts of agricultural land, fertilizers are applied. The main crops that are frequently fertilized with phosphate-based fertilizers are cotton, soybeans, corn, wheat, and coarse grains. Phosphate fertilizers are included as a NORM material because agricultural applications use significant amounts of fertilizer. Radiation levels in the ecosystem and food chain may eventually rise because of the continuous use of phosphate fertilizers. Federal regulations specifically addressing wastes containing NORM are currently nonexistent. Waste production is so great that it is usually impractical to dispose of it in an LLW facility. Implementing a cost-effective solution is necessary to protect the industry from high disposal costs and to guarantee public health and safety.

## 7.4.2. Low-level Radioactive Waste

The most diverse type of radioactive waste is called low-level radioactive waste (LLRW), which is created by a variety of methods. LLW is often defined as what it is not. LLRW is defined as "radioactive waste not classified as high-level radioactive waste (HLRW), TRU waste, spent nuclear fuel, or byproduct material as defined in Section 11(e) of the Atomic Energy Act of 1954" by the Low-level Waste Policy Act of 1980." HLW and spent nuclear fuel are not included in this definition due to their extremely high activity. Due to the time required for them to decay to acceptable levels, TRU wastes (those containing elements heavier than uranium) are not accepted. In conclusion, mill tailings and byproduct material are not included due to their extremely low radioactive concentrations compared to the massive amount of waste present.

Table 7.2. Typical Waste Streams by Generator Category

| Waste Stream | Power Reactors | Medical and Academic | Industrial | Government |
|---|---|---|---|---|
| Compacted trash or solids | X | X | X | X |
| Dry active waste | X | | | |
| Dewatered ion exchange resins | X | | | |
| Contaminated bulk | X | | X | X |
| Contaminated plant hardware | X | | X | X |
| Liquid scintillation fluids | | X | X | X |
| Biological wastes | | X | | |
| Absorbed liquids | | X | X | X |
| Animal carcasses | | X | | |
| Depleted uranium $MgF_2$ | | | X | |

Source: Mary K. Theodore Louis Theodore; environmental management; ISBN-13: 978-1-4200-8908-0.

Nuclear power facilities, healthcare and educational establishments, businesses, and government agencies are among the producers of LLW. LLW can be produced by any process that uses radionuclides. Table 7.2 lists the various waste streams along with potential generators for each.

All the generators mentioned above generate wastes that fit into the LLW category. The waste streams listed in Table 7.2 are classified according to the generation process, though they can occasionally also be listed according to the kind of generating facility. Shallow land burial is the method used for LLW disposal. Typically, this process entails filling individual waste containers into sizable overpacks of concrete. The purpose of the overpack is to minimize the quantity of water that could encounter the waste. Once institutional control of the facility is lost, the overpack also serves as a barrier to prevent trespassers from being exposed to the waste. Before being buried in the landfill, waste that is transported to the facility in boxes, drums, or HDPE liners is first placed in an overpack and sealed with cement.

## 7.4.3. High-Level Radioactive Waste

Spent nuclear fuel, liquid waste from the reprocessing of radioactive reactor fuel, and solid waste from the solidification of liquid HLW are the three types of heavy-level waste (HLW). Reactor fuel used to produce electricity is known as spent reactor fuel. A public utility reactor, a commercial reactor, or a government reactor own the spent fuel. Fuel reprocessing produces waste either produced by the government or the private sector. The amount of liquid HLW that has been produced commercially is quite small. In West Valley, New York, the country's only commercial fuel reprocessing facility generated about 600,000 gal of waste. The government has produced the remaining HLW in American weapon facilities. Reactor spent nuclear fuel is taken out and kept in a nearby pool of water. In addition to cooling the fuel, the water in the spent fuel storage pools protects the environment and the workers from the fission products. One way to measure residual heat from a fuel assembly is as much as 6% of the reactor's operating power. Fuel damage at Three Mile Island resulted from a failure to provide further cooling after the fission reaction had stopped. The spent fuel assemblies will be sealed in a canister and disposed of after a geologic repository is built. Underground storage tanks hold the majority of the liquid HLW. A geologic repository may not be available anytime soon because many of these tanks are getting older. There have been numerous approaches to solidifying wastes for transportation and final disposal studied. The goal is to house HLW in a single, central location across the United States. Yucca Mountain, Nevada, has been selected as the site.

## 7.4.4. Transuranic Waste

TRU wastes contain isotopes that are heavier than uranium, U. Generally, transuranic isotopes are not found in nature. These isotopes are manmade, and produced by the irradiation of heavy elements, such as uranium and thorium. TRU wastes are

$$^{238}92\mathrm{U} + {}^{1}0\mathrm{n} \rightarrow {}^{239}93\mathrm{Np} + {}^{0}(-1)\mathrm{e} \leftarrow {}^{239}93\mathrm{I}$$

where Np and Pu stand for plutonium and neptunium, in that order. They are typically produced by the government, especially through the testing of weapons. The TRU waste is currently being held at several DOE facilities nationwide in preparation for its eventual permanent disposal at WIPP in Carlsbad, New Mexico.

# 7.5. NUCLEAR WASTE DISPOSAL

Nuclear waste, also known as radioactive waste, is defined as material contaminated with radionuclides at activity concentrations higher than the regulatory body's established clearance levels, beyond which no further legal or regulatory purpose is anticipated. The radioactive (nuclear) waste is a byproduct of nuclear energy utilization and the processing of some naturally occurring radioactive materials (NORM), e.g., within oil and gas production and ore beneficiation.

Nuclear energy has many uses, including medical diagnostics and treatment. It also generates a significant amount of power; there are currently 422 nuclear power reactors operating, with a combined capacity of 378,314 MW(e) and 19,399 reactor-years of operation under their belts. Moreover, 57 more nuclear power reactors are being built, bringing the total number of installed nuclear power reactors worldwide to 58,858 MW(e). A portion of spent nuclear fuel (SNF) that is not meant to be reprocessed is among the byproducts that are unavoidably produced when nuclear energy is used. These byproducts can take the form of radioactive or contaminated materials for which no further use is anticipated. When compared to the waste materials produced by non-nuclear activities, the total amount of these unnecessary materials is orders of magnitude smaller. For instance, every year, a typical 1 GW(e) nuclear power plant (NPP) produces roughly 25 tonnes of SNF, which can either be stored for later reprocessing or declared as waste (SNFW). A few hundred cubic meters of low and intermediate radioactive waste (LILW) are also produced by it each year. Compared to this, an average 1 GW(e) coal-fueled power plant generates approximately ~$6.5 \times 10^6$ tonnes of gaseous $CO_2$ per year that is pumped into the atmosphere; more than $300 \times 10^3$ tonnes of solid waste in the form of ash residue that contains approximately 400 tonnes of toxic heavy metals, including radioactive uranium and thorium; and more than $5 \times 10^3$ tonnes of harmful gases. The nuclear industry understands the importance of planning for nuclear waste management (NWM) well in advance. It incorporates NWM at the national level to persuade stakeholders and public to trust it. In many cases, the public views nuclear waste as a problematic issue with either no solutions at all or uncertain solutions. Assuring the sustainability of nuclear energy requires proving to the public that radioactive waste can be managed and does not pose a threat to future generations. Failure to do so would go against the core safety principles of nuclear energy. This entry provides a brief description of NWM's last disposal stage. (Finsterle and Muller, 2021)

### 7.5.1. Nuclear Waste Lifecycle

Nuclear waste typically follows several stages—steps in its life cycle, from generation to disposal–which are as follows:

- Pretreatment. The first step includes all operations carried out with waste before treatment i.e., waste collection, segregation, chemical adjustment, and decontamination. It aims to facilitate the selection and further use of waste processing technologies.
- Treatment. The activities that alter the properties of nuclear waste to increase NWM's economy and safety are included in the second step. Its goals are to: (i) lower the volume of waste, e.g., utilize compaction to do so; (ii) extract radionuclides from the waste, e.g., through cleaning up polluted waters; and (iii) alter the makeup of waste, e.g., by turning the original (raw) burnable waste into an oxide form that cannot be burned, which is represented by the ashes and residues left over after burning. Notably, treatment may produce a waste form appropriate for conditioning and safe handling, e.g., a glass crystalline material that is strong and long-lasting can be produced by plasma treatment.
- Conditioning. This step results in waste packages that are suitable for further safe handling, transportation, storage, and/or disposal. It is typically accomplished by enclosing the waste in containers and providing overpacks when needed. Although not always necessary, this stage may include the conversion of the raw liquid or dispersed waste materials into a solid monolithic waste form via immobilization i.e., using solidification, embedding, or encapsulation.
- Storage. To ensure the retrievability of nuclear waste, this step entails its confinement, isolation, environmental protection, and monitoring in a designated storage facility for predetermined amounts of time.
- Transportation. This step is the purposeful, methodical physical transfer of radioactive waste, i.e., a method created especially to move packages safely from one location to another.
- Disposal. This step, which marks the conclusion of NWM, calls for the placement of radioactive waste in a suitable disposal location with no plans to retrieve it.

Therefore, the final stage of the lifecycle of nuclear waste is disposal, except forthe waste that can be released from regulatory oversight once the amount of radionuclides it contains is less than the exemption levels. Finding the waste's physical, chemical, and radioactive characteristics is a crucial part of the NWM process. Characterization of nuclear waste is one of the steps in the process. It seeks to support the need for additional conditioning, adjustment, and treatment as well as to determine suitability for additional handling, processing, storage, and disposal. Nuclear waste can be properly classified thanks to characterization as well.

### 7.5.2. Nuclear Waste Classification

A universally recognized classification scheme for radioactive waste has been created by the IAEA. This system supports different kinds of nuclear waste and offers generally advised disposal choices that give alternative disposal paths advance thought. However, since such solutions are always based on considerations and safety assessments for each nuclear waste stream, disposal site, and facility used, it does not specifically prescribe a disposal method. The use of NWM end points forms the foundation of the IAEA nuclear waste classification system. First, the IAEA conventionally divides nuclear waste radionuclides, depending on their decay half-lives, into two categories:

- Short-lived;
- Long-lived.

All radionuclides with half-lives of less than 31 years are classified as short-lived, including $^{137}Cs$, which has a half-life of 30.17 years; on the other hand, all radionuclides with half-lives more than 30.2 years are classified as long-lived.

Six classes of waste are also defined by the IAEA. Exempt waste (EW) is the lowest class of waste in terms of radionuclide content. Conventionally non-radioactive waste materials, or waste materials with radionuclide concentrations below exemption thresholds, make up the EW.

The following five classes of nuclear waste are characterized by increased levels of radionuclide content, and are defined as follows:

- Very short-lived waste (VSLW). This is nuclear waste that can be kept to naturally decay over a reasonable amount of time—usually a few months, but no more than a year or so at most. Because the radionuclide content is so low (below exemption levels), the nuclear waste can then be released from regulatory control. VSLW is usually caused using nuclear energy in medicine, primarily in diagnostic procedures. However, it can also arise from research endeavors and occasionally from training and educational activities.
- Very low-level waste (VLLW). This is nuclear waste with radionuclides present at negligible excess over clearance levels. For each waste radionuclide, the activity concentration in this waste usually stays below 100 times the clearance levels. Considering its potential for hazard, this waste does not necessitate a high degree of containment and isolation. Considering this, very little regulatory control can be placed on the disposal of VLLW in facilities akin to near-surface landfills. Usually appearing as dirt and debris, VLLW is frequently disposed of in specially designed disposal sites like earthen trenches.
- Low-level waste (LLW). Long-lived radionuclides are present in small amounts in low-level waste (LLW), but their contents are far above clearance levels. Because of this, LLW must be kept in near-surface disposal facilities (NSDF) under strict isolation and containment for up to a few hundred years before waste radionuclides naturally decay to levels below exemption. LLW is usually the largest by-volume nuclear waste stream resulting from the use of nuclear energy in industry, research, medicine, and nuclear power plants for electricity generation. Most often, LLW is disposed of in engineered nuclear waste disposal facilities (NSDFs), which can have complex or straightforward designs and consist

of concrete vaults and engineered trenches housing nuclear waste containers. Utilizing subsurface disposal facilities is an additional LLW disposal choice.

- Intermediate-level waste (ILW). This is radioactive waste that doesn't require any space for radiogenic heat dissipation because it contains significant amounts of both short- and long-lived radionuclides. Comparing ILW to LLW, a higher level of containment and isolation is necessary due to the higher concentration of long-lived radionuclides. Co-disposing of ILW with SNF and high-level waste is also an effective option; the required disposal depths for ILW are in the range of several tens to a few hundred meters. The IAEA does not specify a precise border between LLW and ILW, although for long-lived radionuclides, it does recommend a limit average content of 400 Bq/g, and up to 4000 Bq/g for any individual package. The necessary limiting levels are specific to the disposal site and facility in question and cannot be applied universally.
- High-level waste (HLW). This is nuclear waste that produces enough radiogenic heat through radionuclide decays to require shielding from radiation emitted by decaying radionuclides. The levels of activity concentration in this waste are also high enough to require shielding. Although they vary depending on the sizes and volumes of waste items, the heat generation levels above which provisions for heat dissipation are required are typically above a few W/m3 at a typical volume of ~1 m3. Long-lived radionuclides make up a larger portion of HLW nuclear waste, so disposal facility designs needs to consider thus. The most widely accepted method of disposing of heavy liquid waste (HLW) is through geological disposal facilities (GDFs), which are situated in stable, deep geological formations several hundred meters or more below the surface of the Earth. This is because HLW needs a disposal endpoint where it will be kept isolated for geological epochs spanning hundreds of thousands or potentially millions of years. Geological formations that are stable and meet certain geochemical and geophysical requirements can offer the right conditions to safely contain and isolate the HLW in the GDFs from the populace and the surrounding environment. Thus, GDFs can shield HLW from potential hazards arising from future surface modifications, such as changes in climate or unanticipated human activities.

It appears that the IAEA's classification scheme for nuclear waste associates different classes of waste with corresponding disposal options. Additionally, a logic diagram for choosing a disposal path based on nuclear waste classes is provided by the IAEA guidance document (Figure 7.10).

The IAEA's classification of nuclear waste classes do not strictly correspond to one another, e.g., While it is generally agreed upon that the half-life of $^{137}Cs$ (approximately 30.2 years) distinguishes long-lived waste radionuclides from short-lived ones, the exact definition of the distinction between LLW and ILW is not well established because the limiting contents of radionuclides (total amount and concentrations) are determined by safety report results, which make them site and facility-specific. Higher radionuclide contents can be used in the same disposal facility with improved (more robust) waste forms and/or containers, all without sacrificing the facility's safety features.

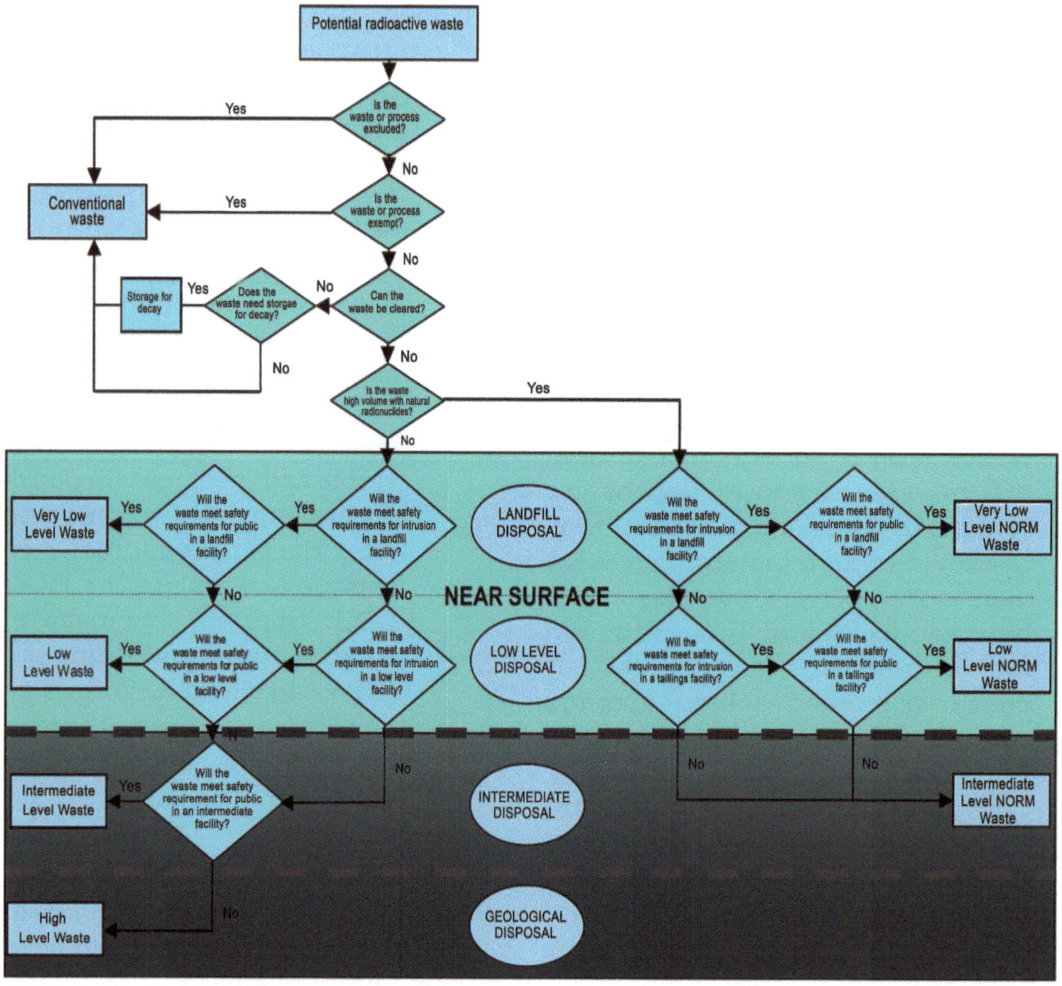


Figure 7.10. Logic diagram aiming to select the NWM end point (disposal option) based on the IAEA nuclear waste classification scheme.

Source: Michael I. Ojovan, Nuclear Waste Disposal, Encyclopedia 2023, 3, 419–429. https://doi.org/10.3390/encyclopedia3020028.

National categorization systems generally adhere to IAEA recommendations (see, for instance, references); however, in numerous other instances, the historically established laws of a specific nation deviate from those suggested within. The IAEA has launched a dedicated predisposal network, IPN, for the preparation and storage of radioactive waste. The most recent information on regulations and disposal practices can be found in the publications of the IAEA networks, DISPONET and URF. The endpoints of NWM according to the IAEA recommendations are summarized in Table 7.3.

Table 7.3. Disposal Options for Various Waste Streams: NSDF—Near Surface Disposal Facility; IDF—Intermediate Depth Disposal Facility; GDF—Geological Disposal Facility; SNFW—Spent Nuclear Fuel Declared as Waste; NORMW—Naturally Occurring Radioactive Material Declared as Waste; NS—Not Possible for Safety Reasons; NTE—Not Possible for Technical and Economic Reasons; A—Acceptable; O—Optimal; P-Preferrable; PTE—Possible, However Requires Assessment from Technical and Economic Aspects

| Nuclear Waste Stream | | End Point | | | | | |
|---|---|---|---|---|---|---|---|
| Decay Storage | | Surface Trench NSDF | Engineered NSDF | IDF | GDF (Deep) | Borehole | |
| VSLW | Low volume | O | A | A | PTE | P | NTE |
| | High volume | O | A | P | PTE | P | NTE |
| VLLW | Low volume | NS | O | A | PTE | NTE | NTE |
| | High volume | NS | O | A | PTE | NTE | NTE |
| LLW | Low volume | NS | A | O | O | A | NTE |
| | High volume | NS | PTE | O | O | A | NTE |
| ILW | Low volume | NS | NS | NS | O | O | NTE |
| | High volume | NS | NS | NS | O | O | NTE |
| HLW and SNFW | | NS | NS | NS | NS | O | O |
| DSRS | Short-lived | A | A | O | A | PTE | O |
| | Long-lived | NS | NS | A | O | O | O |
| NORMW | Low volume | NS | O | A | A | PTE | NTE |
| | High volume | NS | O | PTE | PTE | PTE | NTE |

Source: Michael I. Ojovan, Nuclear Waste Disposal, Encyclopedia 2023, 3, 419–429. https://doi.org/10.3390/encyclopedia3020028.

### 7.5.3. Nuclear Waste Disposal Practices

During the operational phase of nuclear facilities, the nuclear industry systematically implements strategies designed to reduce the amount of waste meant for disposal. These strategies are based on prevention principles and involve preparing waste for reuse, recycling, or other forms of recovery (Figure 7.11).

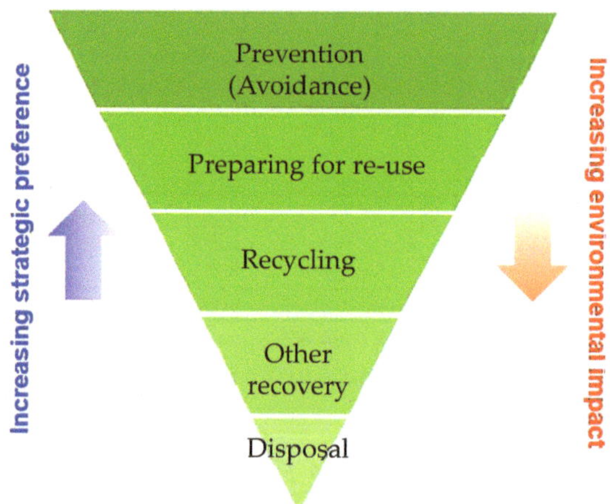


Figure 7.11. Increasing strategic preference versus increased environmental impact.

Source: Michael I. Ojovan, Nuclear Waste Disposal, Encyclopedia 2023, 3, 419–429. https://doi.org/10.3390/encyclopedia3020028.

Achieving the waste disposal stage entails responsible management of nuclear waste. The reduction of nuclear waste volume (as shown in Figure 7.12) enhances the overall cost-effectiveness of NWM. When building disposal facilities, this is also a crucial consideration. By design, nuclear waste disposal facilities can conventionally be classified as follows:

- Earthen trenches classified as NSDF;
- Near-surface engineered structures classified as NSDF;
- Subsurface disposal systems at intermediate depths, which may include boreholes;
- Disposal in deep stable geological formations is classified as GDF, which may include boreholes.

Only waste packages that meet the waste acceptance criteria (WAC), which are always site- and facility-specific and compliant with the operational and post-closure safety cases, will be accepted by disposal facilities. In general, WAC demands that the waste be solid, have stable chemical and physical characteristics, and be as compatible as possible with both natural and engineered barriers. The WACs typically outline the limits on radionuclides and radioactivity, the specifications for waste forms or encapsulation, and the characteristics of waste containers. The disposal facilities are kept under institutional control for a set amount of time after packages are accepted. This is to ensure environmental protection, public safety, and security against any intrusion, including unintentional and intentional ones. The disposal facilities are shut down following the institutional control period, which can last for several decades, based on tested hypotheses that the likelihood of any foreseeable radioactive waste interference with the biosphere is either completely nonexistent or greatly reduced. In 1944, the first nuclear waste disposal took place in an earthen trench in Oak Ridge, Tennessee, in the United States. Since then, there has been a significant development in concepts for disposing of radioactive waste, leading to the current level that aims to maximize the retention capacities of disposal facilities. Several decades of productive operation can be attributed to global disposal practices. There are many different types of disposal facilities in use, such as sub-surface low- and intermediate-level waste (LILW) NSDF in Sweden and Finland, LLW borehole-type disposal facilities in the US, LILW GDF in Germany and the US, and VLLW trenches (the simplest type NSDF) in France, Spain, and Sweden. LLW trench-type NSDF in arid areas of Argentina, India, Iran, South Africa, and the US. LLW engineered NSDF in China, the Czech Republic, France, India, Japan, the Russian Federation, Slovakia, Spain, the UK, Ukraine, and the US. Examples of nuclear waste streams and nuclear waste disposal facilities in various nations are shown in Table 7.4.

Table 7.4. Some Examples of Nuclear Waste Disposal Facilities of Different Types and for Different Waste Categories, Modified After

| Facility Type | Facility | Nuclear Waste | Country |
|---|---|---|---|
| NSDF, trench | CIRES, Morvilliers | VLLW | France |
| NSDF, trench | El Cabril | VLLW | Spain |
| NSDF, engineered | CSA, De l'Aube | Short-lived LILW | France |
| NSDF, engineered | CSM, La Hague | Short-lived LILW | France |
| NSDF, engineered | Dukovany | LLW from NPPs | Czech Republic |
| NSDF, engineered | National Radioactive Waste Repository Mochovce | Solid and solidified low-activity waste from the operation and decommissioning of NPPs, research institutes, laboratories, and hospitals | Slovakia |
| NSDF, engineered | El Cabril Disposal Facility | LLW and ILW from hospitals, research centers, industries, and NPPs | Spain |
| NSDF, engineered | Low-Level Waste Repository, Drigg, Cumbria | LLW from a range of customers, such as the nuclear industry, the Ministry of Defence, non-nuclear industries, educational, medical, and research establishments | UK |

| | | | |
|---|---|---|---|
| Former limestone mine, access via horizontal drifts | Richard repository, Litom⌧⌧ice | Institutional waste from healthcare, industrial, agricultural, and research sectors | Czech Republic |
| Former uranium mine, access via horizontal drifts | National Repository for LILW—Bai⌧a-Bihor | LILW | Romania |
| Intermediate depth, tunnels | Loviisa VLJ repository | LILW generated at the Loviisa NPP | Finland |
| Intermediate depth, silos | Olkiluoto VLJ repository | LILW generated in state-run industry, medical care, and research; in the future also waste from the Olkiluoto NPP | Finland |
| Intermediate depth, vaults and silos | SFR Forsmark | LILW from the operations of Swedish NPPs but also from hospitals, veterinary medicine, research, and industry | Sweden |
| Intermediate depth, boreholes | Planned | Disused sealed radioactive sources (DSRSs) | Ghana, Philippines, Malaysia, Brazil |
| GDF | Waste Isolation Pilot Plant (WIPP, Carlsbad, New Mexico) | Transuranic nuclear waste (clothing, tools, rags, residues, debris, soil, and other items contaminated with small amounts of Pu and other man-made radioactive elements), categorized as 'contact-handled' or 'remote-handled,' based on dose rate measured at the surface of containers | US |
| GDF | Asse II (abandoned salt mine, nuclear waste intended for retrieval) | Waste with negligible heat generation | Germany |

| | | | |
|---|---|---|---|
| GDF | ERAM (Morsleben) (former salt mine, facility to be closed) | Waste with negligible heat generation | Germany |
| GDF | Konrad (former iron mine, under construction) | Waste with negligible heat generation | Germany |
| GDF | Onkalo disposal facility (under construction | SNF | Finland |
| GDF | Forsmark disposal facility (under construction) | SNF | Sweden |
| GDF, deep boreholes | Under consideration | HLW, SNF, DSRS | UK, US, Australia, Russia |

Source: Michael I. Ojovan, Nuclear Waste Disposal, Encyclopedia 2023, 3, 419–429. https://doi.org/10.3390/encyclopedia3020028.

The disposal of nuclear waste has been practised for almost eight decades, demonstrating a high degree of operational safety and environmental protection. The IAEA has assessed the global inventory of nuclear waste disposed of in cubic meters, as follows: VLLW—273,000 (0.36%), LLW—65,192,000 (85,64%), ILW—10,589,000 (13.91%), and HLW—72,000 (0.09%).

## 7.5.4. Safety of Disposal

The collection of technical, scientific, administrative, and managerial justifications and supporting data for the safety procedure is called the safety case, and it is used to assess the safety of disposal. These standards offer proof of the effectiveness of the disposal location, the design of the disposal facility, including the building and maintenance of the facility, and the evaluation of any possible long-term radiation hazards to the populace. It is recommended that the safety case comprise the subsequent elements: (i) background, (ii) safety plan, (iii) facility overview, (iv) safety assessment report (SAR), (v) boundaries, (vi) guidelines and restrictions, (vii) design optimization and iteration, (viii) handling uncertainty, and (ix) incorporating safety arguments. The Safety Analysis Report (SAR) is the central document in the safety case. It describes the expected behavior of the disposal facility in both normal and potentially disturbing event scenarios that occur within expected time frames and causes the radionuclides in nuclear waste to decay below clearance levels. The long-term decay of the artificial barriers and the eventual release of radionuclides into the formation housing the disposal facility are described using complex mathematical models and computer codes. This is primarily due to groundwater encountering the disposed-of waste packages and then slowly moving those radionuclides into the biosphere. Long-term tests of waste form and barrier materials, as well as readily available natural analog materials, subjected to procedures akin to those anticipated to take place in a disposal facility, are commonly used to test and validate models. The results of the SAR are compared with the appropriate national criteria, which are typically set on the basis, and on account, of international standards, such as, which focus mainly on the radiological criteria aiming to:

- limit the exposure doses (typically given in mSv/y) or risks (relative to other risks);
- achieve levels of radiological protection for future generations that are the same (not less) as those currently achieved;
- ensure that the additive impact of the disposal system on the natural radiation background is limited.

The International Commission on Radiological Protection (ICRP) suggests that the population's annual dose limit for radiological criteria should be 0.3 mSv/y, and for occupationally exposed workers, it should be below 20 mSv/y, or 100 mSv in 5 years.

A typical example of calculated (potentially expected) doses for a population from a planned GDF is presented in Figure 7.12, where the expected doses are significantly lower than the conservative dose limit of 0.3 mSv/y.

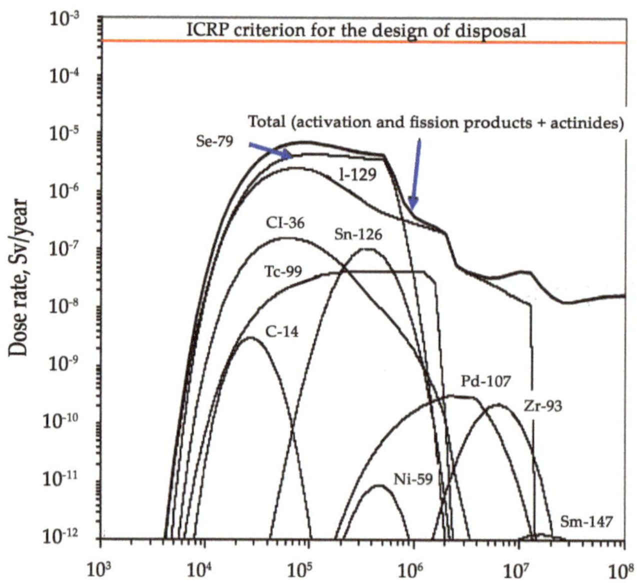


Figure 7.12. An example of expected annual doses ($S_V$/year) that can potentially be caused by nuclear waste radionuclides from a disposal site. The background radiation level is an order of magnitude higher compared with the ICRP recommended level of 0.3 mSv/a. Modified from with permission of Elsevier.

Source: Michael I. Ojovan, Nuclear Waste Disposal, Encyclopedia 2023, 3, 419–429. https://doi.org/10.3390/encyclopedia3020028.

Research on the specific disposal facilities that have been used thus far has demonstrated that nuclear waste disposal is secure and offers sufficient environmental isolation (which can be bolstered by related time frames associated with radionuclide release). For LLW, these can be a few hundred years; for ILW, they can be many thousands of years; and for HLW, they can be hundreds of thousands or even longer. Subsequently, the radionuclides emitted from the disposal sites are insignificant and could only slightly increase the current levels of natural radiation exposure.

# 7.6. CURRENT STATUS OF NUCLEAR WASTE MANAGEMENT

Nuclear or radioactive materials are used in many different applications in modern society. Hospital patients are treated with radioactive materials, which are also used in nuclear power plants to create electricity. The federal government, electrical utilities, the private sector, hospitals, and universities are the main producers of radioactive waste today. Even though these generators all use radioactive materials, the waste that they produce may differ greatly and needs to be managed accordingly. Radioactive waste, or nuclear waste, is defined as any material containing radioactive isotopes in detectable amounts. For the duration of this chapter, radioactive waste and nuclear waste will be used interchangeably.

The field of waste management deals with waste reduction, stabilization, and final disposal. Reducing the quantity of material that needs to be disposed of is known as waste reduction. Among the frequently used techniques for waste reduction are dewatering, compaction, and incineration. Isolating waste from the biosphere and, in the case of radioactive waste, giving it time to decay to sufficiently safe levels are the goals of waste disposal. A timeline of the laws that have impacted radioactive waste management procedures over the past 50 years can be found in Table 7.5.

The disposal of land-level waste (LLW) generated within individual states' borders, or interstate compacts—which Congress forms and dissolves—is required by federal law. These states were initially expected to have the disposal capacity operational by 1993. No state or compact has a facility available to take in waste, even though access to the few remaining facilities is becoming increasingly limited. While some states are moving forward, none of the suggested facilities are currently under construction.

Table 7.5. Major Events Affecting Nuclear Waste Management

| Year | Event |
|---|---|
| 1954 | The Atomic Energy Act is passed |
| 1963 | First commercial disposal of LLW |
| 1967 | DOE facilities begin to store TRU wastes retrievably |
| 1970 | National Environmental Policy Act becomes effective; Environmental Protection Agency is formed |
| 1974 | Atomic Energy Commission divides into the Nuclear Regulatory Commission (NRC) and the Energy Research and Development Administration (ERDA) |
| 1975 | WIPP proposed as unlicensed defense TRU waste disposal facility; West Valley, New York low-level disposal facility closed |
| 1977 | President Carter deferred reprocessing, pending the review of the proliferation implications of alternative fuel cycles |
| 1979 | Three Mile Island, Unit #2 accident; report to the President of the Interagency Review Group on Radioactive Waste Management |
| 1980 | Low Level Waste Policy Act is passed; all commercial disposal of TRU wastes ends |
| 1982 | Nuclear Waste Policy Act is passed; 10 CFR Part 61 issued as final regulation for LLW |
| 1985 | Low Level Radioactive Waste Policy Act Amendments |
| 1986 | The reactor explosion at Chernobyl |
| 1987 | Nuclear Waste Policy Act Amendments provide for the characterization of the proposed HLW repository at Yucca Mountain, Nevada |

Source: Berlin, R. and Stanton, C. Radioactive Waste Management, John Wiley & Sons, Hoboken, NJ, 1989. with permission.

There are designated sites for the TRU and HLW waste programs at the Waste Isolation Pilot Plant (WIPP) in Carlsbad, New Mexico, as well as at Yucca Mountain. The Department of Energy (DOE)'s research and development facility, WIPP, is built to handle 25,000 feet of remotely handled TRU waste in addition to 6 million feet of contact-handled TRU waste. Waste produced by the military will be received by the facility and stored in a geologic repository that can be retrieved. Here, a geologic repository is the salt formations close to Carlsbad. The facility has a 25-year design-based lifespan.

President Obama had promised to fulfill a campaign promise and gain political points in Nevada by cutting off most of the funding for the Yucca Mountain nuclear waste project when this manuscript was submitted in 2009. This presents new concerns regarding the handling of radioactive waste from the country's nuclear power plants, as the federal government may have to pay the utility industry billions more because of the decision. Should the decision be upheld, the majority of the $10.04 billion allocated since 1983 to locate a repository for radioactive waste would have been squandered. Furthermore, if the repository is ultimately abandoned, the country will be left without a solution to an issue it has faced for fifty years. It makes sense that attorneys are projecting damage suits worth tens of billions of dollars from utilities that must pay for waste storage rather than having the government bury them (Ojovan, 2023)

# CASE STUDY

## NUCLEAR WASTE MANAGEMENT

### Background

A fictional country, Xanadu, operates several nuclear power plants that generate electricity for its population. Over the years, these plants have accumulated significant amounts of nuclear waste, ranging from spent fuel rods to contaminated equipment and materials.

### Challenges

- *Storage Capacity:* Existing storage facilities are nearing capacity, and there is an urgent need for expansion or alternative solutions.
- *Safety And Security:* Ensuring the safe containment and management of radioactive materials to prevent environmental contamination and unauthorized access.
- *Public Perception:* Addressing public concerns and misconceptions about nuclear waste, focusing on transparency and education.

### Solutions Implemented

- Long-Term Storage Facility Expansion
    - *Action:* Xanadu authorities decide to expand their deep geological repository (DGR) for high-level nuclear waste.
    - *Rationale:* DGRs provide a stable, isolated environment deep underground, minimizing the risk of radiation exposure to humans and the environment over the long term.
    - *Implementation:* Rigorous geological surveys and safety assessments are conducted to select suitable sites. Construction begins with advanced engineering techniques to ensure structural integrity and containment.
- Research and Innovation in Waste Treatment
    - *Action:* Investing in research and development of advanced nuclear waste treatment technologies.
    - *Rationale:* Innovations like partitioning and transmutation aim to reduce the radiotoxicity and volume of high-level waste.
    - *Implementation:* Collaborating with international research institutions and industry partners to pilot new technologies. Continuous monitoring and evaluation of effectiveness and safety.
- Community Engagement and Education
    - *Action:* Launching a comprehensive public outreach program.

- *Rationale:* Building trust and understanding among citizens about nuclear waste management practices and safety measures.
- *Implementation:* Holding public forums, distributing informational materials, and inviting experts to explain technical aspects in accessible language. Encouraging community involvement in monitoring and oversight.

### Results and Future Considerations

- *Storage Capacity:* The expanded DGR has increased Xanadu's capacity to safely store nuclear waste for several decades, alleviating immediate storage concerns.
- *Safety And Security:* Enhanced containment measures and technological innovations bolster safety protocols, reducing environmental and health risks.
- *Public Perception:* Improved public awareness fosters informed discussions and support for long-term nuclear waste management strategies.

### Conclusion

Through proactive measures in facility expansion, technological innovation, and community engagement, Xanadu demonstrates a commitment to responsible nuclear waste management. Continued research and adaptation of best practices will be crucial in ensuring sustainable solutions for future generations.

# CLASS ACTIVITY

## "NUCLEAR WASTE DILEMMA"

Description: In groups, students will brainstorm and write down the potential risks and benefits of nuclear waste disposal. Each group will then present their findings to the class, focusing on the environmental, economic, and societal impacts of different disposal methods (e.g., storage, reprocessing, deep geological disposal). Finally, the class will discuss and debate the most viable solution for managing nuclear waste in the future.

## SUMMARY

- Nuclear waste, also known as radioactive waste, encompasses materials that contain radioactive isotopes generated from nuclear-related activities such as nuclear power generation, nuclear weapons production, medical treatments, and industrial applications involving radioactive materials. These isotopes emit radiation, which can be harmful to living organisms and the environment if not properly managed.
- Low-level waste (LLW) refers to radioactive waste that contains relatively low levels of radioactivity per unit volume or weight. LLW is generated from various sources, including nuclear power plants, hospitals, research institutions, and industrial facilities that use radioactive materials.
- An atom consists of an extremely small, positively charged nucleus surrounded by a cloud of negatively charged electrons. Although typically the nucleus is less than one ten-thousandth the size of the atom, the nucleus contains more than 99.9% of the mass of the atom! Nuclei consist of positively charged protons and electrically neutral neutrons held together by the so-called strong or nuclear force.
- Radioactive elements are a consequence of nuclear fission both naturally and deliberately in nuclear reactors, particle accelerators, etc. Radioactive elements can be found naturally in the geosphere and can also be produced due to human activities.
- Gamma radiation is an electromagnetic emission at the short-wave edge of an electromagnetic wave spectrum. It originates from the nuclei and X-ray radiation occurs in the electron shells of atoms.
- Nuclear fusion is a process where two atomic nuclei combine to form a heavier nucleus, releasing a significant amount of energy in the process. This is the same process that powers the sun and other stars. In stars, fusion typically involves the fusion of hydrogen nuclei (protons) to form helium nuclei, along with the release of energy in the form of photons (light) and neutrinos.
- Nuclear fission is the process in which a heavy atomic nucleus, typically uranium-235 or plutonium-239, is split into two or more lighter nuclei when bombarded with neutrons. This releases a large amount of energy, along with additional neutrons that can then trigger further fission reactions in a chain reaction.
- Radioactive waste (which is the same as nuclear waste), is material that contains, or is contaminated with, radionuclides, at activity concentrations greater than the clearance levels established by the regulatory body, beyond which no further legal or regulatory purpose is foreseen.

## REVIEW QUESTIONS

1. What is nuclear waste and what are its primary sources?
2. Describe the composition and characteristics of nuclear waste. How does it differ from other types of waste?
3. Explain the process of nuclear fission and how it contributes to the generation of nuclear waste.
4. Discuss the different categories of nuclear waste based on their levels of radioactivity and how they are classified for disposal.
5. What are the current methods used for the disposal of nuclear waste? Evaluate their effectiveness and challenges.
6. Describe the principles behind long-term storage solutions for nuclear waste. What factors are considered in choosing storage sites?
7. What are the technological advancements in nuclear waste management? How do these innovations improve safety and efficiency?

## REFERENCES

1. Ahn, J., & Apted, M. J. (2017). Geological Repository Systems for Safe Disposal of Spent Nuclear Fuels and Radioactive Waste (2nd ed.). Elsevier: Cambridge, UK. 778 pages.
2. Anderson, M., & Woessner, W. (1992). Applied Groundwater Modeling. San Diego, CA: Academic Press Incorporated. ISBN 0-12-059485-4.
3. Attix, F. (1986). Introduction to Radiological Physics and Radiation Dosimetry. New York: Wiley-VCH. ISBN 978-0-471-01146-0.
4. Chapman, N., & Hooper, A. (2012). The disposal of radioactive wastes underground. Proceedings of the Geologists' Association, 123, 46–63.
5. Finsterle, S., Muller, R. A., Grimsich, J., Bates, E. A., & Midgley, J. (2021). Post-closure safety analysis of nuclear waste disposal in deep vertical boreholes. Energies, 14, 6356.
6. IAEA. (2009). Policies and Strategies for Radioactive Waste Management. IAEA Nuclear Energy Series, NW-G-1.1. IAEA: Vienna, Austria. 68 pages.
7. ICRP. (2013). Radiological Protection in Geological Disposal of Long-Lived Solid Radioactive Waste; ICRP Publication 122. Annals of the ICRP, 42(3). ICRP: Ottawa, ON, Canada.
8. Lyman, E. S. (1994). A perspective on the proliferation risks of plutonium mines. Nuclear Control Institute. Archived from the original on November 25, 2015. Retrieved November 25, 2015.
9. National Academies of Sciences, Engineering, and Medicine. (2022). Merits and Viability of Different Nuclear Fuel Cycles and Technology Options and the Waste Aspects of Advanced Nuclear Reactors. The National Academies Press: Washington, DC, USA.

10. Ojovan, M. I. (2023). Nuclear waste disposal. Encyclopedia, 3(2), 419–429. https://doi.org/10.3390/encyclopedia3020028.
11. Ojovan, M. I., Lee, W. E., & Kalmykov, S. N. (2019). An Introduction to Nuclear Waste Immobilisation (3rd ed.). Elsevier: Amsterdam, The Netherlands. 497 pages.
12. Rohlig, K. J. (2022). Disposal and other conceivable strategy 'endpoints' for different types of waste. In K. J. Rohlig (Ed.), Nuclear Waste: Management, Disposal and Governance (Chapter 7, pp. 7.1–7.19). IOP Publishing, Ltd.: Bristol, UK.
13. Rohlig, K. J. (2022). Geologic ('deep') disposal of high-level and other long-lived waste: Host rocks, concepts, current international status. In Nuclear Waste: Management, Disposal and Governance (Chapter 8, pp. 8.1–8.25). IOP Publishing, Ltd.: Bristol, UK.
14. Yudintsev, S. V., Nickolsky, M. S., Ojovan, M. I., Stefanovsky, O. I., Nikonov, B. S., & Ulanova, A. S. (2022). Zirconolite polytypes and murataite polysomes in matrices for the REE—Actinide fraction of HLW. Materials, 15, 6091.

CHAPTER

8

# Electronic Waste

## LEARNING OBJECTIVES

- **Give an overview of electronic waste**
- **Describe electronic waste recycling and disposal**
- **Explain e-waste management practices**
- **Elaborate recovery and recycling**

## KEY TERMS FROM THIS CHAPTER

Contamination
Economic development
Environmental degradation
Leaching process
Polychlorinated biphenyls
Toxic substances
Dumpsites
Electronic waste
Hazardous constituent
Metal dust
Recycling
Transboundary movement

# INTRODUCTION

Electronic waste, commonly known as e-waste, refers to discarded electrical and electronic devices such as computers, smartphones, televisions, and household appliances. As technology rapidly advances and consumption patterns shift towards frequent upgrades and replacements, the volume of e-waste has surged, presenting a significant global challenge. E-waste is a complex mixture of materials, including metals, plastics, and other compounds, some of which are hazardous. These components can contain toxic substances like lead, mercury, cadmium, and brominated flame retardants, posing severe risks to human health and the environment if not properly managed.

The improper disposal of e-waste, such as incineration or landfilling, can lead to the release of harmful chemicals into the soil, water, and air, contributing to environmental pollution and health problems. For instance, heavy metals from e-waste can leach into the ground and contaminate groundwater supplies, while burning e-waste releases toxic fumes and greenhouse gases, exacerbating air pollution and climate change. In many developing countries, informal recycling sectors process e-waste under unsafe conditions, exposing workers to toxic substances and contributing to localized environmental degradation.

Effective e-waste management involves several key strategies, including reducing waste generation through more sustainable design and consumption practices, promoting the reuse and repair of electronic devices, and improving recycling processes to recover valuable materials. E-waste recycling can reclaim precious metals like gold, silver, and palladium, which reduces the need for raw material extraction and supports a circular economy. Additionally, regulatory frameworks and international agreements, such as the Basel Convention, play crucial roles in governing the transboundary movement of hazardous e-waste and promoting safe disposal practices.

Addressing the e-waste challenge requires coordinated efforts from governments, industries, and consumers to foster sustainable electronic consumption and establish robust systems for collection, recycling, and disposal of electronic products. By advancing these efforts, we can mitigate the adverse environmental and health impacts of e-waste and move towards a more sustainable and responsible approach to electronic device lifecycle management.

# 8.1. OVERVIEW OF ELECTRONIC WASTE

Electronic waste refers to different types of electric and electronic devices that are no longer useful to their owners or fulfill their intended function. Electronic waste, or "e-waste," is the term for products that at the end of their useful lives due to redundancy, replacement, or breakage. These products include "brown goods" like televisions, radios, computers, and cell phones, as well as "white goods" like microwaves, refrigerators, and washing machines.

The information and technology revolution has resulted in an exponential increase in the use of new electronic equipment, but it has also led to the production of an increasing amount of obsolete products; one of the waste streams that is growing the fastest is e-waste. E-waste is not the same as regular municipal waste because many of the products contain recoverable precious materials, even though they contain complex combinations of highly toxic substances that are dangerous to human health and the environment. E-waste makes up over 5% of all municipal solid waste worldwide, and its percentage is rising as more people purchase electronic goods in developing nations. Most of the world's e-waste is recycled in developing nations, where it is common to find unofficial and dangerous setups for the extraction and sale of metals. Because of stringent environmental regulations and rising waste disposal costs, recycling companies in developed nations may find it more profitable to export their products to small traders in developing nations than to recycle domestically. Significant illicit cross-border e-waste transfers take place, with wealthy industrialized nations donating large amounts of their e-waste to developing nations. Lax environmental laws, dishonest officials, and low-paid laborers provide opportunities for e-waste profiteers to amass substantial profits. Therefore, policies and strategies for the safe disposal and recycling of e-waste are desperately needed if we are to achieve a sustainable future (Figure 8.1).

Figure 8.1. E-waste.

Source. https://cdn.britannica.com/32/156532-050-6650EB1A/waste-garbage-dump.jpg?w=400&h=300&c=crop.

## 8.1.1. E-waste Characteristics

E-waste can be categorized according to its chemical and physical components. Table 8.1 presents the categorization of its attributes in the references. The variations in the data's physical and chemical composition could result from various data years. E-waste may contain different chemical components because of technological advancement. According to a study done between 1991 and 2008 to evaluate the chemical composition of dynamic RAM (DRAM), from 2008 to 2020, DRAM's copper content would increase by 75%, its palladium content would decrease by 80%, and its gold and silver content would remain stable. The physical makeup of the world's e-waste will be impacted by the decreasing number of DRAM modules due to technological advancements and modular design; however, the amount of precious metals in ICT may stay constant (Charles et al.,2017).

Therefore, although technology may cause changes in the physical composition of global e-waste, the overall makeup of this waste is expected to remain stable.

Table 8.1. Composition of E-Waste

| Constituents | Vats and Singh (2014) | Realff *et. al.* (2004) | Kong *et. al.* (2012) | Hossain *et. al,* (2015) |
|---|---|---|---|---|
| Metals | 60.20% | 49% | 13% | 39.50% |
| - Copper | | | 7% | 20.10% |
| - Iron | | | | 8.10% |
| - Tin | | | | 4% |
| - Nickel | | | | 2% |
| - Lead | | | | 2% |
| - Aluminium | | | | 2% |
| - Zinc | | | | 1% |
| - Silver | | | | 0.20% |
| - Gold | | | | 0.10% |
| - Palladium | | | | 0.01% |
| Plastics | 15.20% | 33% | 21% | 30.30% |
| Metal-plastic mixture | 5% | | | |
| Cables | 2% | | | |
| Screens (CRT and LCD) | 11.90% | 12% | | |
| PCB | 1.70% | | | |
| Others | 1.40% | 1% | | |
| Pollutants | 2.70% | | | |
| Wood | | 5% | | |
| Refractory Oxides | | | | 30.20% |

Source: https://www.researchgate.net/publication/342685930_E-Waste_Current_Research_and_Future_Perspective_on_Developing_Countries/link/60f6f02cfb568a7098c3bace/download?_tp=eyJjb250ZXh0Ij p7ImZpcnN0UGFnZSI6InB 1YmxpY2F0aW9uIiwicGFnZSI6In B1YmxpY2F0aW9uIn19.

## 8.1.2. Generation and Flow

As most sources provide varying estimates for various years, the precise amount of e-waste produced from multiple nations will not be covered. Since the lack of data makes it difficult for developing nations to manage e-waste, this section will cover techniques for predicting the amount of e-waste that will be produced. As a result, these techniques may help forecast the advantages and disadvantages of e-waste in developing nations.

The simplest method involves forecasting using EEE production data and estimating the amount of EOL that is recycled and discarded to determine the total volume of EOL (Bhutta et al., 2011). A different study examined the possible production and flow of e-waste using a questionnaire modified from the United Nations Environment Program (UNEP) (Rimantho and Nasution, 2016). However, the respondents' roles in the EEE or WEEE life cycle determine how accurate the data is, so selecting the wrong respondent will distort the picture of the flow of e-waste.

The Material Flow Analysis and Population Balance Models are two popular models used to estimate the generation of e-waste (Andarani and Goto, 2014; Santoso et al., 2019). scenarios developed per the stakeholders' varying usage durations and transfer coefficients. Since not all the necessary data was available, some assumptions must be made. For instance, the frequency of reuse from the end-user market to the secondhand market and the stock that is available on the market.

Another study that used system dynamics modeling demonstrated how changes in consumption patterns, economic development, and population growth impacted the growing number of EOL EEE. In 2018 Alamerew and Brissaud. According to the simulation results, over the past ten years, there has been an increase in EOL WEEE products. If legislation could be implemented, this would also mean a higher recovery rate. Policymakers may enhance the choice and e-waste management program involving the key stakeholders by understanding the critical elements of e-waste generation and recovery.

In developing nations, the reverse e-waste supply chain network comprises four to six primary stakeholders, i.e., end users, which include households, businesses, or institutions; collectors, which include scavengers and service centers; secondhand markets; refurbishment; and dealers who export e-waste that is incapable of being recycled (Veenstra et al., 2010; Hanafi et al,2011; Mittal and Dwivedy, 2012). The number of stakeholders involved will depend on the flow in the nation under study. Some studies used mixed-integer multi-objective linear programming, the Markov chain model, or Life Cycle Assessment (LCA) to examine the flow of e-waste. According to qualitative studies and literature reviews, up to 90% of electronic waste generated in developing nations ends up in unregulated waste sectors. To enhance their roles in the collection to recovery processes, the informal waste sectors require special attention if the government wishes to implement the Extended Producer Responsibility (EPR) system.

### 8.1.3. Classification

It is possible to classify e-waste according to its components and composition. Glass, plastics, pollutants, ferrous and nonferrous metals, and other materials are six material categories that are reported for the composition of e-waste. Plastics are the second-largest portion of waste electrical and electronic equipment (WEEE) materials, after iron and steel. Third in abundance and having substantial commercial value are nonferrous materials, which include metals like copper and aluminum as well as precious metals like silver, gold, and platinum. The following are examples of toxic materials: mercury in switches and flat-screen monitors; lead and cadmium in circuit boards; lead oxide and cadmium in cathode ray tubes; brominated flame retardants on printed circuit boards; plastic and insulated cables; when these exceed threshold quantities, they are considered pollutants and can harm the environment if disposed of improperly.

One of the most widely accepted classifications is based on European Union directives that divide e-waste into the 10 following categories:

- Large Household Appliances: refrigerators, freezers, washing machines, clothes dryers, dishwashers, electric cooking stoves and hot plates, microwaves, electric fans, and air conditioners.
- Small Household Appliances: vacuum cleaners, toasters, grinders, coffee machines, appliances for haircutting and drying, toothbrushing, and shaving.
- Information Technology (IT) and Telecommunications Equipment: mainframes, minicomputers, personal computers, laptops, notebooks, printers, telephones, and cell phones.
- Consumer Equipment: radios, televisions, video cameras, video recorders, stereo recorders, audio amplifiers, and musical instruments.
- Lighting Equipment: straight and compact fluorescent lamps and high-intensity discharge lamps.
- Electrical and Electronic Tools: drills, saws, sewing machines, soldering irons, equipment for turning, milling, grinding, drilling, making holes, folding, bending, or similar processing of wood and metal.
- Toys, Leisure Equipment, and Sporting Goods: electric trains or racing car sets, video games, and sports equipment with electric elements.
- Medical Devices: radiotherapy equipment, cardiology, dialysis, pulmonary ventilators, nuclear medicines, and analyzers.
- Monitoring and Control Instruments: smoke detectors, heating regulators, and thermostats.
- Automatic Dispensers: for hot drinks, hot or cold bottles, solid products, money, and all appliances that automatically deliver various products.

## 8.1.4. Impacts on Human Health

Human health is negatively impacted by the composition and incorrect handling of e-waste. Concern over the possible harm that e-waste poses to human health has grown because of accumulating epidemiological and clinical data, particularly in developing nations like China and India. The antiquated techniques employed by unlicensed backyard operators (e.g., workers are exposed to various harmful substances when they recover, reprocess, and recycle e-waste materials (the unofficial sector). The use of procedures like component disassembly, wet chemical processing, and incineration exposes workers to dangerous chemicals directly and causes respiratory problems. Worker awareness is frequently low, and there is lack of awareness about safety gear like ventilation fans, gloves, and face masks.

When it comes to potential health risks, burning printed wiring boards outdoors, for example, raises the level of dioxins in the air. Inhaling these toxins increases the risk of cancer for both locals and workers. When small amounts of precious metals are manually extracted and collected, toxic metals and poison can also enter the bloodstream. Additionally, workers are constantly exposed to fumes from highly concentrated acids and toxic chemicals. Burning insulated wires to recover resalable copper results in neurological disorders. Acute exposure to cadmium, which is present in semiconductors and chip resistors, can harm the liver and kidneys as well as cause bone loss. Children are particularly vulnerable to the negative effects of prolonged exposure to lead found on printed circuit boards, computer screens, and television screens, which can harm the kidneys and the central and peripheral nervous systems.

## 8.1.5. Environmental Impacts

Electronics are an essential component of modern life, but their potentially harmful effects on the environment should not be disregarded. During the production, recycling, and disposal of these products, electrical and electronic equipment is released into the environment. Environmental degradation is caused by the release of liquid waste into water and drainage systems, the release of gases, fumes, and particulates into the air, as well as the disposal of hazardous wastes. Policies that extend the accountability of all stakeholders, especially the producers, beyond the point of sale and up to the end of product life are required in addition to stricter regulations regarding the recycling and disposal of e-waste.

Certain ways of recycling e-waste can also harm the environment. Air pollution results from the burning of wires and cables to recover metal, which releases dioxins that are chlorinated and brominated. Toxic chemicals with no commercial value are just dumped during the informal sector's recycling process. The local groundwater quality is severely impacted by the toxic industrial effluent that is dumped into subterranean aquifers, rendering the water unfit for human consumption or agricultural use. Dismantling activities release dust particles into the atmosphere that are laden with heavy metals and flame retardants, which leads to atmospheric pollution. Depending on their size, these particles can either be carried over great distances by long-distance transport or redeposit (dry or wet deposition) close to the emission source. Additionally, the dust has the potential to contaminate soil and water systems by leaching into the ground and combining with substances present in both wet and dry depositions. When materials like lead, mercury, cadmium, arsenic, and polychlorinated biphenyls (PCBs) end up in landfills, the soil becomes poisonous.

Human health and environmental conditions are closely related because there are two ways that e-waste can affect health: directly from recycling areas or through contamination of food chains. Most of the health and environmental research is carried out in developing nations, particularly in Asia, such as China, India, and Vietnam. The main reasons backyard treatment emerged were the economic view of e-waste, inadequate knowledge, and limited resources for e-waste treatment technologies. Improper handling of e-waste can result in chemical contaminants contaminating the air, soil, and water, endangering human health.

The components of electronic waste and their effects on health (Gaidajis et al., (2010); Kishore and Monika, 2010; Kiddee et al., 2013; Grant and others. Garlapati (2016) and, 2013). These materials fall into one of three categories of contaminants. The dangerous components of e-waste, such as heavy metals and halogenated compounds, are known as primary contaminants. Secondary contaminants, such as dioxins, polyaromatic hydrocarbons (PAHs), and poly-halogenated aromatic hydrocarbons (PHAHs), are byproducts or residues created because of improper recycling processes. Tertiary emissions, also known as contaminants, are substances used in recycling that careful management to prevent negative effects on the environment and human health. Examples of these substances include aqua regia, which is the byproduct of leaching metal recycling and contains residues of nitric acid, hydrochloric acid, cyanide, thiourea, and bromide. (Khanna et al., 2014).

#### *8.1.5.1. Soil Contamination*

The presence of dioxin compounds in the soil at the e-waste processing sites was verified by a study conducted in Vietnam. The concentration found exceeded the WHO-recommended maximum allowable concentration, primarily because of open burning and open storage practices (Suzuki et al., 2016). The presence of fire retardants in soil and river sediments demonstrates that improper recycling practices, such as open burning and manual dismantling, pollute the soil and rivers in Vietnam (Someya et al., 2016).

In a Chinese study, soil samples from open burning sites, rice paddy sites, e-waste processing sites, and sediment samples from a nearby river were examined. Seven metallic elements, i.e., High concentrations of calcium, copper, molybdenum, lead, nickel, tin, and zinc were found in soil and sediment samples. The strong correlation between e-waste processing and a high concentration of those elements was validated by multivariate analysis. According to Uchida et al., 2018, soil analysis from open-burning sites revealed Cu accumulation, indicating the influence of open-burning activities. Therefore, to lessen the amount of chemical constituents that contaminate the soil, appropriate treatment procedures and wastewater treatment at e-waste recycling facilities must be implemented.

### 8.1.5.2. Air Contamination

In addition to contaminating the soil, burning e-waste raises the concentration of air pollutants, especially particulate matter. A study conducted in India measured the levels of particulate matter (PM10) and heavy metals (Cu, Pb, Zn, Cr, and Ni) in air samples taken from two residential areas and an open burning site for e-waste. The study revealed that the mean concentration of PM10 and heavy metals was highest in e-waste burning sites out of all the areas investigated. Based on the blood analyses, it was also discovered that the people living near e-waste burning sites had the highest level of exposure. This study demonstrated how the open burning of waste by the unorganized sector led to a high level of air contamination, which impacted the residents' high exposure to heavy metals. (Gangwar et al., 2019). According to research conducted in India, two unofficial e-waste recycling sites in Vietnam's urban area had higher levels of PCBs and BFRs in indoor dust than non-e-waste homes (Tue et al., 2013). According to estimates of human exposure, ingestion of dust led to the intake of most BFRs, whereas inhalation of air led to the intake of PCBs.

#### *8.1.5.3. Water Contamination*

Illegal dismantling of e-waste along with open-air burning was the major cause of high contamination of cadmium in groundwater in four districts in India, namely Rampur, Shahjehanpur, Moradabad, and Bareilly (Idrees et al., 2018). The level of cadmium contamination was higher than the regulatory threshold, indicating serious toxicity problems in the groundwater system.

A contaminant's presence in surface water will have an impact on aquatic life and cause biomagnification. In a contaminated pond, plasticizers, PFRs (organophosphorus flame retardants), and PFR metabolites were investigated in water snake samples. A study was conducted in China. Biomagnification factors (BMFs) less than 1, which indicates a bio-dilution fueled by metabolism, indicate that biomagnification did not occur (Liu et al., 2019).

#### *8.1.5.4. Assessment of Human Health*

Numerous investigations have been carried out for identification of heavy metals and other chemicals from e-waste exposure. A study conducted in South America examined the manual collection of metals by economically disadvantaged children in the region and the relationship between this activity and elevated blood lead levels (BLLs) due to lead exposure. It was discovered that the BLLs of the children and teenagers were higher than those recommended (Pascale et al., 2016).

Hair and urine testing are common bio-indicators to evaluate human exposure to e-waste contaminants, in addition to blood testing. Two studies tracked exposure in e-waste sites using various indicators. Zhang et al, (2019) used resident urine samples to examine the exposure to phthalic acid esters (PAEs), which are common in e-waste recycling areas. Residents of the non-e-waste site served as the reference area for this comparison of samples. The findings demonstrated that compared to those residing in the reference area, the e-waste site residents had a significantly higher concentration of ΣmPAEs in their urine. Hazard index (HI) values greater than 1 are present in about 22% of residents of e-waste sites, indicating exposure above acceptable limits that may have unfavorable consequences (Zhang et al., 2019). The most significant discovery is that 68% of them were between the ages of 0 and 18.

For residents and employees living close to the e-waste recycling area, PAH identification was also conducted using hair and urine as bioindicators. Between residents and e-waste dismantling workers, there was no discernible variation in exposure levels. One carcinogenic metabolite was also found in the respondents' internal hair, according to another finding (Lin et al., 2020).

Acute and long-term effects of heavy metal exposure include respiratory, reproductive, irritated, cardiovascular, and urinary diseases (Awasthi et al., 2018). Furthermore, human DNA may be impacted by exposure to e-waste contaminants. DNA damage and the amount of time spent processing e-waste were found to be strongly correlated in a study done to evaluate possible genomic damages in workers recycling e-waste (Wang et al., 2018). A review of the health of people living in unofficial e-waste recycling sites supports this conclusion. In addition, stillbirths, spontaneous abortions, preterm births, and smaller and lighter babies are all caused by exposure to e-waste (Grant et al., 2013).

Radioactive materials are present in some WEEEs, but they make up less than 1% of all e-waste worldwide. There is no immediate radiological risk to the workers or locals near the municipal waste dumpsites or the e-waste dumpsite in Southwest Nigeria, according to an assessment of radiation exposure at these locations (Jibiri et al., 2014). Since not all e-waste contains radioactive materials, the composition of the waste at the dumpsites may haveimpacted the outcome.

Not all the studies were conducted by direct sampling of human bio-indicators and lab analysis. Photo initiator (PI) intakes daily for Chinese workers dismantling e-waste have been estimated through Monte Carlo analysis. It was discovered that e-waste workers consume more PI daily than city dwellers (Li et al., 2020).

The formal recycling plant has the same level of risk even though earlier Chinese research focused more on the danger of incorrect handling. Three recycling plants were the subject of a Swedish study that included office and direct recycling employees. Both individual air exposure and exposure bio-indicators (blood, urine, and plasma) were tracked. The results of an air sample revealed that recycling workers had greater airborne exposure (10 to 30 times) than office workers. The biomarkers revealed that recycling workers' urine, blood, or plasma had significantly higher concentrations of cobalt, chromium, indium, mercury, and lead than did office workers.' To enhance worker and environmental protection, implementing an automated process is advised (Julander et al., 2014).

## 8.1.6. Economic Assessment

Few studies have been conducted to estimate the potential value of e-waste. A 2015 study used the projected e-waste volume for 2020 and the e-waste volume for that year to analyze the potential recovery value of 14 products. CRT monitors had the highest potential revenue in 2015, of 528 million euros. Depending on the advancement of technology, the quantity of individual products varies as the years pass. Accordingly, smartphones had the largest potential revenue contribution in 2020 (746 million euros) (Cucchiella et al., 2015). According to another study, smartphones came in second place with a total value of 2,306 M€, while cell phones were predicted to generate the highest revenue in 2020 (15,777 M€) (Charles et al., 2017). These studies demonstrate that, when handled appropriately, the precious metals found in e-waste can have positive economic effects. Unfortunately, the assessment only considered the potential revenue and ignored the cost of the recovery procedures.

# 8.2. ELECTRONIC WASTE RECYCLING AND DISPOSAL

Hazardous waste comes in a wide variety, and electronic waste, also known as e-waste or waste electrical and electronic equipment (WEEE), is one of them. Any electronic device that a customer or business plans to discard or that can no longer be used for its intended purpose is included in this waste stream. The exponential increase in e-waste generation, the potential benefit of recycling the waste to recover precious metals and other elements, and the risks to human health and the environment from improperly storing, disposing of, and recycling e-waste have all contributed to the significant level of public and political interest in e-waste. The development of producer "take-back" laws, technological advancements in recycling procedures, and the creation of partnerships to facilitate the transfer of e-waste between the informal and formal recycling sectors have been some of the major responses to the rising generation of e-waste (and growing demand for secondary raw materials that it contains) (Lucier et al., 2016).

E-waste is an incredibly complex waste stream, as it encompasses a wide range of items and the exact composition of many electronic components are trade secrets, meaning they are the confidential information of the manufacturer. In general, "up to 60 distinct elements can be found in modern electronics; many of these elements are valuable, some are hazardous, and some are both. The printed wiring boards (PWBs) typically contain the most intricate combination of materials. For instance, "over 40 elements in the periodic table including base metals like copper (Cu) and tin (Sn), special metals like cobalt (Co), indium (In), and antimony (Sb), and precious metals like silver (Ag), gold (Au), and palladium (Pd)" are among the materials that make up a mobile phone.

For over 40 years, the official recycling industry has recycled electronics that were used in commercial or industrial applications, such as medical equipment. These bulky goods are regularly exported from OECD industrialized nations to specialized facilities where they undergo processing to extract secondary raw materials. In the past, recycling consumer electronics from smaller devices like televisions and cell phones has not been profitable in nations with higher labor costs due to the comparatively small number of valuable materials that can be recovered. To recover gold, copper, and other valuable metals, these items were usually disposed of, stored in consumers' homes, or exported (often illegally) to less developed nations like Ghana, Nigeria, China, and India. There, they are recycled by unofficial recyclers using antiquated techniques like manual dismantling, open burning, and acid leaching. The hazardous materials used in these processes, which are also incorporated into consumer electronics, pose serious risks to both human health and the environment in addition to providing subsistence wages for the workers. This chapter will examine these traditional recycling initiatives and how they are changing in tandem with advancements in the world economy and the launch of novel recycling technologies and processes.

In general, there are five fundamental steps in the recycling of e-waste: collection, removal of toxins, preprocessing, end processing, and disposal. The way these phases are handled varies greatly across the globe. E-waste can be gathered formally through producer "take-back" programs that are either required or voluntary, or more informally through "waste pickers" for a large portion of the world's waste stream. Consumer electronics recycling rates have historically been much higher in areas where e-waste is collected by unofficial collectors than in areas where waste is dropped off through official channels. Once at the recycling location, hazardous materials that need extra care (e.g., Freon, batteries, etc.) are separated. Based on material, the units are then divided into more homogeneous groups. This can be completed mechanically, manually, or by combining the two methods. While mechanical dismantling may involve conveyor belts, enormous shredders, and magnets, manual dismantling uses instruments like screwdrivers, hammers, and labeled containers.

After the stages of separation and disassembly, more uniform material groups (e.g., circuit boards, gold, copper, plastic, and other materials) are subsequently subjected to a refining process, which can be carried out metallurgically, chemically, or by heat. This stage can be as low-tech as acid stripping in a Guiyu, China backyard or as high-tech as a massive smelter in Antwerp, Belgium. Studies have revealed that locations frequently vie for waste by presenting low-cost tactics, which is sometimes referred to as a "race to the bottom" process that results in progressively worse environmental protection and standards. Ultimately, every component that can't be sold or utilized as a secondary raw material is disposed of using landfills or incinerators.

How efficiently e-waste is recycled depends on the procedures used, particularly during the separation and disassembly stages. When disassembling electronics, manual methods are frequently far more efficient than automated ones at obtaining the highest caliber secondary raw materials. Automated return initiatives like the ones in the EU are not nearly as effective as the labor-intensive e-waste collection rates observed in many African nations. Dismantling by hand is also better than shredding by machine, which ruins materials and doesn't fully separate them. For instance, only 26% of the gold in abandoned cell phones can be recovered by mechanical shredding, compared to 90% that can be recovered when the phones are disassembled by hand. But unless labor costs are low, these more labor-intensive options are not cost-effective (Wang et al., 2012).

## 8.2.1. Secondary Raw Materials Recovered in Electronic Waste Recycling

Components found in e-waste have historically been valuable in large quantities when the cost of disassembly has been sufficiently low (Manomaivibool et al., 2009). Table 8.2 shows some of the applications and quantities extracted for various "important" or valuable elements within electronic devices.

Table 8.2. A Sample of Valuable Elements in Electronic Wastes

| Element | Main applications | Total tons/year [2006] |
|---|---|---|
| Silver | Contacts, switches, solders | 6000 |
| Gold | Bonding wire, contacts, integrated circuits | 300 |
| Palladium | Multilayer capacitors, connectors | 33 |
| Platinum | Hard disk thermocouple, fuel cell | 13 |
| Ruthenium | Hard disk, plasma displays | 27 |
| Copper | Cable, wire connector | 4,500,000 |
| Tin | Solders | 90,000 |
| Antimony | Flame retardant; CRT glass | 65,000 |
| Cobalt | Rechargeable batteries | 11,000 |
| Bismuth | Solders, capacitor | 900 |
| Selenium | Electro-optic copier, solar cell | 240 |
| Indium | LCD glass, solder, semiconductor | 380 |

Source: Cristina Lucier & Brian Gareau, 2020. "Electronic Waste Recycling and Disposal: An Overview," Chapters, in: Hosam M. Saleh (ed.), Assessment and Management of Radioactive and Electronic Wastes, IntechOpen.

Apart from these metals, there is another group of elements called rare earth elements that are also essential to the operation of the newest electronics, especially those with touch screens and LED lighting. Globally, rare earth elements are widely accessible, but their extraction can lead to several environmental issues, such as radioactive contamination. Table 8.3 lists the rare elements that are used in various electronics. Although these elements are used in relatively small amounts, it is important to note that their characteristics are closely related to the degree of performance of these technologies (Gradael TE, et al., 2015). An important factor in the high-performance operation of magnets is rare earths. The data presented in Table 8.3 has been modified from data obtained from the U. S. Department of Energy, an analysis ordered by the U.S. Department of Interior and the U.S. Geological Survey in addition to trade journals for the sector. The following rare earths are thought to have the greatest potential for resale value: neodymium (Nd), europium (Eu), dysprosium (Dy), terbium (Tb), and yttrium (Y). They also pose the greatest risk of supply shortages.

Table 8.3. Common uses of Rare Earth Elements in Electronic Devices

| Technology | Rare Earths Used |
|---|---|
| Electric and hybrid cars (NiMH battery) | Neodymium, praseodymium, dysprosium, terbium |
| Computers (magnets in hard disk drive) | Neodymium, praseodymium, dysprosium, terbium |
| Flat panel screens (glass coating to produce colors and brightness) | Yttrium, europium, terbium, gadolinium, praseodymium, cerium |
| MRI machines (magnets) | Neodymium, praseodymium, dysprosium, terbium, yttrium, europium |
| Smart phones (magnets and speakers) | Neodymium, praseodymium, dysprosium, terbium, yttrium, europium |
| Other uses (including "chemicals, military weapons and delivery systems, and satellite systems" | Cerium, lanthanum, yttrium, neodymium, praseodymium, samarium, gadolinium |

Source: Cristina Lucier & Brian Gareau, 2020. "Electronic Waste Recycling and Disposal: An Overview," Chapters, in: Hosam M. Saleh (ed.), Assessment and Management of Radioactive and Electronic Wastes, IntechOpen.

Making e-waste recycling more profitable has been the focus of recent technological advancements, such as modifications to the mechanization process and pilot projects that combine low-tech and mechanized methods. Enhancements to the mechanization procedure are quite simple. On the one hand, the consistency and caliber of the materials collected during the preprocessing phase have improved thanks to modifications made to shredding and sorting machines. Furthermore, to extract additional streams of secondary raw materials that were previously unrecoverable, newly mechanized methods are being developed. New methods for removing different rare earth elements from electronics has been one of this field's biggest breakthroughs. Recently, state-of-the-art rare earth extraction facilities have gone live in France and Japan. Since rare earth materials are crucial for technologies pertaining to defence, communications, and other state objectives, continued investment in recycling technologies is viewed as a strategic priority by industrialized nations. However, most of the rare earth mining occurs in China, a global power that recently placed limits on the amount of these materials it is willing to export. The development of new automated technologies to recover rare earth elements from a variety of e-waste inputs has been spurred by worries about the stability and security of the supply of these materials. The United States is currently developing cost-effective technologies to recover secondary neodymium, dysprosium, and praseodymium from e-waste. S. – based research institutions and recyclers. If they are in the EU, the U.S., or Japan, these recently opened recycling facilities will need a significant amount of e-waste inputs to make a profit. This challenge is keeping a sizable amount of e-waste out of landfills and the unofficial recycling sector in developing nations.

### 8.2.2. The Role of Extended Producer Responsibility in E-waste Recycling

The amount of electronic waste produced worldwide in any given year is estimated to vary greatly. These approximations are based on the volume and number of different electronic items that are bought each year, considering the item's estimated lifespan. Recyclers' surveys regarding the amount of electronics they collect can also be considered. However, it is crucial to remember that a large amount of consumer e-waste is either disposed of in landfills or stored in homes by the consumers themselves. Data on the quantity of e-waste collected for recycling is available in certain places, such as those where manufacturers are required to "take-back" unwanted electronics from customers. Extended producer responsibility (EPR), which maintains that producers of goods containing hazardous materials should be responsible for the financial and logistical costs associated with recycling or disposing of their products in an environmentally responsible manner, gave rise to these regulations.

Manufacturers initially opposed EPR legislation, but support for it has increased because of growing awareness of the strategic significance and potential economic benefits of the secondary raw materials (rare earth elements) found in e-waste. Different EPR "takeback" laws have been implemented globally, including in several U.S. states, throughout the European Union, and in numerous Latin American, Asian, and African nations. These regulations suggest a possible move away from the informal sector's dominance of electronics recycling and toward formalized e-waste recycling sector's expansion.

An increasing number of pilot programs and public-private partnerships have been established to collect and recycle electronics efficiently and economically, in response to the growing number of EPR laws that require manufacturers to assume extended responsibility for the environmentally sound recycling of their products. The Basel Convention on the Control of the Transboundary Movements of Hazardous Wastes and Their Disposal (the Basel Convention) applies to certain projects that involve the transportation of electronic waste across national borders. Technical guidelines for the handling and management of e-waste have been developed because of pilot projects and technical working groups that the Basel Convention has organized over the years.

More pilot programs, overseen by the United Nations University, are being developed to support a globalized e-waste recycling chain that includes labor-intensive preprocessing and dismantling in nations with lower labor costs (e.g., China, India, and African nations), and advanced end-processing in nations with more contemporary infrastructure (e.g., the nations of the EU). Government officials, major recycling companies, and producers of electronics all think that these kinds of collaborations will guarantee a larger supply of raw materials for big, sophisticated smelters and give them access to secondary raw materials that were previously "dumped" or otherwise held in the global south countries where informal recycling is currently practised.

### 8.2.3. Environmental and Human Health Hazards of Electronic Waste Recycling

It is becoming more widely known how dangerous many other materials used in electronics are to the environment and human health. Electronics frequently contain harmful substances like PVC and brominated flame retardants (BFRs), in addition, to lead (Pb), cadmium (Cd), mercury (Hg), polychlorinated biphenyls (PCBs), and polybrominated biphenyls (PBBs). A list of some of the known dangerous parts that are typically found in a desktop computer (with a CRT monitor) can be found in Table 8.4. This table is an adaptation of toxicity data from Ceballos et al., (2015) and information provided by the Silicon Valley Toxics Coalition in their report "Poison PCs and Toxic TVs."

Table 8.4. A Sample of Hazardous Elements in an Older-Model PC

| Element | Main applications | Weight (per 60 lbs.) | Dangers |
|---|---|---|---|
| Lead | Metal joining, radiation shield/CRT, printed wiring board | 3.8 | Human effects: neurological, blood, kidney damage. Brain damage and poisoning/death for children Accumulates in the environment |
| Mercury | Batteries, switches/ housing, printed wiring board | <0.1 | Human effects: long-term brain damage and other neurological effects Concentrates through the food chain. |
| Cadmium | Battery, blue/green phosphor emitter/housing, printed wiring board, CRT | <0.01 | Human effects: acute and chronic damage to the kidneys |
| Plastics (including those containing PVC and BFRs) | Casing, cable coating | 22.99 | PVC effects: developmental toxin, reproductive toxin, endocrine disruptor. Carcinogenic when burned due to the production of dioxins BFR effects: endocrine disruptor, neurotoxin, carcinogen (in humans and animals) |

Source: Cristina Lucier & Brian Gareau, 2020. "Electronic Waste Recycling and Disposal: An Overview," Chapters, in: Hosam M. Saleh (ed.), Assessment and Management of Radioactive and Electronic Wastes, IntechOpen.

Several of the health impacts listed in Table 8.4 have been observed in Guiyu, China, where possibly most of the U.S. exports of e-waste have historically been deposited. Nearly 80% of the kids in this area suffer from respiratory issues, and they are particularly vulnerable to lead poisoning. For the employees and their families, neurological, respiratory, digestive, and bone issues are frequent concerns. Apart from the potential long-term health and environmental consequences associated with toxicological hazards, recycling e-waste can pose even greater risks when specific recycling techniques are used. For instance, burning plastic cables to extract the copper within is an informal recycling method that releases dioxins into the air due to the PVC in the plastic burning. In more advanced processes (usually found in Asian nations), leaching printed circuit boards with acids (such as hydrochloric acid and nitric acid) can result in burns, circulatory and respiratory issues, pulmonary edema, and even death. The goal is to maximize the amount of gold recovered. The hazardous residue left over from acid stripping is frequently dumped in streams, where it can acidify the water and kill plants and wildlife. Dust from heavy metals could spread to more populated areas, contaminating larger populations and food supplies.

Studies on the hazards to occupational health and the environment connected to recycling high-tech e-waste have also been conducted; however, experts point out that much more work needs to be done in this area to obtain a more precise assessment of these risks. Numerous studies draw from or are influenced by field research and experiments that quantify levels of hazardous chemicals in the air, water, and workers' surroundings in the vicinity of advanced recycling facilities. These studies show that some technologies, like better ventilation and the use of face masks, can reduce occupational exposure to various heavy metals and other dangerous chemicals.

Within the U. S. the NIOSH (National Institute of Occupational Safety and Health )recently finished surveying 276 electronic waste recycling facilities. The report's conclusion that "most" of the responding facilities rely on manual dismantling—a method akin to that used in new and pilot facilities in less developed nations—is particularly noteworthy. It's also important to note that most of the facilities that responded had received environmentally sound certification from organizations like the EPA's R2 Solutions, the industry standard RiOS, or the activist standard e-Stewards. As a result, in terms of electronics recycling, these facilities probably represent the "best-case scenario." "E-scrap recycling has the potential for a wide variety of occupational exposures, particularly because of manual processes," NIOSH concluded overall. One of the main issues raised in the report is the possibility of encountering "metal dust" when disassembling something by hand. Since most facilities circulate air within the production area or rely on "natural ventilation," the report notes that it is unclear whether most facilities have installed proper filtration systems to remove metal dust from the air. The use of compressed air for cleaning, which can increase exposure to metal dust, is also mentioned in the report. While acute exposure to heavy metals, like lead, is unlikely, chronic lead poisoning, which is more likely at current occupational exposure levels, may not have symptoms or nonspecific symptoms that are not recognized as being associated with lead exposure, according to the initial NIOSH report.

Further research on on-site occupational exposures in formal e-waste recycling facilities has been finished since the release of the NIOSH report. After reviewing 37 studies on the occupational risks related to formal e-waste recycling, researchers concluded that "formal e-recycling workers and their families may experience unhealthful exposures to metals," although worker and environmental health have improved when compared to informal recycling. The authors suggest conducting additional research "to reduce chemical exposures from formal e-waste recycling," as well as developing electronics components that are safer to disassemble and using fewer hazardous materials in the production process. There is an immediate need to put solutions in place to lower the risks involved with recycling e-waste because it is currently thought to be the hazardous waste stream that is growing the fastest in the world.

# 8.3. E-WASTE MANAGEMENT PRACTICES

## 8.3.1. Current Condition in Developing Countries

The management of e-waste faces comparable challenges in developing nations. Since people in these nations view e-waste as a means of subsistence, e-waste imports from developed countries are typically found in developing nations. The informal waste treatment industry has flourished in the vicinity of residential areas due to the lack of formal recycling facilities. In informal treatment facilities, backyard procedures were carried out by hand without sufficient resources or awareness of the risks that could endanger public health and the environment. Certain developing nations, like Indonesia and Cambodia, do not have laws specifically addressing the management of e-waste. (Sothun, 2012; Rode, 2012). Other barriers to e-waste management include ignorance and a lack of data regarding the material flow of e-waste. (Breiter and Ferse, 2013).

Improvements in developing country management can be made by linking national or local regulations to EPR schemes. The success of implementing the new regulation will be increased if it is combined with an incentive system (Zeng et al., 2016; Shevchenko et al, 2019). A study employing system dynamics modeling has demonstrated the efficacy of legislation. The main factor influencing how much e-waste is collected for product recovery is the way laws are enforced. In 2018 Alamerew and Brissaud.

However, as a stakeholder, the government must also support establishment of formal e-waste treatment plants and urge EEE producers to prioritize Extended Producer Responsibility (EPR) (Wei and Liu, 2012; Ilankoon et al, 2018). Working together, the formal and informal waste sectors (IWS) can lessen the harm that results from improper IWS treatment (Li and Tee, 2012). Technical support is required for the formal recycling plant that is currently in operation, offering technical guidance) and monetary assistance (e.g., investment subsidies) to enhance the treatment procedures (Yoshida et al., 2016).

China provides the best illustration of how the e-waste treatment scenario evolved from antiquated, artisanal methods into a sophisticated system (hydrometallurgical or pyrometallurgical). While hydrometallurgical produces more acid waste streams and pyrometallurgical releases hazardous substances like dioxins and furans, both processes have drawbacks that could be mitigated with system improvements (Ilankoonet et al., 2018).

### 8.3.2. Extended Producer Responsibility (EPR)

EPR designates the obligation of importers and manufacturers to gather and recycle EOL EEE. Although the idea of EPR is widely understood in developed nations, there are two main obstacles to its implementation in developing nations. The first difficulty is identifying the producer, which can be brought about by small-scale shop assembly, imitation product manufacturing, or smuggling. Hidden material flows will arise from this when EPR is implemented. Another difficulty is the potential for inaccurate e-waste collection numbers, which could impact the incentive system subsidies (Kojima et al., 2009). The implementation of EPR will encounter numerous gaps in developing nations lacking treatment infrastructure, legislation, and stakeholder awareness (public, private, and government) (Nguyen et al., 2017). In developed nations, like the Republic of Korea, the Extended Producer Responsibility (EPR) system is incorporated with recycling laws and is reinforced by multiple recycling centers and recycling advocacy (Rhee, 2016).

### 8.3.3. Reuse, Remanufacturing, and Refurbishing

The first level of the waste hierarchy, reuse, offers greater environmental benefits than merely gathering e-waste. A study used a comparative Life Cycle Assessment (LCA) to examine two EOL computer treatment chains in Belgium. The first chain focuses on manufacturers' take-back responsibilities. The second chain, in the meantime, puts manual disassembly and reuse of e-waste ahead of mechanical treatment. Because of the higher recovery rate and reuse of the latter chain, it was discovered to have greater environmental benefits (Gonda and Degrez, 2018).

Reuse scenarios are better suited for expensive or seldom-used products. According to a study, the primary constraint on the reuse scenario's economic viability is collection methods. Major expenses include tasks related to sorting, testing, and repair (Angouria-Tsorochidou, 2018). By addressing the major problems with logistics, process operation (including testing, repairing, cleaning, and refurbishment), quality, and customer acceptance, there are numerous strategies to encourage reuse on a larger scale. Creating standards or quality labels for repurposed goods will address concerns about public trust (Cole et al., 2017). Reuse scenarios will also be encouraged by appropriately differentiating reusable and non-reusable electronic waste (Terazono et al., 2006). The primary concern in this scenario is the loss of reuse potential during waste collection due to handling damages and the flow to the waste system (Cole et al., 2018).

Two more types of reuses are refurbishing and remanufacturing. Current state of photocopier remanufacturing and refurbishing was examined by interviewing two companies (OEM or IR) in three countries about these practices. The interviews took place in Japan, Singapore, and Indonesia. For over ten years, photocopier OEMs in Japan have been remanufacturing their machines. Of these, 80 are returned, with 10-15% undergoing remanufacturing. Remanufacturing allows for a 20 reduction in overall CO2 emissions without sacrificing functionality. For large-lot purchases, remanufactured products are frequently combined with new products. The refurbished product holds a 60–70%

market share in Indonesia. Twenty percent of the new price is charged for used B/W photocopiers that are imported. However, according to Kamigaki et al., (2017) the company did not accept the used goods back.

## 8.3.4. Role of Automation and Cyber-physical System (CPS)

The environment and health of workers can be better protected by automation in the dismantling of e-waste. Nevertheless, different kinds of e-waste make it difficult to classify and disassemble the process, making it difficult to implement a fully automated. The ideal solution is thought to be a collaborative robot that divides the task between human operators and robots (Alvarez-de-los-Mozos and Renteria, 2017). An operator can train a robot to perform low-skilled tasks like cutting and separating by combining manual and robotized operations. This will address the issues with e-waste identification, classification, and disassembly.

To reduce the expense of sorting e-waste in Japan, another treatment utilizing a cyber-physical system (CPS) was suggested (Torihara et al., 2015). E-waste particles that have been roughly crushed are fed onto a conveyor, and a sorting device separates the particles that are rich in metal. By making changes, the sorting machine can be operated remotely from places with lower labor costs or through the internet like a game. The e-waste refurbishment industry can benefit from the Cyber-physical system (CPS) in several key areas, including sorting, disassembly and repair, cost, and routing. When CPS is used, traceability is increased because RFID tags on cores enable control, monitoring, and process-appropriate decision-making. The system lowers process inefficiencies and saves up to 664 hours of operating time. (Sharpe et al., 2018).

To recycle electronic EOL, a different study suggested a cooperative web-based platform that links producers, end users, services, dismantlers, recyclers, or remanufacturers. The algorithm used to make decisions will assess the reuse scenarios' environmental and economic impacts (Marconi et al., 2017). The industrial symbiosis between businesses in the same or different sectors will also be supported by the web platform (Marconi et al., 2018). In summary, automation and cyber-physical systems enhance e-waste management through process sharing to support efficiency and human health, data sharing to enable cooperation in the refurbishment system and/or industrial symbiosis, and remote processes to cut costs.

## 8.3.5. Consumer Behaviour on E-waste

The willingness of customers to take part in e-waste management is influenced by their actions. In addition to attitudes and knowledge, behavior is also recognized as a significant factor that influences support for increasing funding for e-waste management (Pandebesie et al., 2019). According to Hanafi et al (2011), a study which was carried out in Indonesia, 77 of participants were aware of e-waste, and 58 of them indicated that they might be interested in recycling it. Customers will mix e-waste with solid waste and fail to dispose of it in an environmentally friendly manner, making it difficult for the e-waste management system to be implemented without knowledge of e-waste disposal practices and willingness to participate (Bhat and Patil, 2014).

# 8.4. RECOVERY AND RECYCLING

The wide range of material values in e-waste is the primary obstacle to recycling it. Precious and rare metals found in e-waste can be valuable raw material resources (e.g., copper, aluminum, and gold), but low-value plastics make up 30% of e-waste (Realff et al., 2004; Kong et al., 2012). As plastic and metal recycling technologies advance, so do studies aimed at mitigating the shortcomings of the techniques covered in this section.

## 8.4.1. Metal Recovery and Recycling

Metal recycling techniques like burning, dismantling by hand, strong acid leaching, and hydraulic shaking bed separation are inefficient in recovering metal and harm the environment. There will be significant Volatile Organic Compound (VOC) contamination because of incineration. In comparison to the VOCs treated by the electric heating furnace, blower, soldering iron, and mechanical cutting, the total VOCs produced by the incinerator are approximately 40, 139, 180, and 190 times higher. Workers' risk of developing cancer will rise with prolonged exposure to VOCs (An et al., 2014).

Many technologies exist for recycling metal; some, like pyrometallurgical and mild extracting technology, have a high recovery rate for practically all metals but have certain financial constraints owing to reagents or investment costs. Supercritical technology produces waste gas and oil while recovering nearly all metals at a high rate. Although vacuum metallurgy and biometallurgy (bioleaching and bio-sorption) are thought to be environmentally friendly processes, their industrialization and selectivity still require work (Zhang and Xu, 2016).

Enhancing the metal recycling process can be achieved by integrating various technologies or methods into the process, or by optimizing the technology's operating parameters. Torihara et al, (2015) enhanced the sorting procedure by combining electrostatic and magnetic separation. Using a sorting device, metal-rich particles were separated from printed circuit board (PCB) origin particles through pulverization and physical separation.

The recovery of precious metals (silver, Ag and gold, Au) was suggested through the combination of hydrometallurgical and biometallurgical processes. Chemicals are used in hydrometallurgical processes. Cyanide, to extract silver and/or gold from WEEE. The bio-sorption utilizes the biomass of Eicchornia roots and the bio-sorbent, i.e., powdered waste tea (Bath et al., 2012).

Many kinds of research have been conducted to advance current technologies. Consider a study aimed at enhancing printed circuit board (PCB) pyrolysis. PCBs were processed at a high temperature to extract lead and recover valuable metals. The temperature at which PCBs were heated using argon gas for up to 20 minutes was 1150–1350°C. When ceramic impurities precipitated, PCB pyrolysis improved the separation of metals and ceramics (Khanna et al., 2014; Evangelopoulos et al, 2017). Khatri et al, (2018) conducted experiments to determine the kind of chemicals that would produce the best metal leachate results. After four days of reaction time, it was discovered that the 70:30 combination of ferric and ferrous sulfates was the most effective for leaching copper. After 4 and 7 days of reaction time, respectively, ferric sulfate will provide the best dissolution for zinc and nickel.

Using polyaniline, a novel technique for recovering gold from e-waste was investigated. Using the redox characteristics of polyaniline, e-waste was crushed and treated using acidic solutions without the need for large amounts of reagents or outside energy (Wu et al., 2017). Cyganowski et al, suggested using a brand-new resin to adsorb gold in the recycling of metals. Core-shell anion exchange resins are used in the enrichment, leaching, and adsorption processes. The resins exhibit exceptional gold selectivity; they can remove up to 86% of the gold-containing polymers from the leachate and recover 100% of the gold.

Recent research has focused a lot of attention on microbial remediation of metals, or bio-metallurgical, since it is considered an environmentally friendly method. The most prevalent type of bacteria used in bioleaching belongs to the acidophilic group. Pseudomonas putida, Acidithiobacillus ferrooxidans, Acidithiobacillus thiooxidans, Leptospirillum ferrooxidans, Aspergillus niger, Penicillium simplicissimum, and Bacillus megaterium are the most common microorganisms studied in bioleaching. These bacteria have a high copper (65%–100%) and other metal recovery rate, i.e., Zn, Li, Ni, and Al. In the bioleaching process, the temperature range is 25 to 30 °C, with a pH of 1.5 to 9.5. Chromobacterium violaceum has been reported to have a leaching efficiency of over 70% for gold recovery (Liu et al., 2016). Pseudomonas fluorescens and Paeruginosa can increase the efficiency of gold leaching by reinforcing the generation of cyanide during the leaching process.

### 8.4.2. Plastic Recycling

Since the plastics fraction of WEEE is made up of more than 15 different types of polymers, recycling plastic in WEEE is difficult. The recycling process is complicated by the presence of brominated flame retardants (BFRs), such as polybrominated diphenyl ether (PBDE) and polybrominated biphenyls (PBB) (Kong et al., 2012).

There are four categories of plastic recycling technologies currently in use. Primary recycling is the process of mechanically processing plastic waste to create a product with properties similar to the original product. The polymer will change into a product with inferior qualities through secondary recycling in the interim. Chemical components are recovered from plastic waste through tertiary recycling. Quaternary recycling is the process of turning plastic waste into electricity. Notable technologies that can process plastics from e-waste include CreaSolv (tertiary recycling) and KDV (quaternary recycling). Since it can convert plastics containing BFR into value-added goods like silicon carbide, polymer composite, Grenew briquettes for steelmaking, super-capacitors, and 3D printing filaments, a new technology known as the "micro-factory" is regarded as the fifth category (Sahajwalla and Gaikwad, 2018)

### 8.4.3. Glass Recycling

Glass-to-glass and glass-to-lead recycling are two categories of e-waste recycling techniques. Glass was recovered in the first process by using a lead smelter. The amount of lead in CRT glass is between 0.5 and 5 kg, and it serves as a shield for the screen against radiation exposure. High throughput is an advantage of this process, but it might devalue premium glass (Kong et al., 2012). In the latter technique, fresh cathode ray tubes were made using the gathered glass as a raw material. It lowers expenses and boosts the furnace's effectiveness during the recycling process. But because it's hard to pinpoint the precise makeup of recycled glass, there's a risk (Kong et al., 2012)

# CASE STUDY

## ELECTRONIC WASTE RECYCLING INCREASES WASTE DIVERSION

The U.S. Railroad Retirement Board headquarters in Chicago recycled more than 3,200 pounds of electronic waste in one year, a nearly 10 increases over its baseline the previous year. This was accomplished by finding additional recycling or repurposing options for the agency's equipment through the property management process and the Computers for Learning program.

RRB's electronic waste reuse and recycling is driven by two federal programs. The agency began donating excess computers in 1997 upon enactment of Executive Order 12999, Computers for Learning. This order established a registry of schools in need of computer equipment and streamlined the disposition process. Because RRB is a small, independent agency with minimal staff, the Computers for Learning program's efficiency was appealing. The schools receiving the computer equipment were thrilled and expressed their gratitude, which boosted RRB staff morale and provided an incentive to continue and improve the initiative. Since 1997, all surplus desktop equipment in good condition has been donated through the program.

Most other electronic equipment, including televisions, computer servers and networking equipment, is recycled by a certified electronics recycler which ensures that the best management practices are met.

Using certified electronics recyclers offers a way to assess the environmental, worker health and security practices of the recyclers. RRB collects the items, follows property management practices that might identify other agencies that can reuse them, packages them for shipment, and acquires cost proposals from the electronic recyclers to find the most economical solution.

Finally, RRB rounds out its electronics recycling activities by sending reusable mobile devices back to the telecommunications provider where it receives credit on its bill.

RRB tracks disposition results in GSAXCESS and on the Computers for Learning website. Electronic recycling data is uploaded and available to agency personnel to track and download data and create reports required by RRB.

While the agency has been successful in its electronic waste recycling program, the program has not been without challenges.

One challenge was ensuring consistent and complete removal of data from equipment hard drives. To address this, the agency's property management group met with IT departments to establish processes and procedures to ensure compliance with security requirements before placing electronic equipment in surplus storage areas for recycling.

Source: The Railroad Retirement Board is in the Lipinski Federal Building in Chicago.

Another challenge was finding sufficient storage space for temporary housing equipment until recycling occurs. Mobile and utility racks were adapted to use as much vertical space as possible and to facilitate efficient transfer of equipment when retrieved by the receiving party.

Finally, requirements for disposition change. To hear about these changes, trends in electronics recycling, and new tools and resources, RRB property staff attends property management and disposition training offered by GSA.

## CLASS ACTIVITY

Create a magazine piece for a news program by interviewing one another, peers or family about e-waste and recycling, to probe their understanding and attitudes and highlight the importance of acting to re-use or recycle.

## SUMMARY

- Electronic waste, commonly known as e-waste, refers to discarded electrical and electronic devices such as computers, smartphones, televisions, and household appliances.
- Some common models for estimating e-waste generation are the Material Flow Analysis and the Population Balance Model.
- E-waste can be classified based on its composition and components. Ferrous and nonferrous metals, glass, plastics, pollutants, and others are the six categories of materials reported for e-waste composition.
- A growing body of epidemiological and clinical evidence has led to increased concern about the potential threat of e-waste to human health, especially in developing countries such as India and China.
- Environmental conditions have a strong connection to human health since e-waste's impact on health may come in two ways, which are the contamination from food chains and the direct exposure in recycling areas.
- E-waste burning, besides contaminating the soil, also increases the concentration of air pollutants, particularly particulate matter.
- Identification of PAH using hair and urine as bioindicators was also done for workers and residents near the e-waste recycling area.
- E-waste management in developing countries has similar constraints. Import of e-waste from developed countries is usually found in developing countries since e-waste is considered as a kind of livelihood by the residents.

## REVIEW QUESTIONS

1. What do you understand by electronic waste?
2. What are the classifications of electronic waste?
3. `What is the role of extended producer responsibility in e-waste recycling?
4. Explain the environmental and human health hazards of electronic waste recycling.
5. Write short notes on Reuse, Remanufacturing, and Refurbishing for electronic waste.

## REFERENCES

1. Akahori, T., & Hiroshige, Y. (2012). Economical evaluation of recycling system for rare-earth magnets. Electronics Goes Green 2012+ Conference, 9-12 September 2012, Berlin, Germany. pp. 1–5.
2. Alamerew, Y. A., & Brissaud, D. (2018). Modelling and assessment of product recovery strategies through systems dynamics. Procedia CIRP, 69, 822–826.
3. Alvarez-de-los-Mozos, E., & Renteria, A. (2017). Collaborative robots in e-waste management. Procedia Manufacturing, 11, 55–62.

4. An, T., Huang, Y., Li, G., He, Z., Chen, J., & Zhang, C. (2014). Pollution profiles and health risk assessment of VOCs emitted during e-waste dismantling processes associated with different dismantling methods. Environment International, 73, 180–194.

5. Andarani, P., & Goto, N. (2014). Potential e-waste generated from households in Indonesia using material flow analysis. Journal of Material Cycles and Waste Management, 16, 306–320.

6. Angouria-Tsorochidou, E., Cimpan, C., & Parajuly, K. (2018). Optimized collection of EoL electronic products for Circular economy: A techno-economic assessment. Procedia CIRP, 69, 986–991.

7. Awasthi, A. K., Hasan, M., Mishra, Y. K., Pandey, A. K., Tiwary, B. N., Kuhad, R. C., Gupta, V. K., & Thakur, V. K. (2019). Environmentally sound system for e-waste: Biotechnological perspectives. Current Research in Biotechnology, 1, 58–64.

8. Awasthi, A. K., Wang, M., Awasthi, M. K., Wang, Z., & Li, J. (2018). Environmental pollution and human body burden from improper recycling of e-waste in China: A short-review. Environmental Pollution, 243, Part B, 1310–1316.

9. Awasthi, A. K., Zeng, X., & Li, J. (2016). Comparative examining and analysis of e-waste recycling in typical developing and developed countries. Procedia Environmental Sciences, 35, 676–680.

10. Barletta, I., Johansson, B., Reimers, J., Stahre, J., & Berlin, C. (2015). Prerequisites for a high-level framework to design sustainable plants in the e-waste supply chain. Procedia CIRP, 29, 633–638.

11. Barwod, M., Li, J., Pringle, T., & Rahimifard, S. (2015). Utilization of reconfigurable recycling systems for improved material recovery from e-waste. Procedia CIRP, 29, 746–751.

12. Basel Convention Secretariat. Environmentally Sound Management Toolkit. Available from: http://www.basel.int/Implementation/CountryLedInitiative/EnvironmentallySoundManagement/ESMToolkit/Overview/tabid/5839/Default.aspx (accessed on 31 July 2024).

13. Basel Convention Secretariat. Pilot Projects. Available from: http://www.basel.int/Implementation/CountryLedInitiative/EnvironmentallySoundManagement/ESMToolkit/Pilotprojects/tabid/5846/Default.aspx (accessed on 31 July 2024).

14. Bhat, V., & Patil, Y. (2014). E-waste consciousness and disposal practices among residents of Pune city. Procedia–Social and Behavioral Sciences, 133, 491–498.

15. Bhat, V., Rao, P., & Patil, Y. (2012). Development of an integrated model to recover precious metals from electronic scrap—A novel strategy for e-waste management. Procedia–Social and Behavioral Sciences, 37, 397–406.

16. Bhutta, M. K., Omar, A., & Yang, X. (2011). Electronic waste: A growing concern in today's environment. Economic Research International, 2011, 2090–2133.

17. Börner, L., & Hegger, D. L. T. (2018). Toward design principles for sound e-waste governance: A research approach illustrated with the case of the Netherlands. Resources, Conservation & Recycling, 134, 271–281.

18. Budijati, S. M., Wibisono, M. A., & Masruroh, N. A. (2015). A study of consumers' post-consumption behaviour for mobile phone in Indonesia. Industrial Engineering, Management Science and Applications 2015, 563–573.

19. Cahill, T. M., Groskova, D., Charles, M. J., Sanborn, J. R., Denson, M., & Baker, L. (2007). Atmospheric concentrations of polybrominated diphenyl ethers at near-source sites. *Environmental Science and Technology, 41*, 6370–6377. https://doi.org/10.1021/es070844j.

20. Ceballos, D., & Dong, Z. (2016). The formal electronic recycling industry: Challenges and opportunities in occupational and environmental health research. *Environment International, 95*, 157–166. https://doi.org/10.1016/j.envint.2016.07.010.

21. Ceballos, D., Gong, W., & Page, E. (2015). A pilot assessment of occupational health hazards in the U.S. electronic scrap recycling industry. *Journal of Occupational Environmental Hygiene, 12*(7), 482–488. https://doi.org/10.1080/15459624.2015.1018516.

22. Charles, R. G., Douglas, P., Hallin, I. L., Matthews, I., & Liversage, G. (2017). An investigation of trends in precious metal and copper content of RAM modules in WEEE: Implications for long term recycling potential. *Waste Management, 60*, 505–520.

23. Chibunna, J. B., Siwar, C., Begum, R. A., & Mohamed, A. F. (2012). The challenges of e-waste management among institutions: A case study of UKM. *Procedia–Social and Behavioral Sciences, 59*, 644–649.

24. Cole, C., Gnanapragasam, A., & Cooper, T. (2017). Towards a circular economy: Exploring routes to reuse for discarded electrical and electronic equipment. *Procedia CIRP, 61*, 155–160.

25. Cole, C., Gnanapragasam, A., Singh, J., & Cooper, T. (2018). Enhancing reuse and resource recovery of electrical and electronic equipment with reverse logistics to meet carbon reduction targets. *Procedia CIRP, 69*, 980–985.

26. Cucchiella, F., D'Adamo, I., Koh, S. C. L., & Rosa, P. (2015). Recycling of WEEEs: An economic assessment of present and future e-waste streams. *Renewable and Sustainable Energy Reviews, 51*, 263–272.

27. Cyganowski, P., Garbera, K., Le⊠niewicz, A., Wolska, J., Pohl, P., & Jermakowicz-Bartkowiak, D. (2017). The recovery of gold from the aqua regia leachate of electronic parts using a core–shell type anion exchange resin. *Journal of Saudi Chemical Society, 21*, 741–750.

28. Danish Ministry of Environment. (2012). *The Greening of Electronics*. Available from: http://www2.mst.dk/Udgiv/publications/2012/07/978-87-92779-99-1.pdf (accessed on 31 July 2024).

29. Debnath, B., Roychoudhuri, R., & Ghosh, S. K. (2016). E-waste management: A potential route to green computing. *Procedia Environmental Sciences, 35*, 669–675.

30. DOE (U.S. Department of Energy). (2010). *Critical Materials Strategy*. Available from: http://www.doe.gov/sites/prod/files/edg/news/documents/criticalmaterialsstrategy.pdf (accessed on 31 July 2024).

31. Dwivedy, M., & Mittal, R. K. (2012). An investigation into e-waste flows in India. *Journal of Cleaner Production, 37*, 229–242.
32. Elsevier, B. V. (2020). *Science Direct*. Retrieved from: https://www.sciencedirect.com/ (accessed on 31 July 2024).
33. Emerald Publishing Ltd. (2020). *Emerald Insight*. Retrieved from: https://www.emeraldinsight.com/ (accessed on 31 July 2024).
34. Ercan, O., & Bllen, K. (2014). Research on electronic waste awareness and environmental attitudes of primary school students. *Anthropologist, 17*, 13–23.
35. Evangelopoulos, P., Kantarelis, E., & Yang, W. (2017). Experimental investigation of pyrolysis of printed circuit boards for energy and materials recovery under nitrogen and steam atmosphere. *Energy Procedia, 105*, 986–991.
36. Gaidajis, G., Angelakoglou, K., & Aktsoglou, D. (2010). E-waste: Environmental problems and current management. *Journal of Engineering Science and Technology Review, 3*, 193–199.
37. Gangwar, C., Choudhari, R., Chauhan, A., Kumar, A., Singh, A., & Tripathi, A. (2019). Assessment of air pollution caused by illegal e-waste burning to evaluate the human health risk. *Environment International, 125*, 191–199. https://doi.org/10.1016/j.envint.2019.01.007.
38. Gao, Y., Ge, L., Shi, S., Sun, Y., Liu, M., Wang, B., Shang, Y., Wu, J., & Tian, J. (2019). Global trends and future prospects of e-waste research: A bibliometric analysis. *Environmental Science and Pollution Research, 26*, 17809–17820. https://doi.org/10.1007/s11356-019-05095-0.
39. Garlapati, V. K. (2016). E-waste in India and developed countries: Management, recycling, business, and biotechnological initiatives. *Renewable and Sustainable Energy Review, 54*, 874–881. https://doi.org/10.1016/j.rser.2015.10.075.
40. Gnoni, M. G., Mossa, G., Mummolo, G., Tornese, F., & Verriello, R. (2017). Supporting circular economy through use-based business models: The washing machines case. *Procedia CIRP, 64*, 49–54. https://doi.org/10.1016/j.procir.2017.03.123.
41. Gonda, L., & Degrez, M. (2018). End-of-life management of computers in Brussels: Environmental comparison of two treatment chains. *Procedia CIRP, 69*, 968–973. https://doi.org/10.1016/j.procir.2018.01.119.
42. Goonan, T. G. (2011). Rare earth elements—End use and recyclability: U.S. Department of the Interior, U.S. Geological Survey Scientific Investigations Report 2011–5094. Available from: https://pubs.usgs.gov/sir/2011/5094/ (accessed on 31 July 2024).
43. Gradael, T. E., Harper, E. M., Nassar, N. T., & Reck, B. K. (2015). On the materials basis of modern society. *Proceedings of the National Academy of Sciences, 112*(20), 6295–6300. https://doi.org/10.1073/pnas.1312752112.
44. Grant, K., Coldizen, F. C., Sly, P. D., Brune, M. B., Neira, M., van den Berg, M., & Norman, R. E. (2013). Health consequences of exposure to e-waste: A systematic review. *Lancet Global Health, 1*, 350–361. https://doi.org/10.1016/S2214-109X(13)70104-7.

45. Grant, R. (2019). E-waste challenges in Cape Town: Opportunity for the green economy? *Urbani Izziv, 30*, 5–23. https://doi.org/10.5379/urbani-izziv-en-2019-30-01-002.

46. Ha, N., Agusa, T., Ramu, K., Nguyen, T., Murata, S., Bulbule, K., et al., (2009). Contamination by trace elements at e-waste recycling sites in Bangalore, India. *Chemosphere, 76*, 9–15. https://doi.org/10.1016/j.chemosphere.2009.02.056.

47. Hanafi, J., Christiani, A., Kristina, H. J., & Utama, K. P. (2014). Collecting end-of-life mobile phones in Jakarta: A pilot. In *Re-Engineering Manufacturing for Sustainability* (pp. 365–370).

48. Hanafi, J., Kristina, H. J., Jobiliong, E., Christiani, A., Halim, A. V., Santoso, D., & Melini, E. (2011). The prospects of managing WEEE in Indonesia. In *Proceedings of the 18th CIRP International Conference on Life Cycle Engineering* (pp. 1–6). Technische Universität Braunschweig, Braunschweig, Germany, May 2–4.

49. Harler, K. (2018). Rare opportunity to recycle rare earths. Available from: https://www.recyclingtoday.com/article/rare-earth-metals-recycling/ (accessed on 31 July 2024).

50. He, M., Yang, S., Zhao, J., Collins, C., Xu, J., & Liu, X. (2019). Reduction in the exposure risk of farmers from e-waste recycling sites following environmental policy adjustment: A regional scale view of PAHs in paddy fields. *Environment International, 133*. https://doi.org/10.1016/j.envint.2019.105264.

51. Hossain, M. S., Al-Hamadani, S. M. Z. F., & Rahman, M. T. (2015). E-waste: A challenge for sustainable development. *Journal of Health & Pollution, 5*, 3–11. https://doi.org/10.5696/2156-9614-5.9.3.

52. Idrees, N., Tabassum, B., Abd_Allah, E. F., Hasehm, A., Sarah, R., & Hashim, M. (2018). Groundwater contamination with cadmium concentrations in some West U.P. regions, India. *Saudi Journal of Biological Sciences, 25*, 1365–1368. https://doi.org/10.1016/j.sjbs.2018.02.003.

53. Ilankoon, I. M. S. K., Ghorbani, Y., Chong, M. N., Herath, G., Moyo, T., & Petersen, J. (2018). E-waste in the international context–A review of trade flows, regulations, hazards, waste management strategies, and technologies for value recovery. *Waste Management, 82*, 258–275. https://doi.org/10.1016/j.wasman.2018.10.013.

54. International Telecommunications Union. (2017). *Chapter 10: Regional E-Waste Status and Trends*. Available from: https://www.itu.int/en/ITU-D/Climate-Change/Pages/Global-E-waste-Monitor-2017.aspx (accessed on 31 July 2024).

55. Jaiswal, A., Samuel, C., Patel, B. S., & Kumar, M. (2015). Go green with WEEE: Eco-friendly approach for handling e-waste. *Procedia Computer Science, 46*, 1317–1324. https://doi.org/10.1016/j.procs.2015.01.169.

56. Jibiri, N. N., Isinkaye, M. O., & Momoh, H. A. (2014). Assessment of radiation exposure levels at Alaba e-waste dumpsite in comparison with municipal waste dumpsites in southwest Nigeria. Journal of Radiation Research and Applied Sciences, 7, 536–541. https://doi.org/10.1016/j.jrras.2014.04.002.

57. Jowitt, S. M., Werner, T., Weng, Z., & Mudd, G. (2018). Recycling of the rare earth elements. Current Opinion in Green and Sustainable Chemistry, 13, 1–7. https://doi.org/10.1016/j.cogsc.2018.06.002.

58. JSTOR. (2020). *JSTOR*. Retrieved from: https://www.jstor.org/ (accessed on 31 July 2024).
59. Julander, A., Lundgren, L., Skare, L., Grancér, M., Palm, B., Vahter, M., & Lidén, C. (2014). Formal recycling of e-waste leads to increased exposure to toxic metals: An occupational exposure study from Sweden. Environmental International, 73, 243–251. https://doi.org/10.1016/j.envint.2014.08.016.
60. Kamigaki, K., Matsumoto, M., & Fatimah, Y. A. (2017). Remanufacturing and refurbishing in developed and developing countries in Asia–A case study in photocopiers. Procedia CIRP, 61, 645–650. https://doi.org/10.1016/j.procir.2017.03.239.
61. Kang, J. J., Lee, J. S., Wang, W. S., Park, S. W., Alam, M. T., Back, S. K., Choi, H. S., Seo, Y. C., Yun, Y. S., Gu, J. H., Saravana Kumar, A., & Kumar, K. V. (2016). A study on environmental assessment of residue from gasification of polyurethane waste in e-waste recycling process. Procedia Environmental Sciences, 35, 639–642. https://doi.org/10.1016/j.proenv.2016.07.045.
62. Khanna, R., Cayumil, R., Mukherjee, P. S., & Sahajwalla, V. (2014). A novel recycling approach for transforming waste printed circuit boards into a material resource. Procedia Environmental Sciences, 21, 42–54. https://doi.org/10.1016/j.proenv.2014.01.018.
63. Khatri, B. R., Sodha, A. B., Shah, M. B., Tipre, D. R., & Dave, S. R. (2018). Chemical and microbial leaching of base metals from obsolete cell-phone printed circuit boards. Sustainable Environment Research, 28, 333–339. https://doi.org/10.1016/j.serj.2018.01.007.
64. Kiddee, P., Naidu, R., & Wong, M. H. (2013). Electronic waste management approaches: An overview. Waste Management, 33, 1237–1250. https://doi.org/10.1016/j.wasman.2013.01.005.
65. Kojima, M., Yoshida, A., & Sasaki, S. (2009). Difficulties in applying extended producer responsibility policies in developing countries: Case studies in e-waste recycling in China and Thailand. Journal of Material Cycles and Waste Management, 11, 263–269. https://doi.org/10.1007/s10163-009-0295-8.
66. Kong, S., Liu, H., Zeng, H., & Liu, Y. (2012). The status and progress of resource utilization technology of e-waste pollution in China. Procedia Environmental Sciences, 16, 515–521. https://doi.org/10.1016/j.proenv.2012.10.072.
67. Lepawsky, J. (2012). Legal geographies of e-waste legislation in Canada and the US: Jurisdiction, responsibility and the taboo of production. Geoforum, 43, 1194–1206. https://doi.org/10.1016/j.geoforum.2012.05.009.
68. Leung, A., Duzgoren-Aydin, N., Cheung, K. C., & Wong, M. H. (2018). Heavy metals concentrations of surface dust from e-waste recycling and its human health implications in Southeast China. Environmental Science and Technology, 42(7), 2674–2680. https://doi.org/10.1021/es071873x.
69. Li, B., Du, H. Z., Ding, H. J., & Shi, M. Y. (2011). E-waste recycling and related social issues in China. Energy Procedia, 5, 2527–2531. https://doi.org/10.1016/j.egypro.2011.03.432.

70. Li, J., & Zhang, S. (2010). Study e-waste management based on EPR system. In International Conference on E-Business and E-Government (pp. 1–4). IEEE Computer Society. https://doi.org/10.1109/ICEBEG.2010.5582481.

71. Li, J., Li, W., Gao, X., Liu, L., Shen, M., Chen, H., Zhu, M., Zeng, L., & Zeng, E. Y. (2020). Occurrence of multiple classes of emerging photo initiators in indoor dust from e-waste recycling facilities and adjacent communities in South China and implications for human exposure. Environment International, 136. https://doi.org/10.1016/j.envint.2020.105484.

72. Li, R. C., & Tee, T. J. C. (2012). A reverse logistics model for recovery options of e-waste considering the integration of the formal and informal waste sectors. Procedia–Social and Behavioral Sciences, 40, 788–816. https://doi.org/10.1016/j.sbspro.2012.03.217.

73. Li, X., Duan, Y., Sun, H., Zhang, P., Xu, J., Hua, X., Jin, L., & Li, M. (2019). Human exposure levels of PAEs in an e-waste recycling area: Get insight into impacts of spatial variation and manipulation mode. Environment International, 133. https://doi.org/10.1016/j.envint.2019.105274.

74. Lin, M., Tang, J., Ma, S., Yu, Y., Li, G., Fan, R., Mai, B., & An, T. (2019). Insights into biomonitoring of human exposure to polycyclic aromatic hydrocarbons with hair analysis: A case study in an e-waste recycling area. Environment International, 136. https://doi.org/10.1016/j.envint.2019.105514.

75. Little, P. C., & Lucier, C. A. (2017). Global electronic waste, third party certification standards, and resisting the undoing of environmental justice politics. Human Organization, 76(3), 204–214. https://doi.org/10.17730/0018-7259-76.3.204.

76. Liu, R., Li, J., & Ge, Z. (2016). Review on Chromobacterium violaceum for gold bioleaching from e-waste. Procedia Environmental Sciences, 31, 947–953. https://doi.org/10.1016/j.proenv.2016.02.161.

77. Liu, Y., Tang, B., Liu, Y., Lup, X., Mai, B., Covaci, A., & Poma, G. (2019). Occurrence, bio magnification and maternal transfer of legacy and emerging organophosphorus flame retardants and plasticizers in water snake from an e-waste site. *Environment International, 133*. https://doi.org/10.1016/j.envint.2019.105214.

78. Lucier, A. A., & Gareau, B. J. (2019). Electronic waste recycling and disposal: An overview. In *Hazardous Wastes*. https://doi.org/10.5772/intechopen.85983.

79. Lucier, C. A., & Gareau, B. J. (2015). From waste to resources? Interrogating 'race to the bottom' in the global environmental governance of the hazardous waste trade. *Journal of World-Systems Research, 21*(2), 495–520. https://doi.org/10.5195/jwsr.2015.579.

80. Lucier, C. A., & Gareau, B. J. (2016). Obstacles to preserving precaution and equity in global hazardous waste regulation: An analysis of contested knowledge in the Basel Convention. *International Environmental Agreements: Politics, Law, and Economics, 16*, 493–508. https://doi.org/10.1007/s10784-014-9261-6.

81. Manhart, A. (2012). E-Waste Africa Project: Impacts of current recycling practices and recommendations for collection and recycling. In *Pan African Forum on E-Waste* (pp. 14–16 March 2012). Nairobi, Kenya.

82. Manomaivibool, P., Lindhqvist, T., & Tojo, N. (2009). Electrical and electronic equipment in India. *Resources, Conservation and Recycling, 53*(3), 136–144. https://doi.org/10.1016/j.resconrec.2008.10.003.

83. Marconi, M., Favi, C., Germani, M., Mandolini, M., & Mengarelli, M. (2017). A collaborative End of Life platform to favor the reuse of electronic components. *Procedia CIRP, 61*, 166–171. https://doi.org/10.1016/j.procir.2017.03.015.

84. Marconi, M., Gregori, F., Germani, M., Papetti, A., & Favi, C. (2018). An approach to favor industrial symbiosis: The case of waste electrical and electronic equipment. *Procedia Manufacturing, 21*, 502–509. https://doi.org/10.1016/j.promfg.2018.02.080.

85. Mary, J. S., & Meenambal, T. (2016). Incentivization of e-waste and developing a policy-bulk consumer perspective. *Procedia Environmental Sciences, 35*, 643–655. https://doi.org/10.1016/j.proenv.2016.07.084.

86. Mmereki, D., Li, B., Baldwin, A., & Hong, L. (2016). The generation, composition, collection, treatment and disposal system, and impact of e-waste. In *E-Waste in Transition: From Pollution to Resource*.

87. Monika, & Kishore, J. (2010). E-waste management: As a challenge to public health in India. *Indian Journal of Community Medicine, 35*(4), 620–622. https://doi.org/10.4103/0970-0218.74307.

88. Namias, J. (2013). The future of electronic waste recycling in the United States: Obstacles and domestic solutions (Thesis, Columbia University, New York). Retrieved from: http://www.seas.columbia.edu/earth/wtert/sofos/Namias_Thesis_07-08-13.pdf (accessed on 31 July 2024).

89. Nguyen, D., Ha, V., Wiji, Y., & Huynh, T. (2017). Material flows from electronic waste: Understanding the shortages for Extended Producer Responsibility implementation in Vietnam. *Procedia CIRP, 61*, 651–656. https://doi.org/10.1016/j.procir.2017.03.082.

90. Oguchi, M., Sakanakura, H., & Terazono, A. (2012). Toxic metals in WEEE: Characterization and substance flow analysis in waste treatment processes. *Science of the Total Environment, 463*, 1124–1132. https://doi.org/10.1016/j.scitotenv.2012.07.078.

91. Organization for Economic Cooperation and Development (OECD). (2005). EPR policies and product design: Economic theory and selected case studies: OECD report NV/EPOC/WGWPR (2005)9/FINAL. Retrieved from: http://www.oecd.org/fr/env/dechets/factsheetextendedproducerresponsibility.htm (accessed on 31 July 2024).

92. Otto, S., Kibbe, A., Henn, L., Hentschke, L., & Kaiser, F. G. (2018). The economy of e-waste collection at the individual level: A practice-oriented approach of categorizing determinants of e-waste collection into behavioral costs and motivation. *Journal of Cleaner Production, 204*, 33–40. https://doi.org/10.1016/j.jclepro.2018.08.070.

93. Panambunan-Ferse, M., & Breiter, A. (2013). Assessing the side-effects of ICT development: E-waste production and management a case study about cell phone end-of-life in Manado, Indonesia. *Technology in Society, 35*(3), 223–231. https://doi.org/10.1016/j.techsoc.2013.01.001.

94. Pandebesie, E. S., Indrihastuti, I., Wilujeng, A. A., & Warmadewanthi, I. (2019). Factors influencing community participation in the management of household electronic waste in West Surabaya, Indonesia. *Environmental Science and Pollution Research, 26*, 27930–27939. https://doi.org/10.1007/s11356-019-05351-7.

95. Pascale, A., Sosa, A., Bares, C., Battocletti, A., Moll, M. J., Pose, D., Laborde, A., Gonzáles, H., & Feola, G. (2016). E-waste informal recycling: An emerging source of lead exposure in South America. *Annals of Global Health, 82*(3), 489–496. https://doi.org/10.1016/j.aogh.2016.04.007.

96. Perkins, D. N., Drisse, M. B., Nxele, T., & Sly, P. D. (2014). E-waste: A global hazard. *Annals of Global Health, 80*(4), 286–295. https://doi.org/10.1016/j.aogh.2014.10.002.

97. Platform for Accelerating the Circular Economy. (2019). A new circular vision for electronics: Time for a global reboot. In *World Economic Forum, Davos 2019* (pp. 1–24). Switzerland.

98. Priyono, F. X. J. (2017). Law enforcement of electrical and electronic waste smuggling in Batam, Indonesia. *Diponegoro Law Review, 02*. Retrieved from: https://ejournal.undip.ac.id/index.php/dlr/article/view/22931 (accessed on 31 July 2024).

99. Realff, M. J., Raymond, M., & Ammons, J. C. (2004). E-waste: An opportunity. *Materials Today, 7*(4), 40–45. https://doi.org/10.1016/S1369-7021(04)00558-8.

100. Recycling Today. (2012). Solvay opens rare earths metals recycling plants in France. Retrieved from: https://www.recyclingtoday.com/article/solvay-rare-earth-recycling-plants-france/ (accessed on 31 July 2024).

101. Rhee, S. W. (2016). Beneficial use practice of e-wastes in Republic of Korea. *Procedia Environmental Sciences, 31*, 707–714. https://doi.org/10.1016/j.proenv.2016.02.098.

102. Rimantho, D., & Nasution, S. R. (2016). The current status of e-waste management practice in DKI Jakarta. *International Journal of Applied Environmental Sciences, 11*(2), 1451–1468. Retrieved from: http://www.ripublication.com/ijaes16/ijaesv11n2_09.pdf (accessed on 31 July 2024).

103. Robinson, B. H. (2009). E-waste: An assessment of global production and environmental impacts. *Science of the Total Environment, 408*(2), 183–191. https://doi.org/10.1016/j.scitotenv.2009.09.044.

104. Rochman, F. F., Ashton, W. S., & Wiharjo, M. G. M. (2017). E-waste, money, and power: Mapping electronic waste flows in Yogyakarta, Indonesia. *Environmental Development, 24*, 1–8. https://doi.org/10.1016/j.envdev.2017.03.002.

105. Rode, S. (2012). E-waste management in Mumbai metropolitan region: Constraints and opportunities. *Theoretical and Empirical Researches in Urban Management, 7*(2), 90–104.

106. Rosenberg, C., Hameila, M., Tornaeus, J., Sakkinen, K., Puttonen, K., Korpi, A., & Eklund, T. (2011). Exposure to flame retardants in electronics recycling sites. *The Annals of Occupational Hygiene, 55*(6), 658–665. https://doi.org/10.1093/annhyg/mer033.

107. Rotter, V. S., & Chancer, P. (2012). Recycling of critical resources—Upgrade introduction. In *Electronics Goes Green 2012+ Conference* (pp. 1–6). Berlin, Germany.

108. SAGE Publications. (2019). SAGE journals. Retrieved from: https://journals.sagepub.com/ (accessed on 31 July 2024).

109. Sahajwalla, V., & Gaikwad, V. (2018). The present and future of e-waste plastics recycling. *Current Opinion in Green and Sustainable Chemistry, 13*, 102–107. https://doi.org/10.1016/j.cogsc.2018.05.006.

110. Salhofer, S. (2017). E-waste collection and treatment options: A comparison of approaches in Europe, China, and Vietnam. In *The Handbook of Environmental Chemistry* (Vol. 63). https://doi.org/10.1007/978-3-319-45048-5_15.

111. Santoso, S., Zagloel, T. Y. M., Ardi, R., & Suzianti, A. (2019). Estimating the amount of electronic waste generated in Indonesia: Population balance model. *IOP Conference Series: Earth and Environmental Science*. https://doi.org/10.1088/1755-1315/358/1/012044.

112. Schmitz, O., & Gradael, T. (2010). The consumption conundrum: Driving the destruction abroad. Retrieved from: https://e360.yale.edu/features/the_consumption_conundrum_driving_the_destruction_abroad (accessed on 31 July 2024).

113. Sepúlveda, A., Schluep, M., Renaud, F., Streicher, M., Kuehr, R., Hagelüken, C., & Gertsakis, J. (2010). A review of the environmental fate and effects of hazardous substances released from electrical and electronic equipment during recycling: Examples from China and India. *Environmental Impact Assessment Review, 30*(1), 28–41. https://doi.org/10.1016/j.eiar.2009.04.001.

114. Sharpe, R. G., Goodall, P. A., Neal, A. D., Conway, P. P., & West, A. A. (2018). Cyber-physical systems in the re-use, refurbishment, and recycling of used electrical and electronic equipment. *Journal of Cleaner Production, 170*, 351–-361. https://doi.org/10.1016/j.jclepro.2017.09.086.

115. Shen, C., Huang, S., Wang, Z., Qiao, M., Tang, X., Yu, C., & Zhang, H. (2008). Identification of Ah receptor agonists in soil of e-waste recycling sites from Taizhou area in China. *Environmental Science & Technology, 42*(1), 49–55. https://doi.org/10.1021/es071162z.

116. Shevchenko, T., Laitala, K., & Danko, Y. (2019). Understanding consumer e-waste recycling behavior: Introducing a new economic incentive to increase the collection rates. *Sustainability, 11*(15), 4116. https://doi.org/10.3390/su11154116.

117. Solving the E-Waste Problem. (2014). Recommendations for standards development for the collection, storage, transport, and treatment of e-waste. Retrieved from: http://www.step-initiative.org/files/_documents/whitepapers/StEP_WP_Standard_20140602.pdf (accessed on 31 July 2024).

118. Someya, M., Suzuki, G., Ionas, A. C., Tue, N. M., Xu, F., Matsukami, H., Covaci, A., Tuyen, L. H., Viet, P. H., Takahashi, S., Tanabe, S., & Takigami, H. (2016). Occurrence of emerging flame retardants from e-waste recycling activities in the northern part of Vietnam. *Emerging Contaminants, 2*(2), 58–65. https://doi.org/10.1016/j.emcon.2016.03.001.

119. Sothun, C. (2012). Situation of e-waste management in Cambodia. *Procedia Environmental Sciences, 16*, 535–544. https://doi.org/10.1016/j.proenv.2012.10.076.

120. Springer Nature Switzerland AG. (2019). Springer Link. Retrieved from: https://link.springer.com/ (accessed on 31 July 2024).

121. Suzuki, G., Masayuki, S., Matsukami, H., Tue, N. M., Natsuyo, U., Tuyen, L. H., Viet, P. H., Takahashi, S., Tanabe, S., Brouwer, A., & Takigami, H. (2016). Comprehensive evaluation of dioxins and dioxin-like compounds in surface soils and river sediments from e-waste processing sites in a village in northern Vietnam: Heading towards the environmentally sound management of e-waste. *Emerging Contaminants, 2*(2), 98–108. https://doi.org/10.1016/j.emcon.2016.03.002.

122. Terazono, A., Murakami, S., Abe, N., Inanc, B., Moriguchi, Y., Sakai, S., & Li, J. (2006). Current status and research on e-waste issues in Asia. *Journal of Material Cycles and Waste Management, 8*(1), 1–12. https://doi.org/10.1007/s10163-005-0147-0.

123. Torihara, K., Kitajima, T., & Mishima, N. (2015). Design of a proper recycling process for small-sized e-waste. *Procedia CIRP, 26*, 746–751. https://doi.org/10.1016/j.procir.2015.04.027.

124. Tsydenova, O., & Bengtsson, M. (2011). Chemical hazards associated with treatment of waste electrical and electronic equipment. *Waste Management, 31*(1), 45–58. https://doi.org/10.1016/j.wasman.2010.08.014.

125. Tue, N. M., Takahashi, S., Suzuki, G., Isobe, T., Viet, P. H., Kobara, Y., Seike, N., Zhang, G., Sudaryanto, A., & Tanabe, S. (2013). Contamination of indoor dust and air by polychlorinated biphenyls and brominated flame retardants and relevance of non-dietary exposure in Vietnamese informal e-waste recycling sites. *Environment International, 51*, 160–167. https://doi.org/10.1016/j.envint.2012.10.011.

126. Uchida, N., Matsukami, H., Someya, M., Tue, N. M., Tuyen, L. H., Viet, P. H., Takahashi, S., Tanabe, S., & Suzuki, G. (2018). Hazardous metals emissions from e-waste-processing sites in a village in northern Vietnam. *Emerging Contaminants, 4*(1), 11–21. https://doi.org/10.1016/j.emcon.2017.11.003.

127. Vanegas, P., Peeters, J. R., Plessers, F., Cattrysse, D., & Duflou, J. R. (2014). Synergizing industrialized and developing countries to improve resource recovery for e-waste: Case study Belgium-Kenya. *Procedia CIRP, 15*, 283–288. https://doi.org/10.1016/j.procir.2014.06.102.

128. Vats, M. C., & Singh, S. K. (2014). E-waste characteristics and its disposal. *International Journal of Ecological Science and Environmental Engineering*. Retrieved from: http://www.ijesee.org/paper/2014/03/e-waste-characteristics-and-its-disposal (accessed on 31 July 2024).

129. Veenstra, A., Wang, C., Fan, W., & Ru, Y. (2010). An analysis of e-waste flows in China. *International Journal of Advanced Manufacturing Technology, 47*(3–4), 449–459. https://doi.org/10.1007/s00170-009-2241-4.

130. Vuppala, S., & Saritha, V. (2015). Consumer attitudes and perceptions on electronic waste: An assessment. *Pollution, 1*(1), 31–43. https://doi.org/10.22059/pollution.2015.59773.

131. Wang, F., Huisman, J., Meskers, C., Schleup, M., Stevels, A., & Hageluken, C. (2012). The best of 2 worlds philosophy: Developing local dismantling and global infrastructure network for sustainable e-waste treatment in emerging economies.

Waste Management, 32(11), 2134–2146. https://doi.org/10.1016/j.wasman.2012.03.029.

132. Wang, F., Kuehr, R., Ahlquist, D., & Li, J. (2013). E-waste in China: A country report. United Nations University StEP Green Paper Series. Retrieved from: http://isp.unu.edu/news/2013/e-waste_in_china.html (accessed on 31 July 2024).

133. Wang, X., & Wang, L. (2019). Digital twin-based WEEE recycling, recovery, and remanufacturing in the background of Industry 4.0. International Journal of Production Research. https://doi.org/10.1080/00207543.2019.1637876.

134. Wang, Y., Sun, X., Fang, L., Li, K., Yang, P., Du, L., Ji, K., Wang, J., Liu, Q., Xu, C., Li, G., Giesy, J. P., & Hecker, M. (2018). Genomic instability in adult men involved in processing electronic waste in Northern China. Environment International, 117, 69–81. https://doi.org/10.1016/j.envint.2018.04.027.

135. Wei, L., & Liu, Y. (2012). Present status of e-waste disposal and recycling in China. Procedia Environmental Sciences, 16, 506–514. https://doi.org/10.1016/j.proenv.2012.10.074.

136. Wong, C., Dozgoren-Aydin, N., Aydin, A., & Wong, M. (2006). Sources and trends of environmental mercury emissions in Asia. Science of the Total Environment, 368(1), 649–662. https://doi.org/10.1016/j.scitotenv.2005.11.024.

137. World Trade Organization (WTO). (2014). China—Measures related to the exportation of rare earths, tungsten and molybdenum: Dispute settlement: Dispute DS431. Retrieved from: https://www.wto.org/english/tratop_e/dispu_e/cases_e/ds431_e.htm (accessed on 31 July 2024).

138. Wu, Y., Fang, Q., Yi, X., Liu, G., & Li, R. (2017). Recovery of gold from hydrometallurgical leaching solution of electronic waste via spontaneous reduction by polyaniline. Progress in Natural Science: Materials International, 27(5), 514–519. https://doi.org/10.1016/j.pnsc.2017.06.007.

139. Yamasita, T., Sakugawa, T., Akiyama, H., & Hosano, H. (2019). Metal-coated plastics recycling by pulsed electric discharge. Waste Management, 89, 57–63. https://doi.org/10.1016/j.wasman.2019.03.033.

140. Yao, Z., Xie, Z., & Tang, J. (2016). A typical e-waste cathode ray tube glass: Alkaline leaching in the sulphur-containing medium. Procedia Environmental Sciences, 31, 880–886. https://doi.org/10.1016/j.proenv.2016.02.131.

141. Yoshida, A., Terazono, A., Ballesteros Jr., F. C., Nguyen, D., Sukandar, S., Kojima, M., & Sakata, S. (2016). E-waste recycling processes in Indonesia, the Philippines, and Vietnam: A case study of cathode ray tube TVs and monitors. Resources, Conservation and Recycling, 106, 48–58. https://doi.org/10.1016/j.resconrec.2015.10.009.

142. Zeng, X., Duan, H., Wang, F., & Liu, L. (2016). Examining environmental management of e-waste: China's experience and lessons. Renewable and Sustainable Energy Reviews, 72, 1076–1082. https://doi.org/10.1016/j.rser.2016.01.037.

143. Zeng, X., Song, Q., Yuan, W., Duan, W., & Liu, L. (2015). Solving e-waste problem using an integrated mobile recycling plant. Journal of Cleaner Production, 90, 55–59. https://doi.org/10.1016/j.jclepro.2014.11.086.

144. Zhang, B., Zhang, T., Duan, Y., Zhao, Z., Huang, X., Bai, X., Xie, L., He, Y.,

Ouyang, J., Yang, Y., Wu, Y., & Sun, H. (2019). Human exposure to phthalate esters associated with e-waste dismantling: Exposure levels, sources, and risk assessment. Environment International, 124, 1–9. https://doi.org/10.1016/j.envint.2018.12.035.

145. Zhang, L., & Xu, Z. (2016). A review of current progress of recycling technologies for metals from waste electrical and electronic equipment. Journal of Cleaner Production, 127, 19–36. https://doi.org/10.1016/j.jclepro.2016.03.076.

146. Zhong, H., & Huang, L. (2016). The empirical research on the consumers' willingness to participate in e-waste recycling with a point reward system. Energy Procedia, 104, 475–480. https://doi.org/10.1016/j.egypro.2016.12.037.

CHAPTER

9

# Waste Minimization

## LEARNING OBJECTIVES

- Understand the waste minimization
- Illustrate the waste minimization methodology
- Describe the process of recycling
- Discuss on paper and glass recycling

## KEY TERMS FROM THIS CHAPTER

| | |
|---|---|
| Environmental protection | Glass recycling |
| Integrated pollution control | Paper recycling |
| Plastic waste | Product modification |
| Raw material | Recycling |
| Resource recovery | Reuse |
| Source reduction | Source separation |

# INTRODUCTION

A collection of procedures and methods known as waste minimization is meant to lower the quantity of waste generated. Waste minimization aids the promotion of a more sustainable society by lowering or completely stopping the production of hazardous and persistent wastes. Redesigning goods and procedures as well as altering societal consumption and production patterns are all part of waste minimization.

Avoiding the issue entirely is frequently the most economical, environmentally friendly, and cost-effective approach to waste management. For most waste management techniques, managers view waste minimization as the main goal. The benefits of waste minimization can be substantial if done so in an efficient, secure, and sustainable manner because proper waste treatment and disposal can take a long time and huge amount of resources.

Conventional waste management is concerned with handling waste after it is produced, with an emphasis on recycling, reusing, and converting waste into energy. The goal of waste minimization is to prevent waste from being produced during the manufacturing process. A manager must be knowledgeable about the production process, cradle-to-grave analysis (tracing materials from extraction to return to Earth), and the specifics of waste composition to implement waste minimization successfully.

The primary sources of waste differ between nations. In the UK, building construction and demolition account for most of the waste, with mining and quarrying, industry, and commerce following closely behind. The amount of waste generated by households is comparatively small. Industrial waste is frequently linked to supply chain requirements. For instance, a business handling a product might insist on shipping it in a specific way because it meets downstream requirements.

# 9.1. UNDERSTANDING WASTE MINIMIZATION

Traditionally, waste has been seen as an extraneous byproduct of any industry's operations. Waste is a resource that is misplaced and exists at the wrong time. Inefficient use of utilities like fuel, water, and electricity—which are frequently regarded as inevitable overheads—also constitutes waste. Managers typically underestimate the costs associated with this waste. It's critical to understand that the cost of waste includes additional expenses in addition to waste disposal, like:

- Disposal cost
- Inefficient energy use cost
- Purchase cost of wasted raw material
- Production cost for the waste material
- Management time spent on waste material
- Lost revenue for what could have been a product instead of waste
- Potential liabilities due to waste.

Waste minimization can be defined as "systematically reducing waste at source." It means:

- Prevention and/or reduction of waste generated
- Efficient use of raw materials and packaging
- Efficient use of fuel, electricity and water
- Improving the quality of waste generated to facilitate recycling and/or reduce hazard
- Encouraging reuse, recycling and recovery.

Other names for waste minimization include source reduction, pollution prevention, waste reduction, and cleaner technology. To make industrial operations intrinsically pollution-free, it employs managerial and/or technical interventions. Furthermore, it should be made abundantly evident that waste minimization—however alluring—is not a cure-all for environmental issues and might require the assistance of traditional treatment and disposal methods.

Reducing waste generation at its source is the optimal way to minimize waste. Efforts should be made to recycle the waste within the unit after all source reduction opportunities have been exhausted. Lastly, it should be considered to modify or reformulate products to manufacture them with the least amount of waste generation. Table 9.1 lists a few wastes and potential resources.

Table 9.1. Wastes and Possible Resources

| Wastes | Resources |
|---|---|
| Fly ash from power plant | Raw material for cement or brick manufacture |
| Bagasse wastes from sugar manufacture | Fuel for boiler |
| $CO_2$ release from ammonia plant | Raw material for Urea manufacture |

Source: https://beeindia.gov.in/sites/default/files/4Ch13.pdf.

Since waste minimization focuses on resource conservation, environmental protection, and reducing production costs through one or more of the three main strategies of elimination, reduction at source, and recycling, it is a crucial component of sustainable development. Treatment and disposal are not included in the concept of waste minimization. It also doesn't care about waste diluting or waste being transferred from one environmental medium to another.

Elimination is realistically impossible to achieve because any industrial activity will require energy at the very least and cause emissions into the atmosphere, which in turn contributes to waste generation. The generally acknowledged hierarchy of waste management options favors waste reduction at the source over recycling. This is primarily because recycling necessitates the original generation of waste.

## 9.1.1. Basic Principles

The foundation of waste minimization is the balancing of mass and energy. According to the fundamental law of conservation of mass, matter, or mass, cannot be created or destroyed in any non-nuclear process. Therefore, for a process to be steady, the mass entering the system and the mass leaving the system must be equal. If the system unstable, meaning that some or all its components fluctuate over time, then matter may build up within it.

When materials like steel, glass, and plastics are transformed into motorcars, for instance, the substance's makeup and appearance may alter within the system. Chemical compositions may also get altered in biological and chemical processes as they move from raw materials to finished products. The first law of thermodynamics serves as the foundation for the law of conservation of energy, which is closely related to the law of conservation of mass. Energy cannot be created or destroyed in any non-nuclear process. But unlike mass, energy can exist as work and heat as well as a variety of interchangeable forms such as internal, potential, kinetic, and so forth.

An additional principle, particularly applicable to recycling, is that mixing is easy, but separation is always more difficult. This principle should be applied in two main ways:

- The segregation of waste to facilitate recycling.
- To the design of components wherein simple designs with single components are preferred to facilitate recycling unless composite components are required

for structural, rather than aesthetic reasons.

## 9.1.2. Classification of Waste Minimization (WM) Techniques

The techniques of WM (Figure 9.1) include:

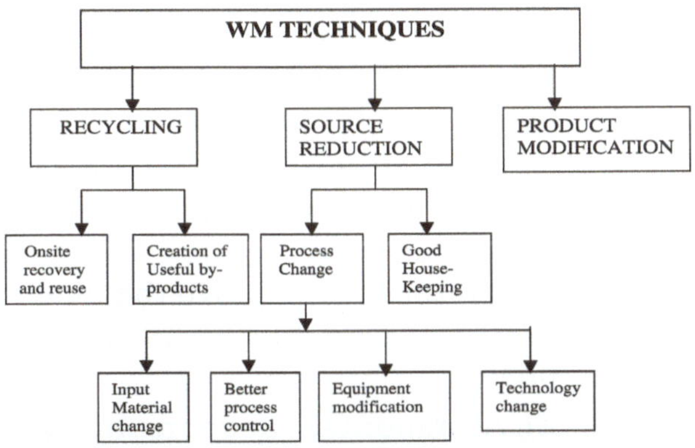


Figure 9.1. Classification of WM techniques.

Source: https://beeindia.gov.in/sites/default/files/4Ch13.pdf.

- Source Reduction: Under this category, four techniques of WM are briefly discussed below:
  - Good Housekeeping: Systems to prevent leakages & spillages through preventive maintenance schedules and routine equipment inspections. Also, well-written working instructions, supervision, awareness and regular training of the workforce would facilitate good housekeeping.
  - Process Change: Under this head, four CP techniques are covered:
  - Input Material Change – Substitution of input materials by eco-friendly (nontoxic or less toxic than existing and renewable) material preferably having longer service time.
  - Better Process Control – Modifications of the working procedures, machine-operating instructions and process record keeping to run the processes at higher efficiency and with lower waste generation and emissions.
  - Equipment Modification – Modification of existing production equipment and utilities, for instance, by the addition of measuring and controlling devices, to run the processes at higher efficiency and lower waste and emission generation rates.
  - Technology Change – Replacement of the technology, processing sequence and/or synthesis route, to minimize waste and emission generation during production.
- Recycling
  - On-Site Recovery and Reuse – Reuse of wasted materials in the same process or for another useful application within the industry.
  - Production of Useful by-Product – Modification of the waste generation process to transform the wasted material into a material that can be reused or recycled for another application within or outside the company.
- Product Modification: Characteristics of the product can be modified to minimize the environmental impacts of its production or those of the product itself during or after its use (disposal).

## 9.1.3. Benefits of Waste Minimization

Industrial organizations can see the benefits of waste minimization by applying the law of conservation of mass. Reduced waste means lower costs for managing and disposing of the waste as well as reduced consumption of raw materials, which saves raw material costs in a system that produces a set number of units, say 100,000 motorcars annually. Additionally, because the system is smaller, it uses less energy to operate, less floor space, and less capital investment. Thus, effective waste minimization initiatives produce "win-win" outcomes, meaning that gains in financial performance and environmental performance occur simultaneously. When it comes to cutting energy use, waste minimization also translates into lower costs for a company when it comes to lowering emissions into the atmosphere, which is one type of waste.

Waste minimization also attracts opportunities, including:

- The promotion of a positive image for stakeholders.
- Reduced risks and liabilities.
- Reduced risks of breaching regulations and compliance.
- A potential reduction in insurance premiums.
- A potential improvement to health and safety.
- A reduced administrative burden.

Waste minimization can protect the environment and often offers positive economic benefits. Waste minimization can improve:

- Efficient Production Practices – waste minimization can achieve more output of product per unit of input of raw materials.
- Economic Returns – more efficient use of products means reduced costs of purchasing new materials, improving the financial performance of a company.
- Public Image – the environmental profile of a company is an important part of its overall reputation and waste minimization reflects a proactive movement towards environmental protection.
- Quality of Products Produced – innovations and technological practices can reduce waste generation and improve the quality of the inputs in the production phase.
- Environmental Responsibility – minimizing or eliminating waste generation makes it easier to meet targets of environmental regulations, policies, and standards; the environmental impact of waste will be reduced.

## 9.1.4. Challenges

Anytime industrial processes are altered, risks inevitably arise. Risk symptoms frequently take the form of obstacles that can be classified into four main areas: cultural, technical, regulatory, economic, and technical.

If it is thought that there is not enough financial incentive to implement waste minimization, then economic barriers may appear. Clearly, if there is any doubt following the initial financial appraisal, then all the less obvious benefits must be considered in economic feasibility studies. Technical barriers may include the following:

- A lack of suitable scientific or engineering information on techniques.

- Concerns about changes to product quality and customer acceptance.
- An interruption to production whilst changes are made.
- The creation of new bottlenecks to the production process.

Environmental performance should be improved by waste minimization, so regulatory obstacles are generally thought to be unlikely. However, modifications to procedures may necessitate adjustments to authorizations for Integrated Pollution Control, licenses, consents, and other regulatory approvals. Planning applications and possibly even environmental impact assessments may be required for larger developments. Cultural barriers might arise from:

- A lack of commitment from senior management.
- A lack of awareness of corporate goals and objectives.
- The resistance to change by individuals.
- Poor internal communication and inadequate training.
- Restrictive employment practices.
- An inflexible organizational structure.
- Bureaucracy, particularly in the generation of cost data.

## 9.1.5. Commitment

Many obstacles to waste minimization can be removed with a strong organizational commitment from senior management. A company policy should be announced along with a plan for putting it into practice. The policy could be included in the environmental policy statement of an environmental management system (EMS), like the European Standard EMAS or the International Standard ISO 14001. The EMS may also include waste minimization targets, goals, and objectives. For new projects and training initiatives, sufficient staff time and financial resources should be provided. To ensure efficient and effective financial appraisals of every waste minimization project, the true and total costs of waste must be attributed, to the greatest extent feasible, to the waste generation sources. If exact allocations cannot be introduced, then the following alternatives can be used:

- Each operating department can be allocated some portion of the fixed costs of the overall on-site service.
- Each operating department can be charged the variable costs of treatment and disposal based on the quantity and nature of wastes arising.
- Various operating departments can be charged transfer prices for the quantities and types of waste treated and disposed of.

# 9.2. WASTE MINIMIZATION METHODOLOGY

Bringing together different industry groups is crucial for an efficient Waste Minimization program because it guarantees implementation. The industry's size, makeup, and waste and emission issues will determine how formalized the program is. The software ought to possess sufficient flexibility to adjust to evolving situations. A systematic, step-by-step process guarantees that all possible waste minimization opportunities are taken advantage of. The following figure illustrates the steps in a typical waste minimization program (Figure 9.2).

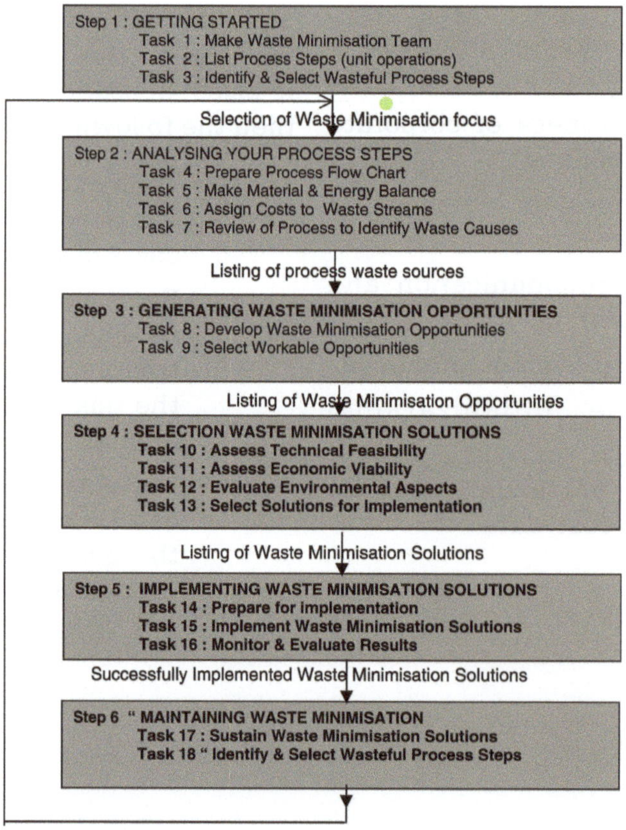


Figure 9.2. Waste minimization methodology.

Source: https://beeindia.gov.in/sites/default/files/4Ch13.pdf.

## 9.2.1. Step 1: Getting Started

### *9.2.1.1. Form a Waste Minimization Team*

The concept of Waste Minimization (WM) applies to every department and division within an industry. The staff of the department in question as well as those of other related departments must assist and cooperate with the activities associated with WM audits. Therefore, the formation of an interdisciplinary and cross-departmental team is necessary for the effective execution of a work management audit. In certain situations, having outside experts would be beneficial for maintaining objectivity and professionalism.

### *9.2.1.2. List Process Steps/ Unit Operations*

The WM team needs to become knowledgeable about the manufacturing processes, including utilities, facilities for treating and disposing of waste, and a list of every step in the process. It would be beneficial to draw out the drainage system, vents, and material-loss areas for the process layout. This makes sure that significant areas are not missed and aids in the establishment of cause-and-effect relationships. Streams of discharge that are intermittent, continuous, and periodic need to be labeled appropriately.

### *9.2.1.3. Identify and Select Wasteful Process Steps*

It could be challenging to begin comprehensive waste minimization initiatives that cover the entire unit in industries with multiple processes. It is best to start with fewer process steps in these situations. The chosen step or steps may be the most inefficient and/or have the greatest potential to reduce waste.

An alternative description of this task would be "preliminary prioritization." The volume of waste, the degree of environmental impact, opportunities for waste minimization, estimated benefits (particularly cost savings), implementation costs, and other factors should all be considered when evaluating the different wasteful steps mentioned in 1.2. Focusing on the process steps and areas for in-depth analysis would be made easier with the aid of this assessment.

## 9.2.2. Step 2: Analyzing Process Steps

### *9.2.2.1. Prepare Process Flow Charts*

This task comes after the one mentioned in 1.2. Flow charts are diagrammatic or schematic representations of production meant to help locate the sources of emissions and waste streams as well as process steps. A flow chart ought to enumerate and describe the input, output, and recycle streams. Even supposedly free or less expensive inputs like sand, water, and air should be considered because they frequently end up being the main source of waste. Chemical equations should be added to the process flow diagram as needed to help with process comprehension. Additionally, materials like catalysts and coolant oil that are only infrequently used or absent in output streams need to be specified. The steps that are batch, continuous, and periodic should all be suitably highlighted. A crucial part of the entire analysis is creating an accurate and thorough process flow diagram, which serves as the foundation for compiling the materials and energy balance.

Create Material and Energy Balances: Identification and quantification of hitherto unidentified losses and emissions are made possible by material and energy balances, which are crucial for any waste minimization program. These balances help assess the advantages and disadvantages of a preventative program as well as tracking advancements made in it. The following lists typical elements of an energy and material balance (see Figures 9.3 and 9.4):

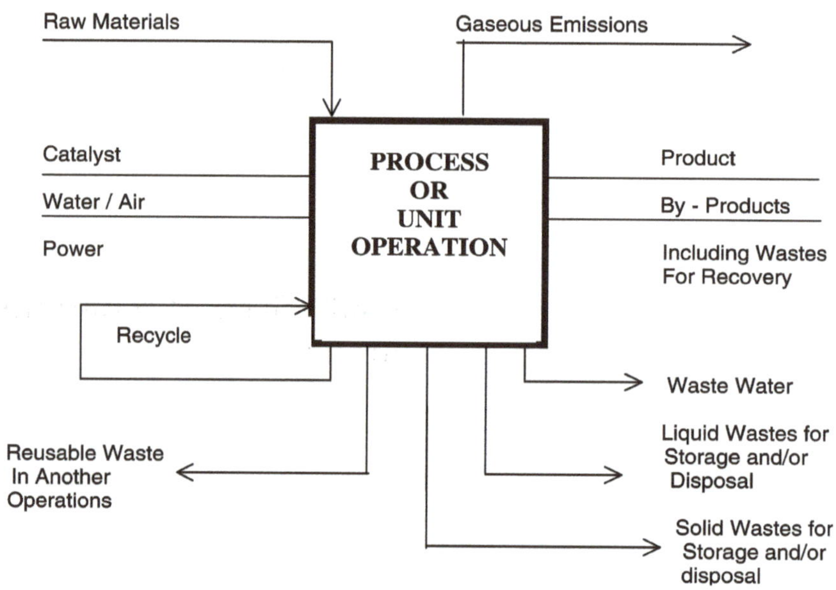


Figure 9.3. Typical components of a material balance.

Source: https://beeindia.gov.in/sites/default/files/4Ch13.pdf.

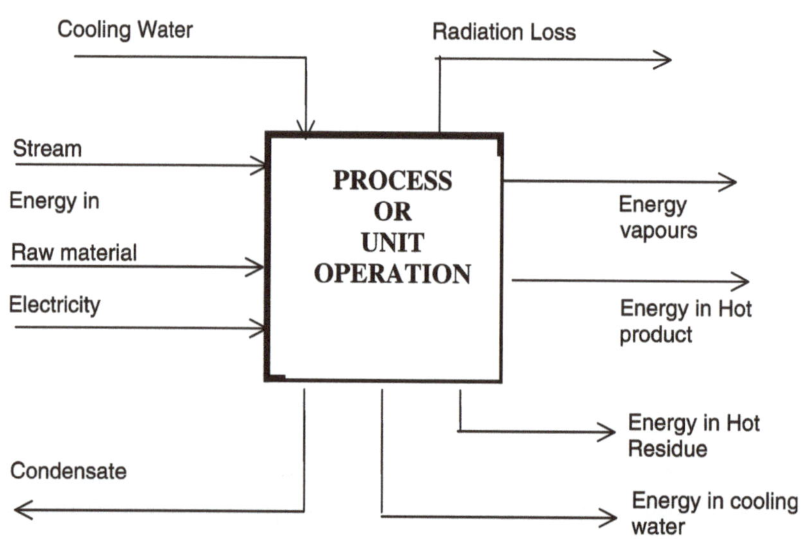


Figure 9.4. Typical components of an energy balance.

Source: https://beeindia.gov.in/sites/default/files/4Ch13.pdf.

Investing more time and money to achieve the ideal material balance is not recommended. Even a rough or preliminary material balance presents opportunities for Waste Minimization that can be turned into a profitable venture.

However, since it is impossible to get a trustworthy estimate of the waste stream by deducting the materials in the product from the raw materials, the accuracy of analytical data and flow measurements is crucial. Under such circumstances, waste streams should be directly monitored and analyzed.

#### 9.2.2.2. Assign Costs to Waste Stream

It would be necessary to assign costs to waste streams to evaluate their potential for profit. This expense represents the money lost because of waste. Except for situations in which there is a direct loss of raw materials or products, a waste stream does not seem to have any quantifiable costs associated with it. A more thorough examination would reveal several direct and indirect cost elements related to waste streams, including:

- Cost of raw materials in waste.
- Manufacturing cost of material in waste
- Cost of product in waste
- Cost of treatment of waste to comply with regulatory requirements
- Cost of waste disposal
- Cost of waste transportation
- Cost of maintaining the required work environment
- Cost due to waste CESS.

Based on this, the total cost per unit of waste (in rupees or kilograms) for each waste stream should be calculated. This number would help determine whether the waste minimization strategies are feasible. The outcome can also be used to rank the waste streams according to importance.

#### 9.2.2.3. Review of Process

It is feasible to perform a cause analysis using the material and energy balances to identify and pinpoint the causes of waste generation. These reasons would then serve as the basis for developing Waste Minimization strategies. Waste generation can arise from a wide range of factors, from basic neglect of hygiene to sophisticated technological factors as listed below.

#### 9.2.2.4. Typical Causes of Waste

- Poor Housekeeping
  - Leaking taps/valves/flanges
  - Spillages
  - Overflowing tanks
  - Worn out material transfer belts
  - Unchecked water/air consumption
  - Unnecessary running of equipment
  - Sub optimal loading
  - Lack of preventive maintenance
  - Sub-optimal maintenance of process condition
  - Ritualistic operation
- Poor Raw Material Quality
  - Use of substandard cheap raw material
  - Lack of quality specification
  - Improper purchase management system
  - Improper storage

- Poor Process/Equipment Design
  - Mismatched capacity of equipment
  - Wrong material selection
  - Maintenance prone design
  - Adoption of avoidable process steps
  - Lack of information/design capability
- Poor Layout
  - Unplanned/ad-hoc expansion
  - Poor space utilization plan
  - Bad material movement plan
- Bad Technology
  - Continuation of obsolete technology
  - Despite product/raw material change
  - High cost of better technology
  - Lack of availability of trained manpower
  - Small plant size
  - Lack of information
- Inadequately Trained Personnel
  - Increased dependence on casual/contract labour
  - Lack of formalized training system
  - Lack of training facilities
  - Job insecurity
  - Fear of losing trade secrets
  - Lack of availability of personnel
  - Understaffing hence works over pressure
- Employee Demotivation
  - Lack of recognition
  - Absence of reward
  - Emphasis only on production, not on people
  - Lack of commitment and attention by top management

## 9.2.3. Step 3: Generating Waste Minimization Opportunities

### *9.2.3.1. Develop Waste Minimization Opportunities*

The process moves into the creative stage once the source and contributing factors of waste and emissions are identified. It is now time for the WM team to start searching for potential ways to cut waste. Your team members' creativity and knowledge will help you identify viable options. The following could be useful in your search for waste minimization opportunities:

- Other personnel from other plants
- Trade associations
- Consultancy organizations
- Equipment suppliers
- Consultants

Opportunities for Waste Minimization should be identified in an atmosphere that fosters ingenuity and independent thought. Moving away from the typical workplace would be advantageous to achieve better outcomes. Different analytical tools and techniques such as "brainstorming" and "group discussions" can be beneficial during this stage.

### *9.2.3.2. Select Workable Opportunities*

Those Waste Minimization opportunities that are deemed unfeasible should be eliminated after screening. The process of discarding should be clear-cut, uncomplicated, and frequently limited to qualitative aspects. Both bias and ambiguity should be absent. Avoid expending needless effort on opportunities for in-depth feasibility analyses that turn out not to be feasible or practicable. Thereafter, a more thorough feasibility analysis is conducted on the remaining Waste Minimization opportunities.

## 9.2.4. Step 4: Selecting Waste Minimization Solutions

When choosing a Waste Minimization solution be practiced, it must be both environmentally friendly and economically and technically feasible.

### *9.2.4.1. Assess Technical Feasibility*

Whether a suggested Waste Minimization option is appropriate for a given application is determined by the technical evaluation. The suggested measure's effect on the rate of production of the product ought to be assessed first. To determine the technical viability, laboratory testing trial runs may be necessary in the event of a substantial departure from the current process practices.

A typical checklist for technical evaluation could be as follows:

- Availability of equipment
- Availability of operating skills
- Space availability
- Effect on production
- Effect on product quality
- Safety aspects
- Maintenance requirements
- Effect on operational flexibility
- Shut down requirements for implementation

### *9.2.4.2. Assess Economic Viability*

The primary factor that determines whether to encourage or oppose waste minimization is frequently economic viability. Therefore, it is crucial that the initial Waste Minimization initiatives be highly profitable to ensure a smooth launch. This kind of approach keeps commitment and generates more interest.

Little investment options that involve more procedural changes (housekeeping measures, operational improvements) can be implemented without a thorough economic analysis using basic techniques like a payback period. However, as Waste Minimization measures become more involved and capital intensive, other profitability analysis methods viz. To obtain the whole picture, return on investment (IRR) or net present value (NPV) is advised.

There may be fixed capital expenses associated with economic investment. E. Costs associated with equipment, working capital, shutdown, O&M, etc. The savings may come from greater production levels and consequently lower specific input costs, lower O&M costs, increased value of byproducts, direct savings of energy and input materials, or a decrease in environmental costs. Transporting and disposing of waste, among other expenses.

### 9.2.4.3. Evaluate Environmental Aspects

It is important to evaluate the waste minimization options considering their environmental impact. If there is a net decrease in the toxicity and/or amount of waste, the environmental benefit will frequently be evident. Changes in the way the waste is treated could have additional effects. Environmental factors might not seem as important as economic ones in the beginning. In the future, environmental considerations will take precedence over economic ones, just as in developed nations.

### 9.2.4.4. Select Solutions for Implementation

The implementation of Waste Minimization measures should be chosen following technical, economic, and environmental assessment. Naturally, those with the greatest financial advantages would be the most alluring, assuming that technical viability is acceptable. However, environmental priorities may end up being the deciding factor in an increasing number of cases, particularly when there is active pressure groups involved.

The completed work ought to be recorded. In addition to serving as a guide for securing implementation approvals, the document would be helpful in securing funding from outside financing sources, informing other agencies of one's status, and setting baselines for performance reviews and evaluations.

## 9.2.5. Step 5: Implementing Waste Minimization Solutions

### 9.2.5.1. Prepare for Implementation

The chosen solutions could be put into practice. A methodical plan of implementation would be necessary for several other measures in addition to basic housekeeping. The team responsible for waste minimization ought to be adequately equipped to tackle the task of execution. In the event of a multi-departmental solution, the preparation would entail setting up links, organizing funds, making technical arrangements, etc.

### 9.2.5.2. Implement Solutions

The work includes equipment fabrication and procurement, layout and drawing preparation, site transportation, installation, and commissioning. Simultaneous training of personnel should be undertaken whenever necessary, as many excellent initiatives have failed miserably due to a lack of appropriately trained personnel.

### 9.2.5.3. Monitor and Evaluate Results

Performance should be tracked with the WM solutions. To determine the cause(s) of any deviations, the results should be compared to those estimated or calculated during technical evaluation. Only after successful commissioning and consistent, stable performance for a reasonable amount of time is the implementation job deemed completed.

## 9.2.6. Step 6: Sustain Waste Minimization

Maintaining Waste Minimization is the largest challenge in waste minimization. The Waste Minimization team's initial enthusiasm gradually fades. Avoiding such tragic ends is a good idea. Some of the frequently experienced causes that one should be aware of and steer clear of are abandoning a commitment, producing more at any cost, not receiving recognition or rewards, and switching priorities.

To encourage the goal of reducing waste, all industry employees should be informed about the monitoring and evaluation of the implemented measures. Encouraging the most employees to participate and providing rewards to those who deserve it will contribute to the long-term sustainability of waste minimization.

After the Waste Minimization team has applied Waste Minimization solutions in the studied area, they should return to Step 2, which involves examining the process steps and deciding which wasteful area to prioritize next. Until every step is completed, this cycle keeps going.

To put it briefly, the organization must adopt a waste reduction mindset. This implies that waste minimization ought to be a crucial component of every business's operation. Until now, every effective waste minimization program has been built around this idea.

# 9.3. RECYCLING

Recycling and the value of products made from scrap must be recognized in an era when solid waste management, energy conservation, and the cost and availability of materials are major concerns for scientists, municipal administrations, and the public. As we shall see, recycling waste has advantages that go beyond the environment to include the economy and aesthetics.

Recycling is the process of gathering and repurposing used materials, which are typically referred to as "waste." These used materials are sorted and processed during this process so they can be utilized as "raw materials" to make new products. "Re-use" and "recycling" are different in that the former simply refers to reusing old products, while the latter uses the essential components of an old product as raw materials to create new products. Frequently recycled materials include plastic, glass, paper, batteries, aluminum, and so forth.

Recycling is not a brand-new idea. Rag pickers were significant recyclers in America as late as the early 20th century. For millennia, agricultural lands have been fertilized with animal manure, plant debris, and "night soil." The 1960s saw the emergence of modern recycling because of growing public awareness of various environmental and public health issues. However, recycling initiatives at the time frequently placed a strong emphasis on separating materials from the waste stream. Regretfully, markets for the acquisition and repurposing of segregated materials were not developed. Many manufacturers lacked the necessary equipment to handle these so-called "secondary materials," and they were hesitant to invest in and use new processing technologies. Many of the separated materials ended up in the landfill as a result. Recycling initiatives failed partly because there were no established markets for the separated materials, but also—and this is more significant—because there was insufficient processing.

By the late 1980s, news of medical waste wash-ups, declining landfill space, potential global warming, and atmospheric ozone depletion had sparked a renewed awareness of the environment. Sanitary landfills were closing at this time, and permitting and siting new ones faced strong opposition from the public and regulatory bodies. Waste disposal expenses also went up in tandem. As a result, there was a noticeable increase in public, industry, and government interest in recycling.

Many neighborhood recycling initiatives in recent years have their roots in attempts to cut down on the amount of waste sent to the nearby landfill to save tax money. Municipalities created materials recovery facilities (MRFs) and recycling drop-off centers in response to public pressure. Both nationally and state-wide, laws that promote MSW recycling have been passed. Certain regulations are directed towards waste generators, be it businesses or individual homeowners; these regulations may take the shape of directives or specifications for prolonging the life of the nearby landfill. Following federal laws enacted since 1990, many states established detailed regulations aimed at lowering the amount of waste that ends up in landfills. A combination of source reduction, recycling, and composting was to be used to meet these quotas. The purchase of recycled materials was covered by other laws. For instance, purchasing paper made with a certain proportion of recycled fibers is now mandated for certain government offices. Many developed countries have developed creative and proactive recycling strategies because of these incentives and pressures.

Source separation and the MRF are the two main methods for segregating MSW for potential recycling. Source separation refers to the division of waste components by the business or homeowner (i.e., at the origin). The specific goods (e.g., aluminum cans, paper, glass, plastics, and so forth) are gathered and brought to a facility to undergo additional processing, like shredding and densification. Brokers or manufacturers purchase and remove these clean, slightly processed materials. The MRF, on the other hand, is a centralized, automated facility that takes raw MSW (commingled) or materials that are separated at the source. The mixed goods are put on conveyor belts, and at predetermined stations, recyclables are taken out by hand or with the help of a specialized mechanical device. The energy consumption, labor costs, capital costs, and separation efficiency of source separation and MRF techniques vary significantly.

## 9.3.1. Recycling Terminology

Terms relating to recycling are often misused; to avoid confusion, it is important at the outset to clarify some of the relevant language.

- Source Separation—Removal of potentially recyclable materials from the waste stream. Conducted by the individual consumer and commercial establishment (Figure 9.5).
- Reuse—Using an item for its original purpose. A common example is refilling a returnable soft drink bottle.
- Recycling—Use of a material in a form like its original use. Newspapers are recycled into cardboard or new newspapers. Plastic is shredded and manufactured into fabric. Aluminum window frames are converted into new beverage containers.
- Waste-to-Energy—The conversion of MSW (preferably the organic fraction only) into energy by combustion in a controlled incinerator. Energy is recovered as heat and can be used directly; however, some facilities convert the heat energy into electrical energy.
- Resource Recovery—Extraction of energy or materials from waste. This term incorporates all the above. Thus, a waste-to-energy facility will incinerate organic waste to generate heat energy. Glass and rubber are separated from waste, processed, and used as road-building materials.

With the above terminology in mind, we can address the overall process of recycling as shown by the universal recycling symbol.

The figure shows source separation as one arrow, which is the removal of materials from the waste stream (e.g., aluminum cans (e.g., putting them aside in the kitchen of the homeowner) and placing them in special containers for pickup at the curb or at a drop-off location.

Figure 9.5. Drop-off centers are one means of segregating MSW components.

Source: John Pichtel. Waste Management Practices: Municipal, Hazardous, and Industrial. Taylor & Francis. ISBN:9781466585188.

The second arrow represents the material's processing. A waste hauler gathers aluminum cans and transports them to a broker or distributor, where they are compressed into substantial bales. The bales are sold and transported to an aluminum smelter off-site under favorable market conditions. The material is pulled into sheets and the bales are melted at the smelter to create new cans. After being filled by a soft drink manufacturer, the cans are shipped to a retail location. At last, the third arrow indicates that the consumer has bought the soft drink kept in the recycled cans.

Because of the cycle, a material cannot be deemed truly recycled until it has passed through all three stages and been bought by the end user. The source separation component of this cycle has grown disproportionately large in comparison to the other two over the last few decades. There have been issues with the second step because some industries have complained that retooling the facilities and equipment required to process recycled stock has cost excessively much. In other circumstances, there might still be financial incentives for the use of raw materials, whether in the form of market pricing or subsidies. The third arrow has also experienced a bottleneck because buyers—especially individual consumers—have not shown enough desire to purchase goods made from recycled materials.

## 9.3.2. Process of Recycling

The recycling process involves three stages.

- In the first stage the old products are collected and processed, where they are sorted, cleaned and made ready for recycling or manufacturing new products.
- The second stage involves the manufacturing of new products from the raw material obtained by the processing of the old products.
- Finally, the process ends with the purchasing of recycled goods by the consumers.

## 9.3.3. Importance of Recycling

- Recycling Saves Energy – A significant amount of energy resources is used in the processing of raw materials. Recycling waste materials lowers the energy needed for numerous manufacturing operations, including mining and refining. Recycled materials, such as glass and aluminum, can significantly lessen the demand for energy resources.

- Recycling Saves Environmental Conditions and Reduces Pollution – Recycling contributes to the reduction of greenhouse gas emissions by minimizing the energy used in industrial production. Coal, diesel, gasoline, and other fossil fuels are some of the main ones used in most industries. All these release toxic gases into the atmosphere, including carbon dioxide, methane, and sulfur dioxide. Toxic materials are produced during the processing of fresh raw materials, contributing to environmental pollution. Recycling minimizes fuel consumption, which lowers the quantity of dangerous pollutants in the environment by lowering the energy required.
  - In a study that involved recycling 35,116 tons of material, it was found that the reduction in greenhouse gas emissions was equivalent to taking 22,140 cars off the road.
  - Recycling a ton of mixed paper or newspaper is equivalent to saving 12 trees.
  - Pollution of the soil and water has increased due to plastic waste. Recycling plastic is a practical way to deal with this issue. During the recycling process, used materials are extracted from plastic waste and subsequently utilized in the manufacturing sector.
  - Recycling Saves Natural Resources – As far as we are aware, recycling entails using and processing an old product's essential components to create new ones. This greatly aids in the preservation of our natural resources. For instance, recycling an old newspaper eliminates the need to cut down another tree to make new paper goods. In this sense, appropriate recycling can assist us in maintaining the natural balance of the environment and protecting our natural resources for future generations.
  - Recycling Saves Space for Waste Disposal – The majority of landfills are overflowing with waste materials that could have been efficiently recycled. Certain waste materials are classified as non-biodegradable, meaning that their breakdown process is prolonged. Recycling allows these waste materials to be used properly and conserves landfill space. Given the rate at which landfills are filling up, if we don't start recycling at home and encourage others to do the same, we may soon run out of landfill space.
  - Recycling Helps Reduce the Amount Of Greenhouse Gas Emissions – Comparatively speaking, producing paper and plastics from recycled materials uses less energy than producing them from virgin or new materials. The extraction and transportation of these virgin materials require more energy. Using vehicles that run on fuels like

gasoline and diesel is known as transportation. These fuels are thought to be the primary cause of emissions of greenhouse gases. Recycling reduces energy use, which leads to a decrease in fuel consumption. As a result, less carbon dioxide will be emitted into the atmosphere.

The advantages of recycling, however, are seen to exceed the dangers associated with improper disposal of waste.

## 9.3.4. Recycling Progress and Statistics

In 2010, Americans produced about 250 million tons of MSW, according to EPA (2011). In 2010, 85 million tons, or 34%, of MSW were recovered through recycling, which included composting. In 2010, the United States reported over 9300 curbside recycling programs and roughly 9000 recyclable drop-off locations (Figure 9.5). To process the materials that have been collected, about 630 MRFs are in use. In all, 2284 yard waste composting initiatives were documented in a Biocycle survey (van Haaren et al., 2010).

Example 9.1. Using the data for the MSW composition of the city of Pristine, IL, in the table below, calculate the maximum contribution of source separation to the city's solid waste disposal program.

| Component | Percent by Weight |
|---|---|
| Paper | 31 |
| Cardboard | 5 |
| Ferrous metals | 10 |
| Nonferrous metals | 1 |
| Glass | 9 |

Solution: These components total 56% of the MSW stream. Of this 56%, however, not all is potentially recyclable.

| Component | Expected Recovery, % | Percent |
|---|---|---|
| Newspaper | 25 | |
| Cardboard | 100 | |
| Ferrous | 75 | |
| Nonferrous | 50 | |
| Glass | 75 | |

MSW comprises about 50% of total solid waste. The remainder is industrial, construction and demolition debris, and so on. Therefore, source separation can handle about 27.6% of 50% or:

27.6% × 50% = 14% of the total solid waste

Unfortunately, no U.S. community has come close to attaining 100% participation in a source separation program. If we optimistically assume 50% participation, then 14% × 50% = 7% is the maximum contribution of source separation to recycling.

## 9.3.5. Recovery and Markets for Components of the Waste Stream

Understanding the various materials involved, the manufacturing processes, and the characteristics of secondary material markets helps develop a comprehensive approach to waste recycling as well as in identifying opportunities and challenges. Questions that recycling professionals must address include the following:

- What is the demand for reclaimed materials?
- What are the specific requirements for a raw material used by an industry?
- How can a recovered material compete, technologically and economically, with virgin feedstock?

- Are new processes or equipment necessary to process and work with waste?
- What are the incentives or disincentives that affect the use of recycled materials by manufacturers?
- Can new uses be identified that will increase the demand for recycled materials?

### 9.3.6. Market Issues

Recycling markets are inherently volatile in terms of price. The cost of recycled materials is determined by the total amount of demand for manufactured goods. The basis for supply and demand for materials is wider issues like raw material markets, international trade agreements, tariffs, and so forth. An effective way to ensure the long-term viability of a recycling program is to manage revenue fluctuations. Certain towns control their swings by generating demand for recycled materials in local manufacturing. Some bargain for contracts with price floors. Recycling initiatives that gather a broad range of materials—including lead automobile batteries, mixed paper, cardboard, glass, metals, and plastic bottles—may have an advantage over those that only gather one or two. If demand for a material declines, it can be hoarded until demand increases, during which time it can be replaced with other, more lucrative materials.

### 9.3.7. Purity of Materials

The homogeneity and contamination-free nature of the materials are usually required by processors and end users of recovered products. A tiny amount of undesirable material could have a detrimental effect on the finished product's quality and, in certain situations, could be dangerous for the workers. Certain industries have stringent composition requirements and won't accept even extremely low levels of contamination. Materials from other industries are frequently processed to remove foreign objects. When comparing raw mixed wastes processed through an MRF to source-separated wastes, there is a noticeable decrease in the amount of foreign material. Nonetheless, a lot of people find it more convenient to send mixed waste to a central processing facility where it will be sorted. Buyers may also require that the materials be compacted or established under a specific condition (e.g., bottles are not to be broken, aluminum beverage cans must be crushed, and high-density polyethylene (HDPE) containers must be baled).

# 9.4. PAPER

Approximately 28.5 of municipal solid waste (MSW) is made up of paper products. S. MSW, which is greater than twice as much as any other part. Paper waste takes up a significant amount of space in collection trucks and landfills, which has an impact on the environment and the economy. Paper manufacturing suggests that additional trees must be cut to meet the ongoing demand for new products. Economic incentives for paper recycling are provided by the enormous volumes of paper waste produced and the related disposal costs.

## 9.4.1. Paper Manufacture

The earliest known method of producing paper was created by the Chinese as early as 100 CE. The paper base was a suspension of bamboo fibers. The first paper mill in North America's English colonies was built in 1690 not far from Philadelphia. Up until the middle of the 1800s, waste paper, cotton, and linen rags were the only materials used to make paper. Papermaking techniques utilizing wood fiber were developed in response to the growing demand for paper and paperboard.

In contemporary papermaking, wood pulp from both conifers and deciduous trees is used. Paper qualities vary depending on the type of wood used, as hardwoods and softwoods have quite different fiber morphologies. Softwoods have stronger and longer fibers than hardwoods, but during the sheet-forming process, softwood fibers tend to tangle and form flocs, which can cause aesthetic issues. Hardwood and softwood fibers are combined to provide the right amount of strength and appearance to mitigate this effect.

Paper does not have a set chemical composition because it is made from fibers that originate in the cells of terrestrial plants. Most plant cells are made up of polysaccharides, or carbohydrates, which are partially combined with lignin, an intricate aromatic polymer. Plant age is typically correlated with an increase in lignin content. The structural polysaccharide cellulose is the main component of the cell's carbohydrate component. There are additional nonstructural polysaccharides with a lower molecular weight called hemicelluloses, which are crucial to the characteristics of pulp and paper. Additionally, there are trace amounts of inorganic materials and water-soluble substances like alcohols, resin acids, and fatty acids (Rhyneretal.1995; Roberts1996).

Whether utilizing recycled or wood fibers as the foundation, the ability of the fibers to be conformable—that is, able to be matted into a consistent sheet—is crucial to produce paper. Additionally, they must be able to create bonds that are strong enough at the point of contact. When the bonds in the wood fibers are broken during the pulping process, proper conformability and bonding start.

Three techniques exist for pulping virgin fiber: (1) mechanical pulping, which involves using mechanical energy to release the fibers; (2) chemical pulping, which involves adding chemicals to dissolve lignin and preserve cellulose; and (3) semi-chemical pulping, which combines techniques 1 and 2.

Beyond pulping, paper manufacture operations are as follows (Smook 1982; Rhyner et al., 1995):

- Sheet Formation. Pulp slurry (1% pulp, 99% water) is guided into a headbox.
- Forming. The pulp is dewatered by about 20%. The fibers are formed into a sheet.
- Pressing. Fibers are pressed together. Another 20% of the water is removed from the pulp.
- Drying. The sheet is dried to about 90–95% solids. Fibers bond together at this point.
- Converting And Finishing. The sheet is pressed between rolls to reduce thickness and increase uniformity and is wound on to reels (Figure 9.6).

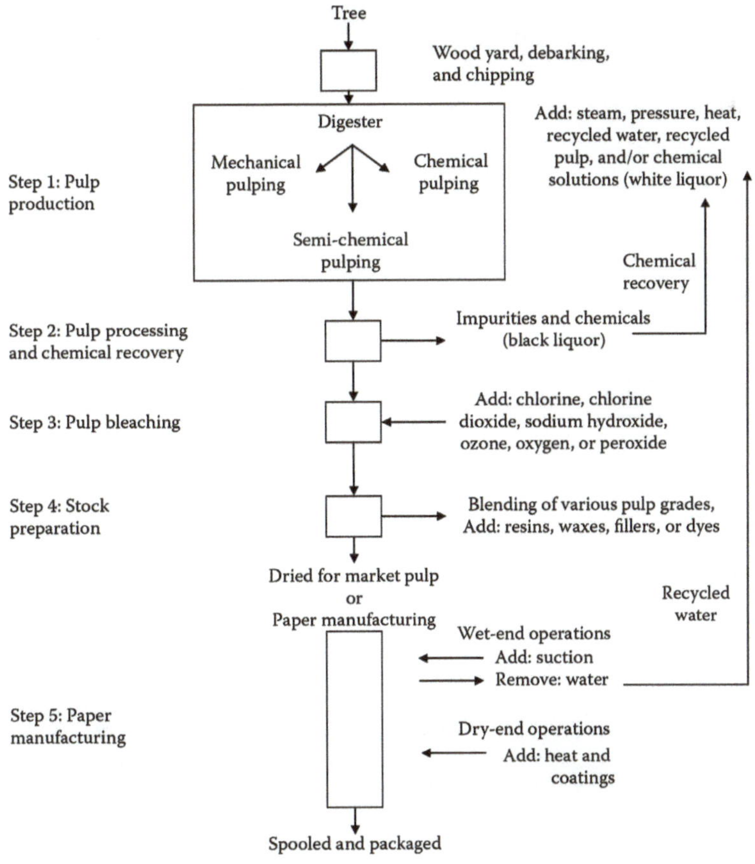


Figure 9.6. Paper making process.

Source: John Pichtel. Waste Management Practices: Municipal, Hazardous, and Industrial. Taylor & Francis. ISBN:9781466585188.

## 9.4.2. Paper Recycling

Recycling paper has a lengthy history. Before World War II, various organizations, like the Boy Scouts, had collection drives. These initiatives grew significantly during the conflict. Out of all the MSW components, paper and paperboard have the highest overall recovery for recycling. In 2010, a total of 85% of corrugated boxes were retrieved for recycling (Table 9.2). Newspapers accounted for 71.6 of the recovered items, followed by high-quality office papers at 74.2, and other papers in smaller amounts. Paper recycling has remained a strong trend over the past few decades. The American Forest and Paper Association reports that the recovery rate for paper and paperboard increased from 28.8% to 65.1% between 1987 and 2012. Out of the 71 million tons of paper products used in the US, only thirty-three percent are made from recycled paper. This figure contrasts with 69% in Western Europe. About seventy-five percent of the raw materials used in paper mills in Mexico are derived from recycled paper, not wood pulp.

Table 9.2. Generation and Recovery of Paper Waste in the Solid Waste Stream, 2010

| | | Recovery | | |
|---|---|---|---|---|
| **Product Category** | **Generation (Thousand of Tons)** | **Thousand of Tons** | **Percent of Generation** | **Discards (Thousand of Tons)** |
| Nondurable goods | | | | |
| Newspapers and mechanical papers | 9880 | 7070 | 71.6% | 2810 |
| Books | 990 | | | |
| Magazines | 1590 | | | |
| Office-type papers | 5260 | | | |
| Standard mail | 4340 | | | |
| Other commercial printing | 2480 | | | |
| Tissue paper and towels | 3490 | | | |
| Paper plates and cups | 1350 | | | |
| Other nonpackaging paper | 4190 | | | |
| Subtotal nondurable goods excluding newspapers and mechanical papers | 23,690 | 10,650 | 45.0% | 13,040 |
| Total paper and paperboard of nondurable goods | 33,570 | 17,720 | 52.8% | 15,850 |
| Containers and packaging | | | | |
| Corrugated boxes | 29,050 | 24,690 | 85.0% | 4360 |
| Gable top and aseptic cartons | 540 | | | |
| Folding cartons | 5470 | | | |
| Other paperboard packaging | 90 | | | |
| Bags and sacks | 1040 | | | |
| Other paper packaging | 1490 | | | |
| Subtotal containers and packaging excluding corrugated boxes | 8630 | 2160 | 25.0% | 6470 |
| Total paper and paperboard of containers and packaging | 37,680 | 26,850 | 71.3% | 10,830 |
| Total paper and paperboard | 71,250 | 44,570 | 62.6% | 26,680 |

Source: John Pichtel. Waste Management Practices: Municipal, Hazardous, and Industrial. Taylor & Francis. ISBN:9781466585188.

Paper waste is categorized as high grade or bulk. Manila folders, hard manila cards, and other paper goods related to computers are rated at the highest level. High-quality waste paper is utilized in place of pulp. Newspapers, corrugated cardboard, and mixed paper waste (unsorted office or commercial paper waste) are the materials that make up bulk grades. Paperboards, construction paper, and other products are made from bulk grades. Paper stock used for repulping in the US and Canada must adhere to standards and procedures set by the Institute of Scrap Recycling Industries. Table 9.3 contains a list of common paper grades.

Table 9.3. Selected Paper Grades for Repulping and Recycling

| No. | Name | Composition |
|---|---|---|
| 1 | Soft mixed paper | Mixture of various qualities of paper not limited as to type of baling or fiber content. |
| 2 | Mixed paper | Baled clean sorted mixture of various qualities of paper containing less than 10% of groundwood content. |
| 3 | — | Grade not currently in use. |
| 4 | Boxboard cuttings | Baled new cuttings of paperboard used in the manufacture of folding cartons, set-up boxes, and similar boxboard products. |
| 5 | Mill wrappers | Baled paper used as outside wrap for rolls, bundles, or skids of finished paper. |
| 6 | News | Baled newspaper as typically generated from news drives and curbside collections. |
| 7 | News, de-ink quality | Baled sorted fresh newspapers; not sunburned; containing not more than the normal percentage of rotogravure and colored sections. May contain magazines. |
| 8 | Special news de-ink quality | Baled sorted fresh newspapers, not sunburned, free from magazines, white blank, pressroom over-issues, and paper other than news, containing not more than the normal percentage of rotogravure and colored sections. This grade must be tare-free. |
| 9 | Over-issue news | Unused overrun newspapers printed on newsprint, baled or securely tied in bundles, containing not more than the normal percentage of rotogravure and colored sections. |
| 10 | Magazines | Baled coated magazines, catalogues, and similar printed materials. May contain a small percentage of uncoated news-type paper. |
| 11 | Corrugated containers | Baled corrugated containers having liners of either test liner, jute, or kraft. |
| 12 | Double-sorted corrugated | Baled double-sorted corrugated containers, generated from supermarkets, industrial or commercial facilities, or both; having liners of test liner, jute, or kraft. Material has been specially sorted to be free of boxboard, off-shore corrugated, plastic, and wax. |
| 13 | New double-lined kraft corrugated cuttings | Baled new corrugated cuttings having liners of either test liner, jute, or kraft. Treated medium or liners, insoluble adhesives, butt rolls, slabbed or hogged medium are not acceptable in this grade. |
| 14 | — | Grade not currently in use. |
| 15 | Used brown kraft | Baled used brown kraft bags free of objectionable liners and original contents. |
| 16 | Mixed kraft cuttings | Baled new brown kraft cuttings, sheets, and bag scrap free of stitched paper. |
| 17 | Carrier stock | Baled printed or unprinted, unbleached new beverage carrier sheets and cuttings. May contain wet strength additives. |
| 18 | New colored kraft | Baled new colored kraft cuttings, sheets, and bag scrap free of stitched papers. |
| 19 | Grocery bag scrap | Baled new brown kraft bag cuttings, sheets, and misprint bags. Prohibitive materials: none permitted. |
| 20 | Kraft multi wall bag scrap | New brown kraft multiwall bag cuttings, sheets, and misprint bags free of stitched papers. |

*Source:* Reproduced with kind permission from ISRI, Washington, D.C. Copyright © 1993 The Institute of Scrap Recycling Industries, Inc.

*Note:* The ISRI Scrap Specifications Circular is subject to change. To find the most recent edition, go to www.isri.org.

Source: John Pichtel. Waste Management Practices: Municipal, Hazardous, and Industrial. Taylor & Francis. ISBN:9781466585188.

When making paper from recovered paper and paperboard, a distinct pulping procedure is needed in contrast to pulping virgin fibers. The continuous pulper is a crucial mechanical element needed for pulping waste paper fibers. After the input material is ground into a smooth pulp, unnecessary materials such as clips, glue, plastic, and metal are taken out. Following recovery, pulps are chemically treated or disintegrated to remove the ink. To further clean the pulp, some plants incorporate a washing step. Using a fine mesh screen, the pulp is cleaned to remove any remaining impurities and ink. Chemicals are added to the pulper in an optional flotation process to produce air bubbles that separate and float ink droplets away from the pulp. Bleaching the pulp is another option. The resulting pulp is thickened and screened in addition to being floated and cleaned. After being processed, the pulp—whether made of recycled or virgin fibers—enters the paper-making process.

Some of the longer fibers in paper are cut shorter each time it is recycled, which usually results in the fibers losing their flexibility and bonding power. The process known as "hornification," which is a composite of multiple partially irreversible physical changes, is to blame for this. To sustain the paper strength necessary for effective runnability at high speeds, virgin pulp is added both at the paper mill and during conversion (e.g., both for the final application and in the printing press). Every time paper is recycled, its brightness also decreases. In conclusion, the quality of waste paper tends to deteriorate during recycling. However, there are some benefits to using recycled fibers: the twice-dried stock may drain more quickly than the virgin equivalent, it needs less refining, it can be co-refined with hardwood pulp or combined with softwood pulp without suffering too much damage, and the fibers improve opacity (Ferguson 2001).

The waste paper market is highly regionally influenced and highly volatile, just like the markets for many other materials. The state of the economy continues to influence the advancement of paper recycling. Mixed paper waste has little value in some places.

The capacity of paper mills is one restriction on the amount of waste paper that can be recycled each year. Since building or altering these mills requires a large amount of capital, investors need to be certain that the mills will have a sufficient supply of waste paper at a competitive price. Much waste paper is shipped to markets along the Pacific Rim (e.g., South Korea), a country with limited timber resources. China has a significant impact on the market for mixed waste paper. In 2010, the United States exported 39% of its recovered paper to foreign markets. In the recipient nation, the mixed waste paper is sorted and processed. Growth in demand is anticipated to continue. Many nations have developed legislative programs in the last ten years mandating a specific percentage of recycled fiber content in office paper, newspapers, and other products. These kinds of programs raise the amount of paper that can be recycled as well as the demand for it. Nevertheless, waste paper recyclers might be reluctant to commit the substantial capital expenditures required to expand plant capacity.

### 9.4.3. Other Paper Markets

Paper waste has been repurposed as a raw material for other products. Old newspapers have been used to prepare insulation for homes. First, an anti-flammability agent is applied to the shreds of cellulosic material, which is then blown straight into wall cavities. Waste paper can also be used for siding, roofing, fiberboard, and insulation boards. Waste paper is used by the molded pulp industry to make packaging materials, meat trays, egg cartons, and plant pots. Waste paper has been utilized as animal bedding in agriculture. In certain establishments, old newspaper bedding has taken the place of straw bedding because it is more readily available and less expensive, especially in years with severe weather. It has been demonstrated that using old newspaper bedding as a component of manure does not negatively impact animals or soil (Rhyner et al., 1995).

RDF, or refuse-derived fuel, is a product made primarily of carbon from MSW. Waste is separated into its noncombustible components, like metals, glass, and stones, usually using machinery. The organic material is then shredded, and the remaining "fluff" is either compressed into pellets or used as fuel directly. RDF is usually burned with coal as a co-fuel, which has several advantages for the environment and the economy.

## 9.4.4. Benefits of Paper Recycling

One ton of paper requires seventeen trees to produce. Every week in the United States, the equivalent of half a million trees is needed to support all Sunday newspapers. On the other hand, significant amounts of energy and trees are preserved when paper is made from waste paper. An estimated 4100 kWh of energy, 7000 gal of water, and 3 yd3 of landfill space are saved for every ton of recycled paper (Liu and Liptak, 2000). Recycling can also reduce the amount of chemicals, such as bleaches, needed to produce paper, which will lead to a reduction in toxic waste production.

# 9.5. GLASS

Glass is typically well-suited for recycling. Glass fragments can be manually selected or mechanically separated where waste goes through shredding, air classification, and other sorts of separation for recovery from municipal wastes or single-stream recycling.

## 9.5.1. Glass Manufacture

Glass for containers was first made by the ancient Egyptians, who shaped a mold out of clay or sand and then wrapped molten glass strands around it. The invention of glass blowing around 50 BCE allowed glass to be used as a container more widely. Over ten years before the Pilgrims' arrival in 1620, the first documented industry in pre-Colonial America was the glass industry across the Atlantic. The nineteenth century saw the development of methods for producing glass containers in large quantities, and in 1903 the first fully automatic bottle machine was created, capable of producing one million bottles every week. The United States has been producing 270–275 million containers of glass annually for the past ten years (Kogel et al., 2006).

Relatively cheap raw materials like silica ($SiO_2$), soda ash ($Na_2CO_3$), and limestone ($CaCO_3$) are needed to produce new glass products. The product's basic building block is silica. Soda ash is occasionally substituted with muriate of potash (KCl). Recycled glass, or cullet, is frequently added. The mixture is liquefied by heating it to between 1480°C and 1570°C (2700°F and 2850°F).

$SiO_4$ is the chemical formula for silica, which is a tetrahedron with a silicon atom in the center and four oxygen atoms symmetrically attached to its corners. Vitreous silica is an amorphous substance that is created when molten silica cools rapidly and forms a randomly arranged network of tetrahedra connected at their corners (Pilkington 2003).

$$Na_2CO_3 + SiO_2 \xrightarrow{1500^\circ C} Na_2SiO_3 + CO_{2(g)} \qquad (9.1)$$

$$Na_2SiO_3 + xSiO_2 \xrightarrow[\text{digestion}]{Na_2SO_4} (Na_2O)(SiO_2)_{(x+1)} \qquad (9.2)$$

The silica can melt at a lower temperature thanks to soda ash. The silicon-oxygen network is penetrated by the sodium-oxygen atoms. Limestone gives the glass its strength; calcium penetrates the network structure and increases its complexity, making it more difficult for the atoms to arrange themselves into crystals during the cooling process.

Glass is melted and then pressed into molds to create jars and bottles. To make holes, air is injected into the containers. The new jars and bottles are strengthened by cooling in an annealing oven. We inspect the cooled product for defects like bubbles. The last stage involves packing food, nonfood items (such as medications, cosmetics, cleaning supplies, and perfume), and other consumer and business goods.

## 9.5.2. Glass Recycling

Except for aluminum drink containers, glass recycling has increased more quickly than any other commodity. From 750,000 tons in 1980 to over 3 million tons in 2010, glass recycling has increased. Increased curbside recycling program collection as well as increased demand from glass manufacturers are to blame for this growth.

Glass makes up about 4.6 of the waste streams in the United States, with an annual disposal volume of about 11.5 million tons. Ninety percent is made up of green or amber bottles, container glass, and clear glass (also known as "flint"). The remaining ten percent is made up of plates and other glass. In 2010, a total of 6.2 million glass containers were recovered for recycling purposes. According to glass generation rates from 2010, an estimated 33.4% of glass containers were recovered for recycling, and the recovery rate for all glass in MSW was 27.1% (Table 9.4). The European Union recycled roughly 70% of its total glass production, which contrasts with estimates for the United States.

Table 9.4. Glass Waste Production and Recycling, 2010

| Product Category | Generation (Thousands of Tons) | Recovery (Thousands of Tons) | Recovery (Percent of Generation) |
|---|---|---|---|
| **Durable goods**[a] | 2,170 | Neg. | Neg. |
| **Containers and packaging** | | | |
| Beer and soft drink bottles | 5670 | 2350 | 41.4 |
| Wine and liquor bottles | 1700 | 420 | 24.7 |
| Other bottles and jars | 1990 | 360 | 18.1 |
| Total glass containers | 9360 | 3130 | 33.4 |
| Total glass | 11,530 | 3130 | 27.1 |

Source: John Pichtel. Waste Management Practices: Municipal, Hazardous, and Industrial. Taylor & Francis. ISBN:9781466585188.

Most of the recovered glass is used to make new containers. A smaller portion goes toward other applications like bricks, glassphalt (a glass and asphalt mixture used as a paving mixture in highway construction), and fiberglass. The glass container industry is the primary buyer of waste glass, or cullet.

## 9.5.3. Glass Recycling Process

Gathering used bottles and other containers is the first step in the glass recycling process. Operators of the collection vehicle, the processing facility, or the consumer themselves separate the glass into three colors: clear, green, and amber (brown). The MRF or another receiving facility crushes the containers into tiny fragments (roughly 3/8 to 3/4 in. across), which are delivered to factories. This stuff is referred to as processed cullet or furnace-ready cullet. Sometimes the containers are sent straight to the manufacturer, not broken.

To make new glass products for consumer, industrial, and other uses, manufacturers buy whole and broken glass cullet and mix it with silica, soda ash, and limestone. Between 10% and 80% of cullet is used by American recyclers to make glass. This contrasts with 70–90% in Germany and Switzerland. The production of glass containers in the modern era needs consistent and clean feedstock.

The four requirements for cullet used in recycling include:

- Be separated by color
- Be contaminant-free
- Meet market specifications
- Be container glass

## 9.5.4. Color

Manufacturers can guarantee the appropriate quality and color consistency of new glass products by sorting glass based on color. It is ideal for businesses or consumers to segregate glass according to color at the source. Additionally, this is the best place to remove impurities like food, labels, and dirt. Mixed glass collection is made possible by numerous community recycling initiatives. While this may seem like a convenient practice for the consumer, it may make the product less marketable.

If color separation is not achieved at the point of generation, contaminants and colors should be sorted out as soon as possible during processing. Color sorting of cullet is almost always done by hand at an MRF or transfer station. Glass benefaction facilities, or intermediate processors, receive glass from recycling programs in some communities and sort it into three color categories using advanced optical sorting machines. Because optical sorting machinery is expensive, hand-picking is usually the only practical sorting method. Unwanted colors are difficult to remove and can ruin a whole load if the cullet is not fully color-sorted and gets thoroughly mixed. There is little market or value for mixed cullet. The amount of mixed cullet that can be used to make new containers is limited by glass manufacturers. Many businesses just would rather not purchase mixed cullet. While there are a few other markets, they are not very large.

## 9.5.5. Contamination

The biggest issue facing glass manufacturers is likely contaminated cullet, which is unsuitable for use in the production of new glass containers. Any stage of the recycling process—at the source, during collection, processing, or shipping—can result in contaminated choplet. Cullet that has been contaminated lowers quality and raises expenses. Glass manufacturers run the risk of having their production interrupted, workers hurt, manufacturing equipment damaged, and subpar product produced because of contaminants. Glass food jars, soft drink bottles, juice containers, beer bottles, wine and liquor bottles, and food jars are almost all recyclables. Glassware made for the home, such as windowpanes, drinking glasses, and lightbulbs, is not permitted to be used to make glass containers. The chemical composition of these products varies greatly, and many of them have distinct melting temperatures. Therefore, combining these products with container cullet may result in flaws in new containers, such as bubbles, cracks, or other weak spots.

Common materials that contaminate cullet include:

- Ceramic cups, plates, and pottery
- Crystal and opaque drinking glasses
- Mirrors
- Windshields and window glass
- Heat-resistant cookware (e.g., Pyrex™)
- Light bulbs
- Clay garden pots
- Laboratory glass

Other contaminants include:

- Ceramic and wire caps for beer bottles
- Metal rings from wine bottles
- Metal caps, lids, and neck rings
- Food and dirt

Glass benefaction centers collect glass from neighborhood recycling initiatives and process it through a series of stages to eliminate impurities (metal caps, stones, and ceramics). Magnetic removal of metals is used. Caps and lids contaminated with nonmagnetic metal are cleaned using eddy current separators. Air classifiers are used to eliminate lightweight materials like plastic and loose paper. Mixed cullet can have some contaminants manually removed, but this is a labor-intensive and potentially hazardous process. The result is a ground glass feedstock that is clean, homogeneous in color, and easily embraced by container makers.Preprocessors, like glass benefaction centers, offer recycling initiatives that don't produce enough material to support direct delivery to a mill, a valuable market. In the industry, however, a clean feedstock at the beginning of processing is strongly preferred.

## 9.5.6. Glass Markets

Recycled glass needs to remain competitive in price to be a commodity on par with raw materials. When compared to most post-consumer recycling materials, crushed glass has a strong and stable market value. It goes without saying that the recovered product's cleanliness and color will have a significant impact on the cullet price.

Cullet, which is sold as clear flint, brown, amber, green emerald, or mixed-color glass, can bring in up to $60 per ton when shipped to the glass factory. Glass scrap that is clean flint cullet is the most sought-after type. The least desirable grade of cullet and one that will fetch the lowest price is mixed-color glass tainted with food particles or ceramic shards (Chen, 2012).

## 9.5.7. Container Glass

Glass containers can be recycled into new ones indefinitely. Glass is a 100% recyclable material. With repeated recycling of cullet, there is no change in chemical or physical properties and consequently no deterioration in quality. Glass collectors, haulers, suppliers, and processors can lower risks and boost profits by adhering to these fundamental glass recycling guidelines, advises the Glass Packaging Institute.

- Contact potential buyers for their specifications and acceptance policies, ability to remove contaminants, transport preference (i.e., truck or rail car), and "furnace-ready" requirements.
- Conduct inspections regularly, especially before adding newly collected glass to stored recyclables and during loading for shipment.
- If stored outdoors, place the cullet on a concrete pad, not on the ground or asphalt, to avoid contamination from dirt or gravel during loading (Figure 9.7). Cover the cullet during inclement weather.
- When storing multiple loads of colored cullet, keep the cullet separated so that no intermingling of colors can occur.
- Prior to loading cullet shipments, wash the truck bed. Inspect the truck bed and the tarp used to cover the previous load for any residue.

Figure 9.7. Cullet stored outdoors and unprotected will inevitably become contaminated.

Source: John Pichtel. Waste Management Practices: Municipal, Hazardous, and Industrial. Taylor & Francis. ISBN:9781466585188.

## 9.5.8. Other Uses for Recycled Glass

Recycled glass is useful in other products and can be used as feedstock to produce new glass containers. A popular substitute market for cullet is fiberglass, which is primarily produced as glass wool for acoustical and thermal insulation. The second-highest volume of post-consumer glass is now made up of recycled glass, which is used to make fiberglass. The industry has very high standards for the consistency and quality of its products. Glassphalt, a road paving material made of asphalt and crushed, low-grade, mixed-color cullet, is another attractive alternative market. Ground glass, sand, gravel, and limestone are all ingredients in some glassphalt mixtures. However, there isn't much of a need for glassphalt given the relatively cheap cost of road-building aggregate. In addition, gravel is utilized in drainage aggregate, sandblasting materials, decorative glass, and highway reflectors and signs. Some wastewater plumbing has been manufactured using cullet. Glass can easily withstand the corrosive agents found in wastewater because it is a relatively inert material. Table 9.5 lists a few of the more inventive uses for recycled glass.

Table 9.5. Other Secondary Uses for Cullet

| | |
|---|---|
| Abrasives | Finely ground container and noncontainer glass used in sand blasting. Such abrasives contain no silica, which is the causative agent of silicosis |
| Aggregate substitute | Container and noncontainer glass utilized as drainage medium and backfill, or for landscaping purposes. |
| Bead manufacturing | Container and noncontainer glass is melted into rounded glass pellets or beads and used in reflective paint for highways. |
| Decorative applications | Ceramic tiles, picture frames, costume jewelry, and some household items may include recycled container and noncontainer glass. |
| Frictionators | Recycled glass is used to make frictionators needed for firing ammunition and lighting matches. |
| Fluxex and other additives | Glass powders used as lubricants, core additives, and fluxes in metal foundry work and fabrication, as well as flux or finders in the ceramics industry. |

Source: John Pichtel. Waste Management Practices: Municipal, Hazardous, and Industrial. Taylor & Francis. ISBN:9781466585188.

### 9.5.9. Benefits of Glass Recycling

Glass recycling—especially that of containers—is a crucial part of the glass production industry, despite the reluctance of certain industrial sectors to use scrap material in regular production due to worries about retooling and exorbitant costs. The standard mixture of silica, soda ash, and limestone has a higher melting point than recovered glass waste, which is 1370°C (2500°F). In the process of making containers, a 10% increase in cullet reduces the melting energy by approximately 2.5. Lower energy consumption in the production of glass goods relative to raw materials will keep production costs low. By using cullet, gaseous emissions are reduced.

The quality and structural integrity of newly made glass containers made from cullet is equal to those made from raw materials alone. By reducing wear on furnaces, recycled glass extends furnace life and lowers maintenance costs. Glass recycling also lessens the amount of solid waste dumped in landfills. Lastly, repurposed glass is typically found nearer bottling facilities than raw material sources. (Smook, 1982)

# CASE STUDY

## MAXIMIZING CULLET RECOVERY REDUCES BATCH COSTS

At Lead Crystal Glass Works, glass was produced by melting a charge of blended raw materials together with a process cullet. However, only about 30% of the cullet produced at the glassworks was of a size appropriate for re-melting. Concern about the significant amounts of valuable raw materials lost in this cullet and being sent for waste disposal, led to the installation of a crushing unit to reduce the cullet to an optimum size for recovery. Operation of the crushing unit was subsequently enhanced by the addition of a vibrating screen and cullet washing system.

The ideal size for cullet pieces, to produce a high quality melt of uniform composition and avoid faults in the blown glass, is 12 – 20 mm. Most of the heavy cullet at the company was present in large pieces that could not be easily broken up manually to the optimum size. Lighter pieces such as those from wine glasses, were also difficult to recycle because they must be broken up manually to obtain a satisfactory charge weight. This generates a lot of fine material, which is unsuitable because it tends to result in air bubbles being trapped in the glass gathered from the furnace pot by the glass blower. Before the waste minimization program, about 560 tons of cullet were disposed for waste disposal each year, costing the company considerably in terms of lost raw materials. The company therefore decided to install a crushing plant capable of producing a consistent output of a size suitable for re-melting and with minimum generation of fine material. Such a plant would allow more cullet to be recycled, leading to a reduction in the cost of both primary raw materials and cullet disposal.

Following discussions with suppliers of crushing plants, the company installed a rotary hammer mill. This resulted in recycling of 74% of process cullet as against 30% previously. Also, alternative uses avoiding waste disposal have been found for the crusher fines and other forms of waste glass. Crushing has also increased the bulk density of the cullet by a factor of three and improved its size distribution.

The benefits of maximizing in-house cullet recovery include:

- Cost savings
- Reduction of 37% in the purchase of primary raw materials
- Improved yield of first quality glass
- Payback period of three weeks

Associated Waste Minimization Measures

In addition to installing the cullet crusher, the company had initiated several other associated waste minimization measures such as segregation by source of in-house cullet, segregating stones from cullet, lead recovery from reject cullet, crusher fines and waste glass prior to disposal.

## CLASS ACTIVITY

A material is not truly considered "recycled" until it has proceeded through several distinct steps and is ultimately purchased by the consumer. True or false? Justify your answer. Students list and discuss the steps.

## SUMMARY

- Waste minimization is a set of processes and practices intended to reduce the amount of waste produced. By reducing or eliminating the generation of harmful and persistent wastes, waste minimization supports efforts to promote a more sustainable society. Waste minimization involves redesigning products and processes and/or changing societal patterns of consumption and production.
- Waste minimization is an important element of sustainable development since it is concerned with the conservation of resources, protection of the environment and the reduction of production costs usually by one or more of three principal means: elimination, reduction at source and recycling. Waste minimization is not concerned with treatment and disposal. Nor is it concerned with actions that only dilute waste and those that simply transfer waste from one environmental medium to another.
- Recycling refers to the process of collecting used materials which is usually considered as 'waste' and reprocessing them. In this process these used materials are sorted and processed to be used as 'raw materials' to produce new products. Recycling varies from 're-use' in the sense that while re-use just means using old products repeatedly, recycling means using the core elements of an old product as raw material to manufacture new goods. Some of the most common items that are recycled are plastic, glass, paper, batteries, aluminum, etc.
- Paper products comprise much of the municipal solid waste stream—about 28.5% of U.S. MSW, more than double that of any other component. Paper waste has its share of environmental and economic costs, as it occupies a substantial volume in collection trucks and landfills. The manufacture of paper implies that more trees must be cut to satisfy continued needs for new products. The massive quantities of paper waste generated, along with the associated costs for disposal, provide economic incentives for paper recycling.
- Traditionally, glass is well recycled. For recovery from municipal wastes or single

stream recycling, glass pieces can be hand-picked or mechanically separated where waste undergoes shredding, air classification, and other types of separation.

- Glass recycling begins with the collection of used bottles and other containers. The glass is separated into clear, green, and amber (brown) colors by the consumer, operators of the collection vehicle, or the processing facility. At the MRF or other receiving facility, containers are crushed into small pieces (approximately 3/8 to 3/4 in. across), which are shipped to manufacturing plants. This material is known as processed, or furnace-ready cullet. In some situations, the containers are not broken but sent directly to the manufacturer.

## REVIEW QUESTIONS

1. What is waste minimization?
2. What are the 3R's in waste minimization techniques?
3. Which would you prefer between recycling or source reduction? Justify.
4. List down a few housekeeping measures by which wastes can be reduced.
5. Explain how modifying a product can help minimize the waste with few examples.
6. Can employees be a factor in reducing wastes? Explain.
7. What is meant by recycling?
8. What are the benefits of paper recycling?
9. What are the four requirements for cullet used in recycling?
10. What are the primary contamination concerns with recycled glass?

## REFERENCES

1. Ackerman, F., & Mirza, S. (2001). Waste in the inner city: Asset or assault? Local Environment, 6(2), 113–120. https://doi.org/10.1080/13549830120044228.
2. Agarwal, A., Singhmar, A., Kulshrestha, M., & Mittal, A. K. (2005). Municipal solid waste recycling and associated markets in Delhi, India. Resource Conservation and Recycling, 44(1), 73–90. https://doi.org/10.1016/j.resconrec.2004.07.003.
3. Al-Salem, S. M., Lettieri, P., & Baeyens, J. (2009). Recycling and recovery routes of plastic solid waste (PSW): A review. Waste Management, 29(10), 2625–2643. https://doi.org/10.1016/j.wasman.2009.06.004.
4. Creason, J., & Podolsky, M. J. (2002). Economic impacts of municipal recycling. Review of Regional Studies, 31, 149–164.
5. Derksen, L., & Gartrell, J. (2000). The social context of recycling. American Sociological Review, 58(4), 434–442. https://doi.org/10.2307/2657365.
6. Donnelly, J. E. (2002). Numbers never lie, but what do they say? A comparative look at municipal solid waste recycling in the United States and Germany. Georgetown International Environmental Law Review, 15, 29–53.

7. Ferguson, L. (2001). Can deinked pulp dare to compete with virgin pulp? In 6th Research Forum on Recycling (p. 181). Magog, QC.
8. Gamba, R. J., & Oskamp, S. (2000). Factors influencing community residents' participation in commingled curbside recycling programs. Environment and Behavior, 26(5), 587–612. https://doi.org/10.1177/0013916500265005.
9. Kogel, J. E., Trivedi, N. C., Barke, J. M., & Krukowski, S. T. (2006). Industrial Minerals and Rocks: Commodities, Markets and Uses (7th ed.). Littleton, CO: Society for Mining, Metallurgy, and Exploration.
10. Liu, D. H. F., & Liptak, B. G. (2000). Hazardous Waste and Solid Waste. Boca Raton, FL: Lewis Publishing.
11. McMurry, J. (1992). Organic Chemistry (3rd ed.). Pacific Grove, CA: Brooks/Cole.
12. Rhyner, C. R., Schwartz, L. J., Wenger, R. B., & Kohrell, M. G. (1995). Waste Management and Resource Recovery. Boca Raton, FL: Lewis Publishing.
13. Roberts, J. C. (1996). The Chemistry of Paper. Cambridge: Royal Society of Chemistry.
14. Rubberecycle. (2001). Tire recycling. Rubber Cycle. Retrieved from: http://www.rubberecycle.com (accessed on 31 July 2024).
15. Russell, C. S., & Vaughan, W. J. (1976). Steel Production: Processes, Products, and Residuals. Baltimore, MD: Johns Hopkins University Press.
16. Salhofer, S., & Isaac, N. A. (2003). Importance of public relations in recycling strategies: Principles and case studies. Environmental Management, 30(1), 68–76. https://doi.org/10.1007/s00267-002-2818-8.
17. Schwarz, S. C., & Brunner, C. R. (1983). Energy and Resource Recovery from Waste. Park Ridge, NJ: Noyes Data Corporation.
18. Smook, G. A. (1982). *Handbook for Pulp and Paper Technologists*. Montreal: Canadian Pulp and Paper Association.
19. U.S. Environmental Protection Agency (EPA). (2001). *Municipal Solid Waste in The United States: 1999 Facts and Figures* (EPA 530-R-01-014). Washington, DC: Office of Solid Waste and Emergency Response, U.S. EPA.
20. U.S. Environmental Protection Agency (EPA). (2003). Municipal solid waste: Recycling. Retrieved from: http://www.epa.gov/epaoswer/non-hw/muncpl/recycle.htm (accessed on 31 July 2024).
21. Van Haaren, R., Themelis, N., & Goldstein, N. (2010). *The State of Garbage in America*. Retrieved from: http://www.biocycle.net/images/art/1010/bc101016_s.pdf (accessed on 31 July 2024).

CHAPTER 10

# Pharmaceuticals and Personal Care Products

## LEARNING OBJECTIVES

- Discuss emerging organic contaminants
- Define the sources and fate of pharmaceuticals and personal care products in the environment
- Describe constructed wetlands: a decentralized wastewater treatment eco-technology
- Understand the removal and behavior of pharmaceuticals and personal care products in constructed wetlands for urban wastewater treatment

## KEY TERMS FROM THIS CHAPTER

| | |
|---|---|
| Advanced oxidation processes (AOPs) | Agricultural pollution |
| Aquatic organisms | Carbamazepine (CBZ) |
| Contaminant candidate list (CCL) | Diclofenac (DCF) |
| Drinking water | Emerging organic contaminants |
| Environmental Protection Agency | Gasoline |
| Horizontal flow (HF) | Hormone ethinylestradiol |
| Millennium development goal (MDG) | Multixenobiotic resistance (MXR) |
| Oxybenzone | People equivalent (PE) |
| Persistent petroleum products | Personal care products (PCPs) |

# INTRODUCTION

People who live in dry areas are forced to rely on contaminated sources of drinking water. The risk of contracting waterborne illnesses (such as cholera, typhoid fever, etc.) can rise when using water of poor quality). Nearly 2 million child fatalities are attributed to dirty water and inadequate sanitation each year. The situation is getting worse even though the seventh Millennium Development Goal (MDG) seeks to cut in half the percentage of the population without sustainable access to basic sanitation and safe drinking water by 2015. Water scarcity is becoming a bigger issue as urbanization, population growth, and rising domestic and industrial water demands all contribute to increased water needs. Protecting water resources, ensuring sustainable development, and distributing the benefits of water use requires well-funded infrastructure and strong governance structures at both the national and local levels. Globally, initiatives are being launched to address this issue. Reuse, reclamation, and wastewater treatment are therefore essential components of that strategy. The uses of accessible renewable freshwater on Earth for municipal, industrial, and agricultural purposes account for more than one-third of the total. Thousands of synthetic and natural chemical compounds are eventually introduced into freshwater systems across the globe because of the majority of activities. Approximately two million tons of waste are produced by all industrial and human activities combined (such as synthetic compounds used in consumer and industrial products, fertilizer and pesticide-related agricultural pollution, oil and gasoline components from unintentional spills, etc.). ) are daily discharged into receiving waters after receiving little to no treatment (Figure 10.1) (UNESCO, 2003; Schwarzenbach et al., 2006).

Figure 10.1. Ratio of treated to untreated wastewater discharged into water bodies.

Source: https://upcommons.upc.edu/bitstream/handle/2117/85009/Chapter7-POPs-v1.pdf;jsessionid=1F1E493C10B51024F2F00074596B3B5D?sequence=1.

The existence of organic micropollutants in the environment has drawn a lot of attention in the past ten years due to widespread concerns about the potential negative effects that many of these pollutants may have on the environment and on living things (Cunningham et al., 2006; Kümmerer et al., 2009).

Known as "emerging organic contaminants" (EOCs), these trace pollutants are primarily man-made substances like pesticides, plasticizers, antiseptics, and pharmaceutical and personal care products (PPCPs) that are released into the environment regularly due to waste disposal, consumer activities, unintentional releases, and intentional introduction (Daughton, 2004a). The release of effluents from wastewater treatment plants (WWTPs), where the removal of EOCs is typically not complete, is one of the primary sources of EOCs into the environment (Heberer, 2002). A plethora of advanced treatment technologies, including ozonation, chemical advanced oxidation, and UV radiation, have recently surfaced to eliminate EOCs (Kim et al., 2009; Liu et al., 2009; Rosal & associates, 2010) and have been demonstrated to improve their elimination, the cost of these treatments is frequently excessive and unjustified in ligconsideringresent standards and concepts for wastewater treatment and sanitation. Additionally, they might cause the generation of transformation products that, in certain circumstances, are more hazardous or persistent than the original compound (Fatta-Kassinos et al., 2011).

In this context, decentralized, extensive low-cost treatment technologies—like constructed wetlands—become an excellent substitute for treating and reusing wastewater because they practically never need maintenance or energy. To increase the natural wastewater treatment systems' treatment capacity, constructed wetlands, or CWs, highlight the processes that occur in natural wetlands (Kadlec and Wallace, 2009). They provide a more affordable option to traditional WWTPs, especially for smaller communities with fewer than 2000 people equivalent (PE), (Puigagut et al., 2007). Apart from the removal of conventional wastewater pollutants (e.g., nutrients such as nitrogen, phosphorus, total suspended solids, and biochemical oxygen demand) and heavy metals, CWs have demonstrated a high potential for EOC removal. Studies on the occurrence and behavior of pharmaceuticals at horizontal flow (HF) wetlands at the micro and meso-scale are the majority of the literature currently in publication (Matamoros et al., 2008a; Ávila et al., 2010; Hijosa et al., 2010b; Dordio et al., 2010; Hijosa et al., 2011b; Zhang et al., 2011; Reyes-Contreras et al., 2012; Zhang et al., 2012b; Ávila et al., 2013a), as well as pilot-scale (Matamoros and Bayona, 2006; Matamoros et al., 2005). Moreover, several studies have been carried out in vertical flow (VF) systems (Matamoros et al., 2007; Matamoros et al., 2009a; Song et al., 2009) and others at full-scale tertiary treatment systems, which include the use of a free water surface (FWS) wetland (Matamoros et al., 2008b; Llorens et al., 2009; Hijosa et al., 2010a; Matamoros and Salvadó, 2012). Additional examples include the use of a CW following an anaerobic digester or a traditional WWTP (Reyes-Contreras et al., 2011; Verlicchi et al,. 2013), and the application of a three-phase hybrid constructed wetland systems that draw inspiration from various wetland types (Ávila et al., 2014 (a, b). An aerial photo of a surface-constructed wetlands system serving as a tertiary treatment unit in Spain following a traditional activated sludge unit, as assessed by Matamoros et al., (2008b) is shown in Figure 10.2.

Figure 10.2. Aerial view of the activated sludge WWTP and tertiary treatment constructed wetland system.

Source: https://www.researchgate.net/publication/270758256/figure/fig3/AS:614370379849728@1523488721511/Aerial-view-of-the-activated-sludge-WWTP-and-tertiary-treatment-constructed-wetland.png.

To shed more light on the prevalence of pharmaceuticals and personal hygiene products in the environment, their potential negative effects, and the technologies that could prevent their release into aquatic environments.

# 10.1. EMERGING ORGANIC CONTAMINANTS

Most of the time, unregulated pollutants whose emissions have become an issue for the environment are known as emerging organic contaminants, or EOCs. Thousands of chemical compounds, including pharmaceuticals, hormones, pesticides, flame-retardants, plasticizers, personal care products (PCPs), surfactants, flame-retardants, industrial additives and agents, and their transformation products, are included in this diverse collection (Farré et al., 2008). The production and consumption of these compounds have been steadily rising (Campbell, 2007), and new analytical methods have made it possible to detect them (González et al., 2007; Farré et al., 2012), and about ten years ago, the scientific community became aware of it and expressed concern (Halling-Sorensen et al., 1998). Since then, numerous writers have conducted in-depth studies on the presence, distribution, and ecotoxicological consequences of these contaminants in the environment (Kolpin et al., 2002; Cunningham et al., 2006; Barceló and Petrovi⊠, 2007; Brooks et al., 2012). Europe is a leader in this field, with numerous programs and projects underway that investigate the occurrence, fate, and ecotoxicology of these pollutants in all the various aquatic environment compartments as well as possible ways to get rid of them from wastewater.

Some examples of finalized projects are:

- POSEIDON (2001–2004): Assessment of Technologies for the Removal of Pharmaceuticals and Personal Care Products in Sewage and Drinking Water Facilities to Improve Indirect Potable Water Reuse
- P–THREE (2002–2005): Removal of Persistent Polar Pollutants through Improved Treatment of Wastewater Effluents
- EMCO (2004–2007): Reduction of Environmental Risks, Posed by Emerging Contaminants, through Advanced Treatment of Municipal and Industrial Wastes
- ERAPharm (2004–2007): Environmental Risk Assessment of Pharmaceuticals
- KNAPPE (2007–2008): Knowledge and Need Assessment on Pharmaceutical Products in Environmental Waters
- PILLS (2009–2012): Pharmaceutical Input and Elimination from Local Sources

With funding from the European Commission, the NORMAN network (Network of Reference Laboratories for Monitoring of Emerging Environmental Pollutants) was established in September 2005 to serve as an independent network of research centers, reference laboratories, and associated organizations throughout Europe for the biomonitoring and monitoring of EOCs. To meet identified needs, the goals are to improve information exchange, harmonize methodologies, and encourage coordinated, interdisciplinary projects on problem-oriented research. In a similar vein, the Environmental Protection Agency (EPA, USA) has worked hard to draw attention to the possible issues brought about by the widespread use of pharmaceuticals in the environment. In addition to urging scientists to conduct more research in this area, the EPA has also called attention to policymakers, the medical community, pharmaceutical companies, and patients (Daughton and Ternes, 1999; Daughton, 2004b). The final objective is to create mitigation strategies to address this environmental issue and reduce the risk brought on by these pollutants (Barceló, 2003).

Even though there have been significant advancements in recent years in our understanding of the subject, one of the key conclusions at this point is that our understanding of the fate and effects of emerging contaminants on the environment and living things is still incomplete, frequently addressing only a portion of the problem, leaving much to be understood (KNAPPE, 2008).

Out of the rich array of EOCs, the ones receiving the greatest attention up to date are the following: antibiotics, due to the increase in antibiotic resistance genes in aquatic microbiota and potentially human pathogens (Díaz-Cruz et al., 2003; Knapp et al., 2010; Patra et al., 2012; Rodríguez-Rojas et al., 2013); then, X-ray contrast media, which have shown high persistence in the environment; cytostatic drugs, which, being majorly used in the chemotherapy of oncological patients, are designed to prevent the growth and proliferation of cells, and thus are considered highly hazardous compounds due to their genotoxic properties (Zhang et al., 2013); and finally endocrine disrupting compounds (EDCs) such as estrogens, PCBs and organochlorine pesticides (e.g., DDT), among others, which have been found to induce disruption of the reproductive system, female and male reproductive health in humans and wildlife, sex ratio (in wild fish and mollusks), neurodevelopment in children and wildlife, as well as to cause hormone-related cancers, bone disorders, metabolic disorders, and even immune function, diseases and disorders in humans and wildlife (Bergman et al., 2013).

To increase awareness of the exposure to and risk of chemicals in ecosystems and human health, Daughton (2004b) led an excellent discussion about the universe of chemicals that occur in the environment, both known (regulated or not) and unknown. Due to limitations in the range of analysis tools available, a significant portion of the chemicals present in water samples under analysis are either not identified at all or cannot be identified. Only a very small portion of commercially used contaminants are being studied, even though every day, and for decades now, an array of new drugs with mechanisms of action never encountered by biological systems, are released into the environment. Undoubtedly, the advancements in industrial and technological processes related to chemical production present significant obstacles to the development and structuring of regulatory frameworks.

It is challenging for authorities to control the disposal of EOCs and set concentration limits due to the lack of knowledge regarding their occurrence, fate, and effects. Certain chemical groups, like pesticides and other priority pollutants, are regulated (EC, 1976). Additionally, to prioritize a more thorough review of these 553 endocrine-disrupting compounds (EDCs), the European Commission is working on creating a candidate list of them. Among the many new pesticides and other industrial chemicals included in this list are the industrial chemicals bisphenol A (monomer of hard plastics and epoxy resins) and the estrogenic steroid hormone 17⊠-ethinylestradiol (EC, 2000b). This list may be updated in response to advancements in science or modifications in the ways that chemicals are used. Conversely, the United States Environmental Protection Agency (EPA) maintains a list of chemicals known as the Contaminant Candidate List (CCL). This list of contaminants is not covered by any drinking water regulations, but it is being tracked in case it is added to in the future. The final one has some flame-retardants and multiple EDCs (CCL3, 2008).

However, there are no laws, acts, or regulations of any kind that specify the permissible limits for the concentrations of EOCs in the environment, drinking water facilities, or WWTPs. Unknown effects on the environment and public health result from the unrestricted and widespread release of EOCs into the environment due to a lack of regulations. To achieve more fruitful results, gain a better understanding of the issue, and take action to reduce EOCs' presence in the environment, efforts should be coordinated and integrated amongst the various actors involved in their lifecycle, as well as researchers and public authorities (Schwarzenbach et al., 2006; Daughton, 2009; KNAPPE, 2008).

### 10.1.1. Pharmaceuticals and Personal Care Products

Because of how widely used they are in society, a lot of EOCs are constantly released into the environment. Pharmaceuticals and personal care products (PPCPs) are a subclass of this class that includes an incredibly broad range of chemicals used in veterinary medicine, agriculture, human health, and cosmetics. Prescription and over-the-counter drugs, dietary supplements, diagnostic tools, and other consumer goods like sunscreen, fragrances, disinfectants, and cosmetics are all considered part of the PPCPs. Pseudo-persistent pollutants are those that are found in trace concentrations, are polar, and have a short half-life in water (Barceló and Petrovi⊠, 2007). This is because they are widely used by both humans and animals, which results in a constant release of pollutants into the environment that frequently originates from inadequate or untreated wastewater treatment. PPCPs have large, complex molecular structures that vary greatly between substances and contain functional groups like amine, ketone, carboxyl, and hydroxyl. It is crucial to understand that the way pharmaceuticals are categorized within subgroups based on their active ingredients does not mean that these substances behave in a particular way chemically. Small alterations in the chemical structure may have a big impact on other characteristics like polarity and solubility, which will then affect how they are distributed in the environment in the air, water, sediments, soils, and animals (Kümmerer, 2009). One of the physicochemical properties of organic compounds most applied in the field of environmental chemistry is the n-octanol water partition coefficient ($K_{ow}$), which is defined as the ratio of the compound's concentration in a known volume of n-octanol to its concentration in a known volume of water at equilibrium and at a constant temperature. $K_{ow}$ values are constant for a given compound and reflect the lipophilicity of a compound. Hence, the higher the $K_{ow}$ of a chemical, the higher its lipophilicity. $K_{ow}$ values are generally directly proportional to molecular weight.

$K_{ow}$ values can be measured in the laboratory with shake-flask systems or can be estimated by models. $K_{ow}$ values have been measured for a wide variety of chemicals and range from about 0.001 to over 10,000,000. Since the range of $K_{ow}$ values encompasses ten orders of magnitude, log $K_{ow}$ values are used instead, and range -3 to 7. A substance is usually considered lipophilic at log $K_{ow}$ values ranging from 3-4 to 6.5 ($K_{ow}$>100) (Connell, 2005).

Nonetheless, $K_{ow}$ assays are carried out by modifying the aqueous phase's pH to unionize the compound's predominant form. Thus, while some compounds with log $K_{ow}$ values between 3 and 4 may be predicted to be relatively lipophilic, the presence of carboxylic groups or other polar groups in their structures increases their hydrophilicity by increasing the likelihood that these groups will ionize. Log $K_{ow}$ may not always be a reliable indicator of a compound's hydrophobicity because many medications can be ionizable at ambient pH levels. Specifically, at ambient pH, most non-steroidal anti-inflammatory drugs (NSAIDs) ionize and exhibit highly polar properties. Therefore, the hydrophobicity of weak acids is also dependent on their $pK_a$ and the pH of the media in which they are present (Eq. 1).

The concentration of ionized molecules divided by the concentration of unionized molecules yields the ionization constant ($K_a$), which is a measure of (weak) acid strength. In this sense, stronger acids have higher $K_a$ values and lower $pK_a$ values.

$$\log [A^-]/[AH] = pH - pKa \quad (1)$$

Hence, for a given $pK_a$ of a compound, the higher the pH, the higher the ionization of the compound.

In this sense, a more accurate descriptor for ionizable compounds is the octanol-water distribution coefficient ($D_{ow}$), which is the ratio of the sum of all forms of the compound (ionized + unionized) in each of the two phases. In this instance, the compound is added without substantially changing the pH of the aqueous phase because the pH has been buffered to a particular value. Since the log $D_{ow}$ depends on pH, the pH at which the parameter was measured must be specified. However, for many pharmaceuticals like NSAIDs (e.g., ibuprofen, diclofenac), the molecule occurs primarily in its ionized form and thus holds a hydrophilic behavior at environmental pHs (log $D_{ow}$ = 1.1 for both substances at pH =8). For unionizable compounds, log $K_{ow}$ = log $D_{ow}$ at any pH.

Fragrances found in personal care products (PCPs) are typically organic compounds that are semi-volatile and hydrophobic. These substances have been used for many centuries for a variety of purposes, including perfumes, deodorants, washing and cleaning agents, and cosmetics. According to Cadby et al., (2002) the typical range of fragrance concentrations is 0.3–5% in face creams and up to 20% in perfume extracts. Nowadays, the majority of fragrances are manufactured by the fragrance industry, with nitro and polycyclic musks being the two primary types. These latter are the main types of musks used today, making up nearly two-thirds of global production, with an estimated volume of 6000 t year-1 produced globally (Rimkus, 1999). One of the most popular polycyclic musks in the world, tonalide (AHTN) is found in detergents and cosmetics. Its high octanol-water coefficient (log Kow = 5.9) indicates that it is poorly soluble in water due to its highly lipophilic design, which allows it to sorb to organic materials like skin. Furthermore, an organic UV filter found in sunscreen and other cosmetics is called oxybenzone (OXY). Triclosan (TCS), an antibacterial and antifungal agent found in soaps, deodorants, toothpastes, shaving creams, and other consumer goods, is another PCP that is frequently discovered (Medline Plus, 2010). Unlike medications, PCPs can be released into recreational waters or volatilize into the air in large quantities, which is how they differ from pharmaceuticals. (e.g., musks).

### 10.1.2. Transformation Products

Metabolites are the substances that arise from the structural modification of pharmaceuticals in the human body. Like metabolism, a substance's introduction into the environment or a treatment plant can cause it to undergo a range of biotic (like biodegradation and biotransformation) and non-biotic (like phototransformation and hydrolysis) structural changes that can result in partial or total transformation of the original compound. Their physicochemical characteristics alter as a result, which has an impact on their toxicity and behavior. While it is reasonable to anticipate that parent compound transformation will lessen toxicity, there are situations in which transformation products may be more persistent and/or hazardous than the parent material. This has been observed during advanced oxidation processes (AOPs) and both technical and natural photolytic processes (Kümmerer, 2009; Fatta-Kassinos et al., 2011; Benítez et al., 2013).

Little is known about the degradation pathways and products of most PPCPs in the environment, and less is known about their intermediates of degradation, which just started to be explored (Miao and Metcalfe, 2003; Farré et al., 2008). Gómez et al., (2012) just recently found transformation products of various PPCPs in the Henares River basin (central Spain) at concentrations much higher than those of their parent compounds. Various transformation products of ibuprofen and other pharmaceuticals were identified by Quintana et al., (2005) in municipal wastewater treated by a membrane bioreactor.

These results emphasize how crucial it is to incorporate degradation products into research and monitoring initiatives to gain a deeper comprehension of the behavior and possible toxicology of PPCPs in the environment. It is crucial to remember, though, that it would be impossible, if not infinite, to consider every chemical that is released into the environment daily for research projects or routine monitoring, given the vast array of chemicals and the cost and difficulty of analyzing them. Rather, to find a compromise, well-balanced efforts based on estimated risks associated with each contaminant must be made.

# 10.2. SOURCES AND FATE OF PHARMACEUTICALS AND PERSONAL CARE PRODUCTS IN THE ENVIRONMENT

A multitude of sources, the components of which are shown in Figure 10.3, allow contaminants to enter the environment.

PPCP consumption is high globally. As a case study of human consumption, consider the rise in non-steroidal anti-inflammatory drug (NSAID) consumption in Spain between 1992 and 2006 (defined daily doses/1000 inhabitants day) from 26.30 DHD to 45.81 DHD (AEMPS, 2006). When it came to NSAIDs, ibuprofen was the one with the biggest increase in consumption, going from 0.29 DHD in 1992 to 21.30 in 2006. On the other hand, the NSAID DCF also referred to as Voltaren, demonstrated relatively consistent consumption over time, averaging 7 DHD. 942 tons of the anticonvulsant carbamazepine and 877 tons of DCF were sold in 2007 in 76 major countries, according to data from IMS Health. These sales are estimated to account for 96% of the global pharmaceutical market. Spain's antibiotic consumption did not vary significantly over the past ten years, according to the AEMPS report (2009). Most of the quinolones under study had values between 0.1 DHD, with the penicillin amoxicilin having the highest value at 6.6 DHD. According to the OECD report from 2013, many countries worldwide have experienced a notable rise in pharmaceutical consumption over the past ten years. Considering this, Daughton and Ruhoy (2013) have emphasized the significance of giving patients the "lowest effective dose" of medications as a tactic to reduce PPCP concentrations in the environment and ultimately prevent any potentially harmful environmental effects. Pharmaceuticals are excreted and eventually make their way to wastewater treatment facilities through the urban sewer system after being consumed and partially metabolized. Because PPCPs are frequently not entirely removed by the current treatment methods used to treat municipal wastewater, they are subsequently released into terrestrial and aquatic ecosystems through disposal and reuse practices.

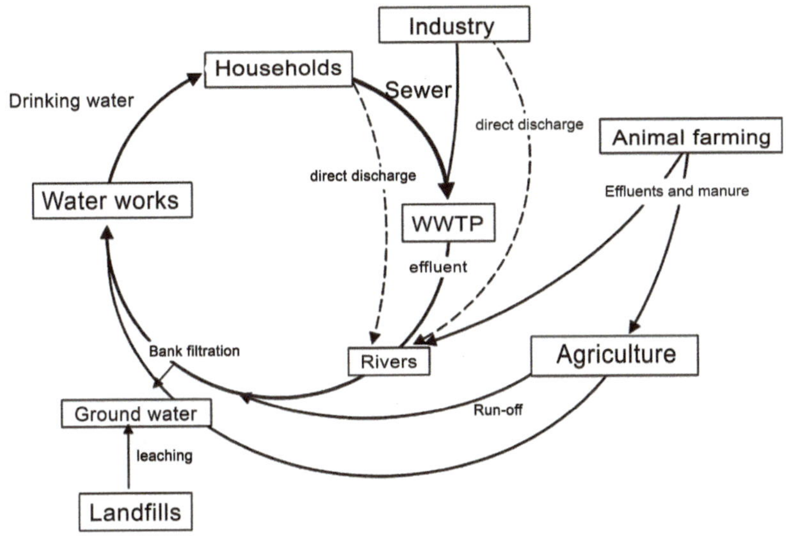


Figure 10.3. Sources and fate of contaminants in the (partially) closed water cycle with indirect potable reuse.

Source: https://upcommons.upc.edu/bitstream/handle/2117/85009/Chapter7-POPs-v1.pdf;jsessionid=1F1E493C10B51024F2F00074596B3B5D?sequence=1.

On the other hand, terrestrial run-off from the agrifood (and particularly meat) industry is the second major source of PPCPs in the environment, following sewage effluent discharge. To give you an idea of the levels of antibiotic consumption on livestock farms, consider that although the United States limits the amount of antibiotics used for human use to less than 8 billion pounds annually, the use of antibiotics on livestock farms has been rising and currently accounts for nearly four-fifths of all antibiotics used in the country (FDA, 2011). Municipal landfill leaching is a major source of concern for PPCPs in groundwater. Another way that these pollutants enter the environment is through the improper disposal of used or expired medications, which can happen in a landfill or down the toilet. PPCPs can either remain retained in the soil and eventually accumulate, impacting the ecosystem and humans through the food chain, or they can reach groundwater and contaminate aquifers, depending on the physicochemical characteristics of the organic contaminant (and their transformation products) and the characteristics of the soil. PPCPs have been found in freshwater environments for decades in trace amounts (HallingSorensen et al., 1998; Kolpin et al., 2002; Loos et al., 2009, 2010; Jurado et al., 2012). As a particular example, Kim et al., (2007) documented the frequent detection of many pharmaceuticals, hormones, antibiotics, as well as flame retardants in surface waters in South Korea. Various endocrine disrupting compounds (EDCs), as well as associated toxicological effects over caged male fathead minnows, were found in 90% of sediment and water samples of 11 studied lakes across Minnesota, USA (Writer et al., 2010). Stackelberg et al., (2004) found pharmaceutical compounds and other organic wastewater contaminants in a conventional drinking-water-treatment plant before and after treatment. Pharmaceuticals were also found at the ng $L^{-1}$ level in the tap water of Berlin (Heberer, 2002). The authors attribute this phenomenon to being characteristic of conurbations like Berlin, with high municipal wastewater discharges and low surface water flows, when groundwater recharge is used in drinking water production.

The range of PPCP concentrations in groundwater (from filtration through soil) and surface waters (from incomplete elimination in WWTPs) is µg $L^{-1}$ to ng $L^{-1}$. In-depth explanations of the various pathways that PPCPs can take from resource up to drinking water level can be found in the study by Halling-Sørensen et al., 1998. PPCPs are mainly dispersed through food chain dispersal and aqueous transport, as their low volatility and polar nature generally prevent them from escaping the aquatic realm (Daughton and Ternes, 1999). Even though the majority of PPCPs have a comparatively short half-life in aquatic environments (Yamamoto et al., 2009), some materials can build up in the soil and become resistant, with concentrations as high as g $kg^{-1}$. For instance, clofibric acid, a substance used to lower blood cholesterol levels, has been discovered to have a half-life of 21 years in soil (Hernando et al., 2006). Applying sewage sludge in this regard may eventually lead to the contamination of soil, surface, and groundwater, either through the gradual build-up of these contaminants in the soil or through potential leaching and migration of contaminants through the soil zone (Chen et al., 2011; Stuart et al., (2012)). The interaction between groundwater and surface water caused by infiltration from rivers' declining reaches, particularly in the vicinity of industrial facilities and sewage treatment plants, is another significant source of groundwater contamination (Stuart et al., 2012). The safety of the drinking water supply may be seriously jeopardized by groundwater contamination. According to a recent study by the GAO (2011), most drinking water samples collected in the USA contained pharmaceutical contaminants, primarily antibiotics and estrogen-based hormones.

To evaluate the possible risks resulting, irrigation must undoubtedly be considered as an additional exposure route for PPCPs in terrestrial ecosystems. Due to the significant water scarcity, water reclamation irrigation is being widely used in Europe for a variety of uses, primarily agricultural, because of the benefits associated with nutrient recovery and the socioeconomic implications. Reusing wastewater without adequate management can result in pathogens and chemical pollutants, which could be harmful to both human health and the environment (Muñoz et al., 2009; Fatta-Kasinos et al.,2011).

## 10.2.1. Toxicity and Evaluation of Risks Associated to Pharmaceuticals and Personal Care Products

Because PPCPs are so common in the environment, scientists are very concerned about the potential toxicological effects these substances may have on ecosystems and public health. Many studies have been conducted in recent years to assess the potential toxicological effects of PPCPs and other newly discovered organic contaminants, primarily on aquatic organisms (Fent et al., 2006) as well as risk management (Enick and Moore, 2007) with to assist stakeholders in making decisions and ultimately formulating policy. Even though naturally occurring PPCP concentrations in the environment are frequently extremely low (ranging from ng $L^{-1}$ to µg $L^{-1}$), and they may not appear to represent a significant risk to human health, these pollutants are continuously released into the environment, making them what are known as "pseudo-persistent" contaminants. While the potential effects of their presence in the environment are often poorly understood in other cases the effects appear clear and concerning. For instance, the use of the anti-inflammatory drug diclofenac on cattle in these countries has resulted in the reduction of 90% of the vulture population in India, Pakistan, and Nepal since the 1990s; as a result, the use of this drug for veterinary purposes was outlawed in India in 2005 (Oaks et al., 2004; Balmford 2013).

Furthermore, it has been discovered that musk scents are pervasive, persistent, bioaccumulative, and occasionally extremely toxic pollutants in aquatic organisms (Daughton and Ternes, 1999; Luckenbach et al., 2004, Zhang et al., 2013). Products of amino musk transformation have important toxicological implications. Fragrances with polycyclic musks have been observed to accumulate in a variety of aquatic species, including fish and mussels, as well as in aquatic organisms at high trophic levels and human breast milk (Kannan et al., 2005; Yin et al., 2012). It is crucial to remember that drugs are made to be biologically active to fulfill their intended therapeutic roles. This is achieved by altering the immune, endocrine, and cellular signaling pathways. As a result, biological systems and ecosystems physiological and biochemical processes may be impacted. Furthermore, some of them are estrogen hormones, which fall under the category of EDCs and exhibit toxicity at ng $L^{-1}$ levels (ethynilestradiol, for example) (Larsson et al., 1999; Ingerslev et al., 2003). But because there is a great deal of diversity among PPCPs, their effects on ecosystems could differ greatly (Halling-Sorensen et al., 1998). PPCPs are disposed of widely, over an extended period, and at lower doses than pesticides, although many of them are EDCs of concern and have thus far received less attention than the latter. Aquatic organisms are therefore exposed for a longer period. More information is available for antibiotics than for any other therapeutic class, even though little is known about the presence and effects of pharmaceuticals in the environment. This is a result of their widespread application in aquaculture, plant agriculture, animal husbandry, and human therapy. It is now obvious that the widespread and excessive use of antibiotics has led to a rise in the prevalence of highly resistant pathogenic bacteria. The effects on aquatic microbiota and the resulting changes to the microbial community's structure, along with the development of resistance to possible human pathogens, are of serious concern. Even though most bacteria found in freshwater sources are not harmful, there may be genetic material exchanges between harmful and non-pathogenic bacteria that give the latter enhanced resistance to antibiotics (Witte, 1998; Barbosa and Levy, 2000; Cabello, 2006; Schlüter et al., 2007; Kelly et al., 2009; Zhang et al., 2009; Middleton and Salierno, 2013).

Research efforts that document contaminant uptake, modes of action, biological endpoints, and exposure are necessary to shed light on the possible toxicological significance of PPCPs in ecosystems and ultimately in public health. Much of the published research is centered on risk and hazard assessments as well as acute effects on aquatic organisms. To extrapolate chronic (long-term) effects, assessment factors are applied to the data on acute effects. Given that the ratio of acute to chronic toxicity varies depending on the PPCP, this approach is frequently unrealistic (Cunningham et al., 2006; Fent in collaboration. 2006). PPCPs have low acute toxicity; however, little is known about their chronic toxicity and possible subtle effects.

As an illustration, Cleuvers (2003) discovered that the anticonvulsant carbamazepine (CBZ) and the non-steroidal anti-inflammatory drugs (NSAIDs) ibuprofen (IB) and diclofenac (DCF) were toxic to algae.

They also discussed how, despite the extremely unlikely occurrence of acute effects of individual substances in aquatic environments due to their small concentrations, combinations of several compounds and/or their transformation products can have an impact.

Brausch and Rand (2011) assessed the acute and long-term toxicity information that was available for personal hygiene products, such as tonalite (AHTN), a musk fragrance, and antiseptic triclosan (TCS). TCS and Triclocarban were the only two that posed a risk. Additionally, Breitholtz et al., (2012) evaluated the ecotoxicity of different micropollutants on macroalgae and crustaceans in a Swedish free-water surface wetland and found that the effluent water had good quality, similar to that treated by sophisticated tertiary treatment methods.

Conversely, Fent et al., (2006) found that a few substances such as DCF had the highest acute toxicity within the class of NSAIDs, and average concentrations found at the effluent of conventional WWTPs were in the range of those being toxic for fish.

What is more, Lai et al., (2002), using a food-web model found that natural and synthetic estrogens, such as the hormone ethynilestradiol (EE2), could be bioaccumulated in fish in river systems, although to a lesser extent than other EDCs such as DDT (Lai et al., 2002). Likewise, AHTN and other fragrances, which are made hydrophobic to adsorb to tissue might accumulate in aquatic biota (Rimkus, 1999). Although acute toxicity in the aquatic systems due to the occurrence of PPCPs is unlikely to occur, there is a lack of chronic toxicity data and effects from long-term, low-level environmental exposures should be further addressed, to have a better understanding of the environmental risk of these compounds. What is more, since pharmaceuticals are not present alone in the environment, but as multi-component mixtures, an accurate prediction of the chronic mixture toxicity is indispensable for an environmental risk assessment. Significant synergistic toxic effects may occur through additive exposures; therefore, they should not be handled in isolation.

The effects and interaction of a cocktail of commonly used pharmaceuticals, including CBZ, IB and the antibiotic sulfamethoxazole were studied using in vitro tests on human and zebrafish cells by Pomati et al., (2008), and they found that the mixture of drugs at ng $L^{-1}$ can inhibit cell proliferation on aquatic life. However, the research in this new field of ecotoxicology is just starting and much remains to be learned (Kümmerer, 2009; Stuart et al., 2012).

Inhibition of Mult xenobiotic resistance (MXR) in aquatic organisms is another recent concern; it has been specifically observed in environmental samples from polluted areas. This system serves as the "first line of defence" against potentially harmful endogenous and exogenous xenobiotics, which are removed from the cell by different transmembrane transport proteins.

The question of whether the substances referred to as chemosensitizers or MXR inhibitors (Smital and Kurelec, 1998) are synthetic or natural remains unanswered. Furthermore, it is unclear if the inhibition is brought on by many chemicals present in contaminated water or if it is made up of a small number of potent MXR inhibitors. The possibility that these compounds could be important in enhancing the effects of other xenobiotics is also discussed (Daughton and Ternes, 1999). Initiated research on this subject includes the investigation of musk fragrances' suppression of MXR (Luckenbach et al., 2004) and there is still several issues on this subject that need to be resolved (Smital et al., 2004,). The lack of obvious ecological effects caused by PPCPs does not negate their existence. These drugs may have gradual, subtle effects over time (Daughton and Ternes, 1999).

Traditional standard ecotoxicological assays are preferred over target- or biomolecule-oriented, or mode-of-action-based investigation activities to gain a better understanding of the mechanisms of action of these contaminants in the environment. Evaluating the effects of persistent petroleum products (PPCPs) on aquatic systems is still very difficult. It calls for better analytical and modeling tools, the creation of new and existing chemical classification schemes based on how likely they are to affect ecosystems and people, and the development of attenuation technologies and other mitigation measures to reduce the amount of these chemicals that are released into the environment (Schwarzenbach et al., 2006).

### 10.2.2. Occurrence and Fate of Pharmaceuticals and Personal Care Products in Conventional Wastewater Treatment Plants

Numerous traditional wastewater treatment technologies, like activated sludge systems, have been thoroughly examined for the presence and fate of particulate matter pollutants (PPCPs). These systems are primarily responsible for the pollution of the environment because they are frequently incapable of breaking down PPCPs. The most popular secondary treatment method utilized in municipal wastewater treatment facilities across the globe is activated sludge. It is made up of a two-stage biological treatment process called suspended growth that is intended to mostly remove organic matter. A mixed microbial population in the first stage of the process eliminates organic matter in an aerated reactor; in the second stage, solids (activated sludge) are separated from water using a settling tank or clarifier. A portion of the activated sludge is recycled back into the reactor, and the remainder is wasted.

The average solids retention time for conventional activated sludge is 5 to 10 days, whereas the hydraulic retention time (HRT) of the water is approximately 8 hours. An intensive wastewater treatment technology is the conventional WWTP. Despite taking up comparatively little surface area can treat large volumes of water (Tchobanoglous and Burton, 1991).

Since non-steroidal anti-inflammatory drugs (NSAIDs) are a common therapeutic class these days, acetaminophen and ibuprofen are the ones that are found in raw wastewater at higher concentrations.

It is important to remember that even at low concentrations, estrogens can have detrimental long-term effects like endocrine disruption (Ingerslev et al., 2003). Numerous studies have also documented the presence of antibiotics in WWTPs (Yang and Carlson, 2004). Removal rates from activated sludge WWTPs are highly variable and compound-dependent because they are not specifically built to remove PPCPs.

An analysis of current research on wastewater treatment technologies and their capacity to eliminate 16 newly discovered organic contaminants, many of which are PPCPs, is presented in a recent report from the EPA. Based on activated sludge, average removal efficiencies were computed from a sample of up to 41 full-scale WWTPs; Table 10.1 shows the removal rates for the different PPCPs. It should be noted that activated sludge systems that reported design modifications from the traditional activated sludge WWTPs (such as those that aim to remove nutrients) were not included in the treatment systems used to obtain the data presented in the table. Even though there are numerous variations of this process, it was not practical to further divide the categories of activated sludge (EPA, 2010).

Table 10.1. Minimum, Maximum and Average Removal Efficiencies of Various PPCPs at Full-Scale Activated Sludge WWTPs Treating Municipal Wastewater

| Compound | Removal efficiency (%) | | | |
|---|---|---|---|---|
| | Min | Max | Avg | # of WWTPs used to calculate removal |
| Ibuprofen | 43 | 100 | 90 | 32 |
| Acetaminophen | >90 | 100 | 97 | 4 |
| Diclofenac | 7 | >99 | 44 | 23 |
| Tonalide | 13 | 97 | 67 | 20 |
| Oxybenzone | >8 | >96 | 76 | 6 |
| Triclosan | >67 | 100 | 89 | 22 |
| Ethynilestradiol | >1 | >99 | 66 | 13 |

Source: https://upcommons.upc.edu/bitstream/handle/2117/85009/Chapter7-POPs-v1.pdf;jsessionid=1F1E493C10B51024F2F00074596B3B5D?sequence=1.

It was discovered that the traditional WWTPs under investigation differed in terms of treatment performance, primarily due to the features of the wastewater that was being fed into them. The average removal efficiencies varied greatly and were dependent on the compound. PPCPs can be eliminated from wastewater through the processes of biodegradation and/or adsorption to the particulate matter in activated sludge treatment. A few substances degrade quickly (e.g., in traditional WWTPs (acetaminophen) (Yu et al., 2006; Miège et al., 2009; Kosma et al., 2010; Rosal et al., 2010). Likewise, ibuprofen and triclosan are consistently removed in most treatment plants (Clara et al., 2005; Lishman et al., 2006). However, some compounds exhibit variable removal, (e.g., diclofenac), while others, such as carbamazepine, have shown to be very recalcitrant (Kosma et al., 2010).

Desorption and/or deconjugation processes are thought to be responsible for higher concentrations that have occasionally been discovered in the effluent than in the influent (Kasprzyk-Hordern et al., 2009; Zorita et al., (2009).

The less soluble materials could adhere to activated sludge or suspended solids. Carballa et al., in this regard. (2005) discovered that in coagulation-flocculation tests involving various doses and coagulant types at various temperatures, compounds with high sorption properties—like tonalide and diclofenac—were significantly eliminated, whereas hydrophilic compounds. This did not impact ibuprofen or carbamazepine. In contrast, Jeli⊠ et al., (2011) found that sludge samples from three conventional WWTPs did not accumulate diclofenac. WWTP-treated wastewater can be recycled for a variety of uses or dumped in surface waterways. The waste sludge could also be used as fertilizer on the land after it has been dried and treated. Soil and surface waters may absorb any leftover PPCP concentrations.

### 10.2.3. Removal of Pharmaceuticals and Personal Care Products through Tertiary and Advanced Treatment Technologies

As PPCPs and other specific and priority substances listed in the Water Framework Directive (EC, 2000a) have proven to be difficult for conventional activated-sludge WWTPs to completely remove, several advanced (or tertiary) treatment technologies that can remove these substances even further are currently being researched and developed.

Table 10.2. Min, Max and Average Removal Efficiencies of Various Pharmaceuticals and Personal Care Products Under Different Advanced Treatment Technologies for Municipal Wastewater Treatment at Full-Scale

| Compound | Technology | Min | Max | Avg (%) | # systems used to calculate removal |
|---|---|---|---|---|---|
| Ibuprofen | Chlorine disinfection | >43 | 100 | 78 | 5 |
| | Ozonation | >90 | 100 | 95 | 2 |
| | Reverse Osmosis | - | - | 72 | 1 |
| | UV disinfection | >81 | 100 | 90 | 6 |
| Acetaminophen | Chlorine disinfection | >90 | >99 | 95 | 2 |
| | Reverse Osmosis | - | - | 90 | 1 |
| | UV disinfection | - | - | 90 | 1 |
| Diclofenac | Chlorine disinfection | >18 | >90 | 66 | 3 |
| | Reverse Osmosis | - | - | 90 | 1 |
| | UV disinfection | >86 | >91 | 89 | 3 |
| Tonalide | Chlorine disinfection | 64 | 93 | 79 | 2 |
| | UV disinfection | - | - | 52 | 1 |
| Oxybenzone | Chlorine disinfection | >8 | >95 | 51 | 2 |
| | Reverse osmosis | - | - | 95 | 1 |
| | UV disinfection | >89 | >96 | 92 | 3 |
| Triclosan | Chlorine disinfection | >67 | >99 | 83 | 4 |
| | Ozonation | 99 | 100 | 99 | 2 |
| | Reverse Osmosis | - | - | 67 | 1 |
| | UV disinfection | >71 | >99 | 90 | 5 |
| Ethynilestradiol | Chlorine disinfection | 1 | 72 | 42 | 4 |

Source: https://upcommons.upc.edu/bitstream/handle/2117/85009/Chapter7-POPs-v1.pdf;jsessionid=1F1E493C10B51024F2F00074596B3B5D?sequence=1.

There is already a wide range of frequently highly standardized unit processes available to remove PPCPs through chemical and physical transformation or removal techniques like adsorption or filtration. Activated carbon is one of these (Ternes et al., 2002; Snyder et al, , 2007), ozonation according to Rose et al., 2010), image-Fenton (Rodríguez-Gil and others). 2010), ultraviolet light (Kim et al., 2009), low-frequency ultrasound (Naddeo et al., 2009), chemical AOPs (hydroxyl radicals), which typically employ a blend of oxidizing agents (e.g., $H_2O_2$, $O_3$, $TiO_2$, etc.). ), UV or ultrasonic radiation, and catalysts (Klavarioti et al., 2009; Liu et al., 2009), oxidation of chlorine dioxide (Huber et al., membrane technologies (Yoon et al., 2005; Snyder et al,. as well as membrane bioreactors (Quintana et al, 2005; Shariati et al, 2010). For instance, diclofenac, an anti-inflammatory medication, has continuously demonstrated good photo degradation when exposed to UV light (Andreozzi et al., 2003; Zhang et al., 2008; Kim et al., 2009). Under ozonation, this material also becomes well-mineralized (Rosal et al., 2010) and the disinfection of chlorine (Kosma et al., 2010). Table 10.2 shows the average removal efficiencies of different PPCPs under various tertiary treatment technologies (EPA, 2010).

Even if the technology is improved, it is highly unlikely that conventional tertiary or advanced treatment technologies—which have been demonstrated to improve the elimination of these contaminants—will be implemented in the context of wastewater treatment for small communities due to energy and financial constraints. Considering this, decentralized, extensive, low-cost treatment technologies—like constructed wetlands—emerge as a fantastic substitute for wastewater treatment and reuse, needing very little upkeep and energy. Their ability to degrade a wide range of contaminants, from conventional parameters to heavy metals or pathogens, has been demonstrated by their successful full-scale implementation. Their potential to remove PPCPs has only recently come to light. The properties, benefits, drawbacks, and effectiveness of artificial wetlands in the treatment of urban wastewater.

# 10.3. CONSTRUCTED WETLANDS: A DECENTRALIZED WASTEWATER TREATMENT ECO-TECHNOLOGY

The natural wastewater treatment systems known as constructed wetlands (CWs) are made up of a shallow basin that has been thoughtfully designed and planted with aquatic vegetation on its substrate. Other elements, like bacteria and aquatic invertebrates, grow organically. To treat wastewater, these systems are designed to replicate the physical, chemical, biological, and microbiological processes that take place in a natural wetland but in a more controlled setting (Kadlec and Wallace, 2009).

These natural treatment systems constitute a cost-effective and technically feasible approach for the treatment of wastewater for several reasons:

- Can be less expensive to build than other treatment options.
- Low operational and maintenance (O&M) expenses (energy and supplies). Easy to maintain (do not require highly qualified staff).
- Able to tolerate fluctuations in flow and load.
- Promote biodiversity. As an example, restored wetlands are becoming common in agricultural landscapes of northern Europe to increase nutrient retention as well as increasing species diversity (Thiere et al., 2009; Matamoros et al., 2012a).
- Can be easily integrated into the natural and rural landscape, and provide aesthetic, commercial and habitat value (Llorens et al., 2009).
- Possible commercial revenue from flower harvesting and/or aquaculture (Jana, 1998).

- However, they have some limitations:
- Generally high land area requirement: they may be more economical than other technologies, but only if land is available and affordable. While activated sludge technology requires approximately 0.06 m2 /PE (Cooper, 2005), vertical subsurface, horizontal subsurface and free water surface CWs require 2, 5 and 8 m2 /PE, respectively to attain the same contaminant removal efficiencies (Cooper, 2005; Vymazal, 2005; Kadlec, 2009).
- Poor maintenance or design can lead to excessive clogging, which can affect hydraulics and treatment performance (Knowles et al., 2011).

Following the adoption of the Water Framework Directive 2000/60/EC (EC, 2000a) and the Directive 91/271/CEE on urban wastewater treatment (EC, 1991), several European nations have had to adhere to strict regulations, particularly when it comes to the treatment of small communities. Due to this, there is a growing need for affordable, aesthetically pleasing, and environmentally sustainable treatment systems, which has focused attention on decentralized wastewater treatment eco-technologies.

In this way, CWs and other biological treatment technologies—like waste stabilization ponds—have minimal to no energy requirements and are simple to use. They therefore provide a good substitute for treating wastewater in small communities. Constructed wetlands have been widely utilized on a large scale for the treatment of various types of wastewater such as storm water, landfill leachate, tannery wastewater, and other industrial wastewaters, all over the world since the 1980s (Schulz et al., 2001; Behrends et al., 2007; Nivala et al., 2007; Vymazal, 2009).

They have been widely implemented in countries like France (Molle et al., 2005), UK (Weedon, 2003; Cooper, 2005), Denmark (Brix and Arias, 2005) or Czech Republic (Vymazal, 2002), and their implementation is increasing in countries like Spain (Vera et al., 2011) or Italy (Masi et al., 2007). Numerous studies have shown their capability to maintain hydraulic, technical, economic, environmental and ecological benefits (Vymazal, 2002; Dixon et al., 2003; Zhou et al., 2009). There are several types of constructed wetlands. They can mainly be divided into subsurface flow (horizontal or vertical), surface flow constructed wetlands, and hybrid systems, which are a combination of different types.

### 10.3.1. Horizontal Subsurface Flow Constructed Wetlands

This kind of wetland is made up of gravel beds covered in wetland vegetation. The wastewater is meant to percolate through the roots and rhizomes of the plants, staying below the surface of the substrate. These wetlands are always flooded, and water depth usually remains between 0.3 and 0.9 m (between 0.05 and 0.1 m beneath the surface of the bed). The recommended organic loading rate (OLR) is around 6 g of biochemical oxygen demand (BOD5) $m^{-2}$ $d^{-1}$ (García and Corzo, 2008). One of the main mechanisms for eliminating pollutants from wastewater is the biofilm that forms on the granular media as well as the roots and rhizomes of the plants. The components of horizontal subsurface flow (HF) constructed wetlands usually consist of the following: filter media, emergent vegetation, an impermeable liner, inlet piping, and outlet piping. Figure 10.4 shows a diagram of a traditional HF wetland.

They work well for denitrification, total suspended solids (TSS), BOD5, and bacteria removal. However, due to their low oxygen transfer capacity (OTC) (they perform poorly in nitrification (Cooper, 1999; Tyroller et al., 2010). The risk of encountering pathogenic organisms is reduced because the water is not exposed, and they also don't offer mosquitoes a suitable place to live (Kadlec and Wallace, 2009).

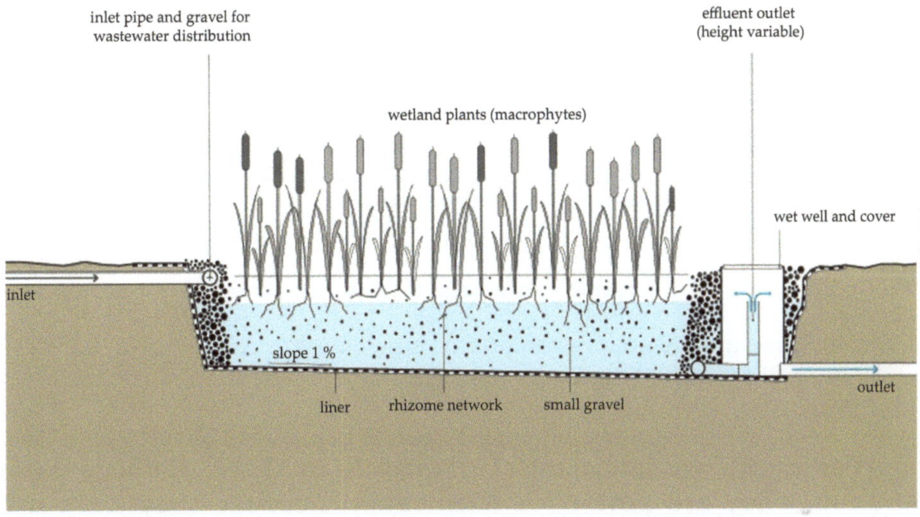


Figure 10.4. Horizontal subsurface flow constructed wetland schematic.

Source: https://sswm.info/sites/default/files/inline-images/TILLEY%20et%20al.%202014.%20Schematic%20of%20the%20Horizontal%20Subsurface%20Flow%20Constructed%20Wetland_0.png.

## 10.3.2. Vertical Subsurface Flow Constructed Wetlands

This kind of wetland collects wastewater from the bottom of the media, where water is freely draining, and distributes water over the surface of a sand or gravel bed covered in emergent vegetation. As the water percolates through the plant root zone and filter media, it is treated. There are various types of wetlands with vertical subsurface flow (VF). In the most popular configuration, the bed is made up of multiple layers with varying grain sizes and has a system of perforated pipes for draining. They typically receive sporadic feedings and frequent loading and resting intervals. In many French systems, two or three beds can function in this way alternately. The wastewater is fed into French VF systems without first undergoing primary treatment (Molle et al., 2008), resulting in the build-up of a solid layer on the top of the bed that serves as a filter. During resting phases, the solid deposits are more likely to mineralize when the cycles of feed and rest alternate (Molle et al., 2005). When compared to HF beds, a higher organic loading rate can be applied in VF wetlands due to their unique operating conditions, which include intermittent dosing and an unsaturated bed. VF wetlands are typically composed of a single bed. Applied OLR values range from 20 g BOD5 $m^{-2}$ $d^{-1}$ in Denmark (Brix and Arias, 2005) to 180 g BOD5 $m^{-2}$ $d^{-1}$ in French VF systems (simply screened wastewater), with significant regional variations. In Figure 10.5, the elements of a normal VF wetland are displayed.

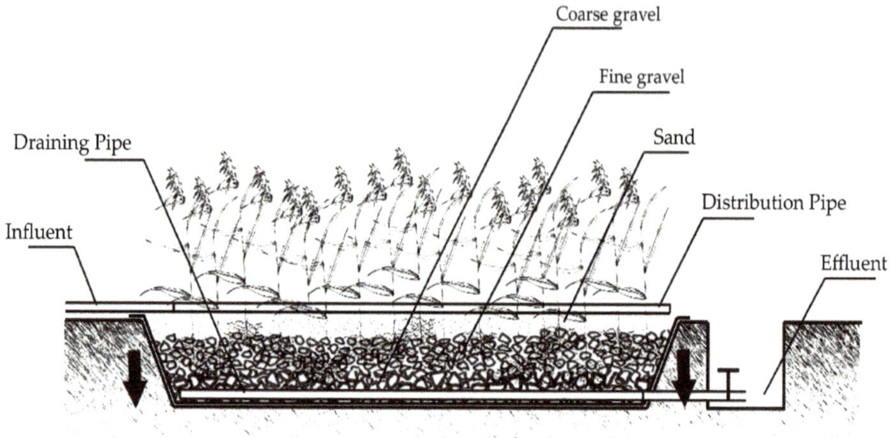


Figure 10.5. Basic elements of a vertical subsurface flow constructed wetland.

Source: https://www.researchgate.net/profile/Cristina-Avila-4/publication/270758256/figure/fig5/AS:614370379837440@1523488721792/Basic-elements-of-a-vertical-subsurface-flow-constructed-wetland-adapted-from.png.

The properties of HF wetlands differ greatly from those of these systems due to their configuration. They are better for nitrification because they have a significantly higher OTC than HF beds. The diluted oxygen found in wastewater, convection during sporadic loading, and diffusion processes taking place in between doses is how oxygen transfer is accomplished (Torrens et al., 2009). In addition to effectively removing BOD5 and COD, high OTC can also effectively remove certain bacteria (Headley et al., 2013) and are likewise significantly smaller than HF systems (roughly 5$m^2$ /PE) (Cooper, 2005; Vymazal, 2005). However, they perform poorly in terms of removing suspended solids, and for single VF systems in which there are no permitted rest periods, clogging of the media may happen quickly if the hydraulic loading rates and granular material selection are incorrect (Platzer and Mauch, 1997; Cooper, 1999; Kayser and Kunst, 2005).

### 10.3.3. Free Water Surface Wetlands

Wastewater flows above the top of the media in these wetlands, which can have a variety of impermeable layers underneath them, such as clay, soil, gravel, etc., as it is not meant to filter the wastewater through it. Open water areas, floating vegetation, and emerging plants are all present in these wetlands, either intentionally or as an inevitable result of the way the design was created. In Figure 10.6, a typical FWS wetland schematic is shown.

These kinds of wetlands frequently draw a wide range of wildlife, including insects, molluscs, fish, amphibians, reptiles, birds, and mammals, because the water is exposed during the treatment process (Kadlec and Knight, 1996). Furthermore, they are frequently used for the tertiary treatment of secondary or tertiary treatment effluents (such as activated sludge systems, waste stabilization ponds, SSF wetlands, etc.) because they pose a risk to the public's health due to human exposure to pathogens). They are particularly suitable for attenuating agricultural runoff at the tertiary treatment level or restoring deteriorated areas (Thiere et al., 2009; Matamoros et al., 2012a; Matamoros and Salvadó, 2012).

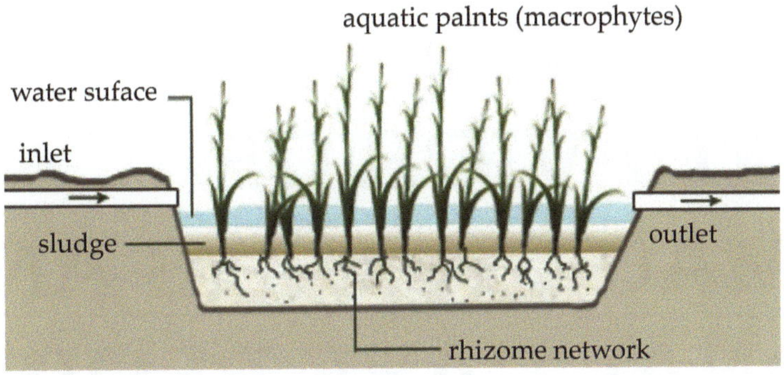


Figure 10.6. Free water surface constructed wetland schematic.

Source: https://www.researchgate.net/profile/Orodi-Odhiambo/publication/286417697/figure/fig4/AS:669459060641792@1536622886033/Free-Water-Surface-Constructed-Wetland.png.

## 10.3.4. Hybrid Systems

To achieve higher treatment efficiency, particularly for nitrogen and pathogens, different types of constructed wetlands can be combined. VF and HF systems combined in a variety of ways typically make up these hybrid systems. Due to a lack of oxygen, nitrification cannot occur in HF wetlands; however, VF wetlands can offer favorable conditions for nitrification, but denitrification does not take place in these systems. From this point on, the benefits and drawbacks of each kind of system cancel each other out, making it feasible to produce an effluent with low concentrations of total nitrogen (TN) and BOD (Vymazal, 2005). According to Brix and Arias (2005), various combinations can be used, such as HF followed by VF wetlands, VF followed by HF wetlands, and other filter stages that involve water recirculation between stages.

# 10.4. REMOVAL AND BEHAVIOR OF PHARMACEUTICALS AND PERSONAL CARE PRODUCTS IN CONSTRUCTED WETLANDS FOR URBAN WASTEWATER TREATMENT

When multiple removal mechanisms, such as biodegradation, adsorption, photodegradation, plant uptake, or hydrolysis, occur simultaneously, pharmaceuticals and personal care products (PPCPs) are eliminated from constructed wetlands (CWs). Certain wetland configurations will favor certain removal mechanisms over others (photo-degradation, for example, occurs only in free water surface (FWS) wetlands). It has been demonstrated that the three types of CWs (HF, VF, and FWS) can easily be freed of the anti-inflammatory medication ibuprofen (IB). Results by Matamoros et al., (2005) and Matamoros and Bayona (2006) show high removal efficiencies in HF wetlands (60 – 80%), observing that the elimination rates are higher in shallower (0.27 m) than in deeper beds (0.5 m). Matamoros et al., (2008a) found the prevalence of OH-IB over Ca-IB metabolite in an HF wetland, which would be explained by its predominant aerobic biodegradation. A comparison between different seasons by Hijosa-Valsero et al., (2011b) in HF beds shows a seasonal variability, with lower removal efficiencies in winter (90%). Moreover, two injection experiments carried out at that same experimental plant in Barcelona, Spain, showed how its removal highly depended on the existence of a high redox status, which would promote aerobic biodegradation of this drug (Ávila et al., 2010, 2013a). The removal of IB is also high in FWS wetlands, as documented by Llorens et al., (2009) and Matamoros et al., (2008b), where efficiencies of over 90% are detected. High elimination rates were also reported for VF beds, with efficiencies of around 99% for unsaturated vertical wetlands (Matamoros et al., 2007).

Conversely, although diclofenac (DCF) has been reported to be recalcitrant in activated sludge wastewater treatment plants (WWTPs) (Heberer, 2002), this substance shows a variable removal in CWs, ranging from values as low as 21% in HF beds (Matamoros et al., 2009a; Matamoros and Bayona, 2006) up to values of 99% at experimental scale HF wetland under an injection experiment (Ávila et al., 2010) or a value of 73% in a VF wetland in Denmark (Matamoros et al., 2007). In FWS wetlands average removal efficiencies have been reported to be as high as 96% and 73-96% in studies carried out by Llorens et al., (2009) and Matamoros et al., (2008b), respectively. This could be attributed to its large removal through photodegradation, as reported by Andreozzi et al., (2003). This substance is hydrophilic and was not detected in the particulate matter associated with the gravel matrix of CWs, as reported by Matamoros and Bayona (2006). Variability in the elimination of diclofenac was also observed by Hijosa-Valsero et al., (2011b) in various experimental meso-scale HF wetlands, who showed that its degradation was favored by high temperatures (summer VS winter) and found variable removal efficiencies in Barcelona (around 70%) and in León (<12%) wetlands. Although reductive dehalogenation has been claimed as a possible important removal mechanism for this compound (Ávila et al., 2010), other studies show how its removal seems to be enhanced by more oxidizing conditions (Ávila et al., 2013a).

In a full-scale FWS wetland (750 $m^2$) with a depth of 0.5 m and a hydraulic retention time of 84 h, Gray and Sedlak (2005) found a 41 removal of the estrogen ethynilestradiol in CWs after spiking it in concentrations slightly above background concentrations. They also reported sorption as a major mechanism for its removal in the aqueous phase. Song et al, (2009) assessed the impact of saturated versus unsaturated conditions (and consequently more oxidized conditions) were superior for the removal of this type of pollutant. This was discovered when investigating the removal of estrogens in VF units.

Furthermore, several authors, including Matamoros et al., have previously investigated the fate of the musk fragrance tonalide in artificial wetlands. In a full-scale FWS wetland, (2008b) reported elimination rates of 88–90%. These rates did not vary with the season, most likely because of their low biodegradability (i.e., major removal mechanism through interaction with organic matter and sediment). Matamoros and Bayona (2006) found this fragrance and galaxolide to be the most abundant on the gravel bed and the suspended particulate matter in another study, this time in two pilot HF wetlands at Les Franqueses del Vallès (Barcelona). They attribute the removal of these musks to hydrophobic interactions with organic matter and the biofilm. Furthermore, it was discovered that this compound had no dependence on water depth. Conversely, Hijosa-Valsero et al., (2011b) suggested that this compound might be temperature-dependent because summer removal efficiencies were much higher than winter ones. This is likely because winter releases hydrophobic compounds (like tonalide and galaxolide), which are released when plant and biofilm activity is lower. Removal efficiencies in VF wetlands were also reported by Matamoros et al., (2007) to be moderately removed with elimination rates ranging from 75 – 82%. The behavior of the sunscreen agent oxybenzone in CWs has been reported by Matamoros et al., (2007) in a VF wetland, where it was very efficiently removed (>95%), and by Matamoros et al., (2009a), where the removal was also >95% in HF wetlands and 90% in VF wetlands. The authors suggest that it could be partially removed by sorption onto the particulate matter, although it is majorly removed by biodegradation (Ávila et al., 2013a, 2014b).

Regarding the elimination of triclosan, Park et al., (2009) reported removal efficiencies in FWS wetlands in South Korea of above 60%. Matamoros et al., (2012a) found triclosan to be the substance achieving the highest removal efficiency (>40%) among all studied compounds in a restored wetland in Denmark.

## 10.4.1. Design and Operational Factors Affecting the Removal of Pharmaceuticals and Personal Care Products in Constructed Wetlands

Microbial and physico-chemical interactions play a complex role in the removal of contaminants from CWs. The depth of the bed, the kind and size of the media, the rates of hydraulic and organic loading, the feeding plan, and artificial/external aeration are just a few of the design and operational variables that affect these processes' rates. The black box concept is frequently used in the design of CWs, and if wetlands are built without taking these factors into account, treatment efficiency may suffer. Much of the research that is currently available on the impact of operational and design parameters on treatment performance concentrates on traditional water quality parameters. It has only been studied in the last ten years how these parameters affect the removal of personal care and pharmaceutical products (PPCPs). Those include very few studies evaluating the influence of water depth (Matamoros et al., 2005; Matamoros and Bayona, 2006); primary treatment (Ávila et al., 2013a), operation under cycles of saturation/unsaturation (Ávila et al., 2013a); type of organic matter (Matamoros et al., 2008a); type of granular media (Dordio et al., 2009, 2010) and other different design parameters and modes of operation (Hijosa-Valsero et al., 2010b, 2011b; Zhang et al., 2012a,b; Ávila et al., 2014b). Therefore, the optimization of the performance of CWs in terms of PPCP removal stands as a necessity that can be achieved through the identification of the optimal design and operational factors of these treatment systems.

### *10.4.1.1. Behavior of PPCPs in Horizontal Subsurface Flow Constructed Wetlands*

Seasonal effects are seen in the removal of PCP from HF wetlands; lower removal efficiencies are seen in the colder months at lower water temperatures. Seasonal patterns have been observed at various CW systems at mesoscale in Barcelona, NE Spain (Hijosa et al., 2011b; Ávila et al., 2013a) and in León, NW Spain (Hijosa et al., 2010b; Reyes-Contreras et al., 2012), being removal efficiencies consistently better in summer than in winter for almost all the studied PPCPs. It is well known that temperature has a significant impact on microbial activity. Warm temperatures (15–25°C) are generally when microorganisms in wetlands are most active, especially nitrifying and proteolytic bacteria (Truu et al., 2009). It has been discovered that denitrifying bacteria follow the same pattern (Spieles and Mitsch, 2000; Boulêtreau et al., 2012). It has been discovered that water temperature affects the removal of PPCPs, and this is particularly true for biodegradable compounds (like ibuprofen and diclofenac). Additionally, summertime vegetation may help eliminate PPCPs through plant assimilation and, to a greater extent, through the rhizosphere's transfer of oxygen (Brix, 1993; Tyroller et al., 2010). This should allow the creation of microenvironments with different redox conditions, which in turn would promote the development of microbial biofilms with functionally different respiration processes (Wiessner et al., 2005; Kadlec and Wallace, 2009; Ávila et al., 2013a).

Furthermore, it is well known that microbial activity plays a major role in the removal of most pollutants, including PPCPs, from CWs (García et al., 2010). PPCP removal efficiency in CWs is frequently higher than in conventional wastewater treatment plants (WWTPs). This is probably because microbial diversity in CWs is higher than in WWTPs (Kadlec and Wallace, 2009). By favoring a broad range of redox conditions, CWs can be engineered to enhance multiple biological processes and remove multiple contaminants from a single CW bed. The optimization of CW technology can be achieved by identifying the factors that impact its performance.

For example, running HF wetlands in batch (or fill and drain) cycles appears to be a critical component in encouraging the removal of most contaminants. The transition between the saturation and unsaturation phases encourages the presence of a higher redox status when operating in a saturated state, which in turn greatly improves the removal of PPCPs (Zhang et al., 2012b; Ávila and others. 2013a). Materials whose primary elimination process is believed to be aerobic biodegradation (e.g., ibuprofen) exhibit the biggest variations (Ávila et al., 2013a). By encouraging the fluctuation of redox conditions where robust aerobic facultative biofilms can operate, operating under cycles of saturation and unsaturation generally appears to be a good strategy to enhance the treatment performance of HF CWs (Stein et al., 2003; Faulwetter et al., 2009).

Other design factors for HF wetlands are water depth, applied hydraulic loading rate, grain size and type of media. Water depth seems to be a determining factor in this type of wetlands, which affects the redox conditions, oxygen supply and thus the treatment performance of the wetlands. Shallower wetlands appear to have higher redox potential and thus promote more variate and energetically favorable reactions, which in turn enhance the degradation of PPCPs (Matamoros et al., 2005). Other authors have shown experimental evidence, which supports this hypothesis (García et al., 2003a, 2005; Headley and Davison, 2005; Huang et al., 2005; Matamoros and Bayona, 2006). Regarding the granular media of HF beds, in general, a wide range of materials and sizes have been applied around the world, and its selection is in many cases dictated by the availability, price and local practices of a certain region. Some studies indicate the importance of using hard, durable and homogeneous materials that do not contain fine grains, which might clog the media (García and Corzo, 2008; Pedescoll et al., 2009). In general, finer materials ($D_{60}$ = 3.5 vs. 10 mm) have been recommended in HF beds when low organic loads are applied since they provide a greater surface area for microbial biofilms. Finer gravel promotes a higher growth of the vegetation and in conjunction a higher removal of pollutants such as ammonia (García et al., 2005). However, it is important to bear in mind that the finer the material, the greater the risk of clogging and hydraulic problems. Moreover, although usually granite gravel is used in HF wetlands, other alternative sorbents have emerged to increase the adsorption capacity of these wetlands, such as light expanded clay aggregates (Dordio et al., 2009, 2010). This is expected to be especially important for hydrophobic compounds such as fragrances. The influence of granular size in HF wetland PPCPs removal has not to date been assessed, though it should majorly show effects on hydrophobic substances that get attached to particulate matter.

Lastly, it is commonly known that in HF wetlands, the oxygen required by wastewater degradation of pollutants is greater than the oxygen present in the wetland bed. For traditional HF wetlands, subsurface oxygen limitation is generally one of the primary rate-limiting factors (Brix and Schierup, 1990; Kadlec and Wallace, 2009; Tyroller et al., 2010). As a result, different variations—like the use of active aeration—are being developed and used globally and have been demonstrated to improve the effectiveness of treatment for this kind of CW (Wallace, 2001; Nivala et al., 2013). However, this approach has high operating and maintenance costs, which are only justified if the wetland's increased performance more than makes up for the energy and equipment costs (Kadlec and Wallace, 2009; Nivala et al., 2012). As far as we are aware, no research has been done to evaluate the removal of PPCPs from actively aerated HF wetlands.

Seasonal variations in the environment are important considerations for both the design of HFCW and its operation and maintenance (O and M), particularly in locations with high seasonality. Adopting strategies that support a higher redox status of the system has generally been found to improve the treatment performance of HF CWs on PPCP removal. By encouraging a higher redox status of the wetland bed, operating HF wetlands in cycles of saturation and unsaturation stands as a very relevant strategy that greatly enhances PPCP removal. Another important factor affecting performance is water depth; shallower wetlands, measuring between 0.5 cm, are more effective at removing PCP than deeper HF CWs.

#### *10.4.1.2. Behavior of PPCPs in Vertical Subsurface Flow Constructed Wetlands*

It has been found that vertical subsurface flow (VF) wetlands remove PPCPs more effectively than horizontal subsurface flow (HF) wetlands. This is consistent with findings from other laboratory and experimental studies that indicate aerobic pathways are generally more effective than anaerobic ones at degrading these substances (Zwiener and Frimmel, 2003; Matamoros et al., 2007; Conkle et al., 2012; Ávila et al., 2013a). However, depending on the organic contaminant being treated, the configuration of the wetland, and certain design and operational parameters, the relative significance of a given degradation process in VF CWs can change. For example, diffusion and convection processes in unsaturated intermittently loaded VF wetlands (typically) are dependent on the loading strategy and are influenced by the number of pulses, the pulse volume, and the surface filter dewatering duration (Molle et al., 2006; Torrens et al., 2009). It is also very important to choose the filter media (i.e., grain size, grain material, and layer depth) in a way that will meet treatment requirements and prevent bed clogging while preserving media oxygen renewal (Cooper, 2003; Kayser and Kunst, 2005).

Because of mass convection and diffusion, a larger batch dosed at a lower feeding frequency should promote oxygen transfer into the media with respect to loading frequency. However, because there will be less time for the water and biomass to come into contact, an excessive volume may result in less removal efficiency. However, loading too frequently may limit the amount of oxygen that can regenerate between pulses, which could have an impact on removal processes like nitrification. In this regard, long-term variations might be more pronounced (Molle et al., 2006; Torrens et al, 2009). More research in this area is required. Loading frequency is generally regarded as a critical operational parameter for ensuring this kind of treatment system operates well, wherein adequate oxygen renewal and organic matter mineralization should be permitted. Particularly in non-rested VF wetlands, applied hydraulic loading rates (HLRs) and possible solid deposit accumulation on the bed surface should be closely monitored and controlled to prevent surface ponding and filter clogging, which could ultimately shorten the treatment system's lifespan and efficiency (Platzer and Mauch, 1997; Langergraber et al., 2003).

Furthermore, grain size is yet another crucial consideration. A recent study assessed the impact of this factor on the removal of PCP and found that sand-based (1-3 mm) VF wetlands performed better than gravel-based (4-8 mm) VF wetlands (Ávila et al., 2014c). The findings imply that decreased treatment efficiency in gravel-based VF wetlands would be caused by reduced contact as well as a decreased capacity to retain solids because of larger pores between grains. Indeed, it has been noted that a swift downward flow is not conducive to ammonia sorption onto the gravel bed, which could ultimately lead to a reduction in nitrification activity (Molle et al., 2006; Torrens et al, 2009). Because of their larger grain sizes, gravel-based VF beds have less surface area per unit volume than sand-based VF beds. This lower surface area may result in insufficient contact time between the biofilm and pollutants, as well as reduced oxygen diffusion and solids entrapment. However, the larger surface area for biofilm attachment and increased oxygen availability in sand-based VFs might encourage the growth of more diverse microbial communities and larger biomass. Indeed, it has been discovered that the upper portion of a VF wetland—roughly the first 10 cm of a 50 cm VF—has higher microbial biomass and activities due to the presence of a higher redox status (Tietz et al., 2007,). Although higher elimination rates have been found at sand-based systems (Ávila et al., 2014c), their long-term operation may make them more vulnerable to clogging. In practice, grain size, as well as, loading frequency should be carefully selected, as well as HLRs controlled (Matamoros et al., 2007), to ensure desired removal of PPCPs, as well as enough oxygen renewal and mineralization of organic matter so that no clogging develops in the filter beds. Nevertheless, in French-style VF wetlands, where various beds alternate cycles of feed and rest, it has been observed that very high hydraulic overloads can be applied without observing a decrease in the infiltration capacity or treatment performance of the system (aprox 0.4-1.8 m $d^{-1}$). Nevertheless, case scenarios, the grain and type of media used should be adapted to available local materials, and the use of gravel could constitute an acceptable solution.

## 10.4.2. Behavior of Pharmaceuticals and Personal Care Products in Hybrid Constructed Wetland Systems

A hybrid constructed wetland system consisting of three stages of VF, HF, and FWS wetland (in this study followed by an open-air water tank indicated for water reuse) was designed, constructed, and monitored at experimental and full-scale for the removal of conventional water quality parameters and PPCPs. The goal of this system is to evaluate the treatment capacity and potential performance enhancement when various CWs of different types are combined (Ávila et al., 2013b, 2014a, b).

The VF, HF, and FWS wetlands have surface areas of 317 $m^2$, 219 $m^2$, and 240 $m^2$, respectively. The results obtained in a full-scale hybrid system with the cited characteristics showed to perform remarkably well in the removal of PPCPs, achieving very high overall removal efficiencies for most of the studied compounds (above 80 removals for all compounds) (Ávila et al., 2014a). Figure 10.7 shows the total removal efficiencies for each treatment unit.

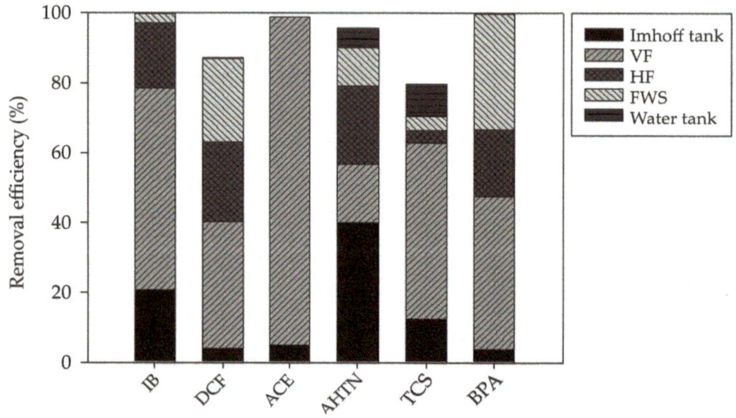


Figure 10.7. Accumulated removal efficiencies of various pharmaceuticals and personal care products at the different units of the pilot hybrid treatment system.

Source: https://upcommons.upc.edu/bitstream/handle/2117/85009/Chapter7-POPs-v1.pdf;jsessionid=1F1E493C10B51024F2F00074596B3B5D?sequence=1.

The target PPCPs final effluent concentrations were incredibly low, falling below the acetaminophen detection limit. The rest were in the ng $L^{-1}$ order (20-100), which is in the range of those found in the environment, such as those reported by Matamoros et al., (2009b) in small ponds or lagoons. These concentrations are also in the range of those obtained in advanced treatment technologies applied at full-scale, such as ozonation or membrane filtration (Snyder et al., 2007; Rosal et al., 2010).

Because different CW configurations would allow for the combination and synergy of different physicochemical and biological removal mechanisms (e.g., biodegradation, sorption, volatilization, hydrolysis, and photodegradation) to occur, the high removal efficiencies can be explained by the differing existing physico-chemical conditions (Faulwetter et al., 2009; Imfeld et al., 2009). In this way, anaerobic biodegradation and sorption would be more common in HF beds whereas aerobic metabolic pathways and solids retention are improved in VF wetlands. Finally, the FWS wetland would oversee any possible photodegradation of chemicals, as well as, less crucially, adsorption onto organic matter and plant material uptake (Matamoros and Salvadó, 2012).

Because of the hydrolysis and energetically favorable aerobic microbial reactions occurring in this type of wetland and made possible by the VF's design and operation strategy, which produces high dissolved oxygen concentrations and effluent redox potentials, it may have performed better in treatment than the other treatment units. Nevertheless, the combined nitrification-denitrification activity seen in the VF bed of the referenced (Ávila et al., 2014a) and an earlier investigation at the same treatment facility (Ávila et al., 2013b) indicate that the presence of both anaerobic and aerobic microenvironments in this wetland bed would permit the occurrence of both processes. This could be partially explained by the presence of a saturated layer beneath the draining pipes at the bottom of the wetland bed. Similarly, this research suggests that while sorption onto organic matter and aerobic biodegradation may be the primary removal mechanisms causing PPCP reduction in VF wetlands, anaerobic metabolism-based alternative processes may also be taking place at anoxic microsites or micropores within the wetland bed (i.e., lower layers) (Cooper et al., 1996; Ávila et al., 2010), which aid in its eradication.

While substances like the musk fragrance tonalide show a great reduction within the HF presumably due to sorption processes, results for triclosan would exhibit little sorption capacity. The behavior of triclosan within the three different wetland configurations indicates aerobic biodegradation as the major removal mechanism involved in the elimination of this compound (Singer et al., 2002; Ying et al., 2007). Although triclosan has been detected in plants and sediments of an HF CW and its concentration generally decreased from inflow to outflow (Zárate et al., 2012), sorption and plant uptake do not appear to constitute a principal mechanism of triclosan removal in constructed wetlands. The similarly high removal efficiencies achieved for diclofenac in the current study in the HF bed if compared to the VF wetland suggest that various alternative mechanisms may determine the elimination of this compound, and to that respect, anaerobic biodegradation through reductive dehalogenation could constitute a predominant degradation pathway of diclofenac when anaerobic conditions prevail (Park et al., 2009; Ávila et al., 2010; Hijosa-Valsero et al., 2010a).

Diclofenac was especially well-absorbed by the FWS wetland. According to Buser et al., these findings validate photodegradation as the main removal mechanism causing diclofenac attenuation in water bodies. 1998; Andreozzi et al, 2003; Matamoros & Salvadó, 2012, 2013), in addition to less common mechanisms (i.e., plant uptake, aerobic and anaerobic biodegradation). Even though there was very little triclosan removed, there appeared to be some additional reduction in the water reuse tank, suggesting that photooxidation processes might have played a tiny role in the removal of the chemical (Mezcua et al., 2004; Ávila and others., 2014b; Salvadó and Matamoros, 2012, 2013). Further reduction was achieved at the water reuse tank, suggesting photodegradation through sunlight exposure as one of the principal mechanisms of the fragrance tonalide removal within this type of wetland configuration, even though the reduction of tonalide concentrations in the FWS could be attributed to sorption onto particulate matter and sediment. At other FWS wetlands functioning as a tertiary treatment step, comparable high removal efficiencies were achieved (Matamoros et al., 2008b; Llorens et al., 2009; Salvadó and Matamoros, 2012).

For wastewater treatment and reuse in small communities, hybrid CW systems—which combine various CW configurations such as a VF CW, an HF wetland, and an FWS wetland—have proven to be an extremely reliable eco-technology (Ávila et al., 2014a, b). Combining various wetland configurations has been demonstrated to optimize several key treatment processes, including conventional water quality parameters (>90), high disinfection capacity, and excellent overall removal efficiency of PPCPs (above 80 for all substances studied) (Ávila et al., 2013b). This has been made feasible by the occurrence of complementary abiotic/biotic removal pathways at wetlands with various configurations that occur under various physico-chemical conditions.

# CASE STUDY

## INNOVATIVE WASTEWATER TREATMENT: SOIL BIOTECHNOLOGY

One of the largest integrated chemical companies in India, which is headquartered in Vaisad, Gujarat. With 140 brands and over 900 products, they serve clients in 83 countries across 30 industries, reflecting a strong commitment to excellence in all operations.

They were looking for efficient wastewater management for one of their plants in Gujrat. The company wanted to ensure compliance with the pollution board standard.

### Problem

Untreated industrial wastewater is a huge problem in India. According to the UN, only 60% of industrial wastewater is treated in the country. This water can cause several problems, such as contaminating the environment and processing streams, and exposure to the wastewater can cause health problems.

Industrial wastewater plants face the following challenges:

- *Footprint and Facilities:* Newer regulations and stricter compliance may require additional space and investment. Additionally, older technologies require more space and are more labor-intensive.
- *Cost of Treatment Chemical:* Plants often require chemicals, and supply chain and price issues can cause problems. Operators have to be creative and look for alternatives.
- *Sludge:* Using inorganic coagulants, while cheaper, can lead to higher sludge volumes and disposal costs due to their inability to dewater effectively.
- *Energy Consumption:* Operating equipment, especially equipment like aeration blowers (to apply oxygen) and pumps for filtration, consumes a significant amount of energy.
- *Operator Skill:* Recruiting skilled operators who can manage sophisticated automated wastewater treatment systems can be challenging.

### Types of Wastewater Treatment

Wastewater treatment plays a vital role in protecting our environment and public health. This process involves various methods, each with its own strengths and applications. Let's delve into some of the most common technologies utilized:

### Soil Biotechnology

The Indian Institute of Technology (IITB) developed Soil Biotechnology or SBT. This innovative technology was developed after two decades of in-depth research and offers a unique and environmentally friendly way to treat wastewater.

SBT harnesses the fundamental reactions in nature—respiration, photosynthesis, and mineral weathering in a specially designed environment. This controlled geological media" houses carefully selected micro- and macro-organisms that work together to purify the wastewater. You can think of SBT as a self-sustaining biological engine that supplies oxygen.

It is designed to address the purification of contaminated municipal and industrial wastewater. The treated water quality surpasses the statutory requirements of CPCB and state PCBs.

### Activated Sludge Process (ASP)

ASP or Actively sludge process (ASP) treats oxidise organic waste present in wastewater. It works by injecting air or oxygen into the raw sewage through an aeration system. This stimulates the growth of bacteria that consume organic matter, releasing $CO_2$ and water, leaving behind a brown floc (clump) of settled matter.

### Sequencing Batch Reactors (SBRs)

Sequencing Batch Reactors (SBRs) are an activated sludge process used for wastewater treatment, focusing on treating wastewater in batches. They involve a sequence of steps in a tank reactor, promoting microbial removal, settling suspended solids, and drawing clear effluent. SBRs are flexible and can operate continuously by having multiple reactors, making them suitable for areas with large flow variations. They are a popular aerobic treatment technology used to treat municipal wastewater and wastewater from various industries, including refineries and petrochemical plants. They can enhance the removal of nitrogen, phosphorus, ammonia, total suspended solids (TSS), and biochemical oxygen demand (BOD).

### Membrane Bioreactor

Membrane bioreactors (MBR) combine biological treatment with microfiltration/ ultrafiltration membranes, offering superior wastewater treatment. There are two configurations:

- *Submerged:* Membranes reside within the biological reactor, directly immersed in water.

- *Side Stream:* Membranes act as a separate stage, treating wastewater after the biological process.

MBRs use the membrane to separate solids from salt, creating reusable effluents. MBR manufacturers commonly use polymeric membranes, especially PVDF, because of the durability and resilience of the material.

### Moving Bed Biofilm Treatment

Moving Bed Biofilm Treatment (MBBT) actively treats wastewater, serving municipal and industrial needs economically and efficiently. It operates as an activated aeration system, utilizing recycled plastic carriers to gather sludge. These carriers boast extensive internal surfaces, ensuring optimal interaction with water, air, and bacteria. MBBT excels in removing organic substances, facilitating nitrification, and promoting denitrification.

### Solution

The client opted to go into soil biotechnology to treat the wastewater at their plant. The plant could effectively meet the pollution norms.

### How Does it Work?

The SBT system is based on the principle of trickling filtration using a natural ecosystem within a controlled environment. It combines physical processes like sedimentation and infiltration with biochemical processes to remove suspended solids, and organic and inorganic components from the wastewater.

The SBT system comprises several key components:

- *Raw Water Tank:* Holds the untreated wastewater before it enters the treatment process.
- *Bioreactor Containment:* This is the heart of the system, housing the media layer and facilitating the biological treatment.
- *Treated Water Tank:* Stores the cleaned water after it exits the bioreactor.
- *Piping and Pumps:* These components are essential for transferring water between different tanks and ensuring proper flow through the system.

The system houses RCC, stonemasonry, or earthen embarkment (soil buds) to create a water-tight container. The container has distinct layers, and each layer plays an important role:

- *Underdrain Layer:* This bottom layer allows the treated water to drain away after the cleaning process.
- *Culture Layer or Media Layer:* The layer resembles soil and contains culture

(microbial and macroorganisms) and bioindicator plants; it is also called the constructed soil filter.

- *Additive Layer (Optional):* This layer, if present, can be used to pre-treat the wastewater.

### Treatment Process

- The wastewater enters the system through the raw water tank.
- The water may pass through the additive layer for processing in the media layer.
- The media layer is where the microorganisms and plants work through photosynthesis, respiration, and mineral breakdown, and remove contaminants from the water.
- The water passes through the underdrain layer for further processing before being stored in the storage tank, ready for reuse or further treatment. Depending on your requirements, you can opt for a single-stage or multi-stage cleaning.

# CLASS ACTIVITY

### Objective

To understand the ecological importance and functions of free water surface wetlands.

### Materials Needed

Images or diagrams of free water surface wetlands, Paper and markers, Access to a wetland area (if possible);

### Assessment

- Begin with a brief explanation of what free water surface wetlands are. Discuss their characteristics, such as being shallow bodies of water with a slow flow rate.
- Divide the class into groups. Assign each group a role: wetland biologist, environmental engineer, wildlife photographer, or park ranger. Each group should discuss the responsibilities and tasks associated with their role in managing or studying free water surface wetlands.

## SUMMARY

- Emerging organic contaminants (EOCs) are, in most cases, unregulated contaminants whose emissions have emerged as an environmental problem. They comprise a diverse collection of thousands of chemical substances, such as pharmaceuticals, personal care products (PCPs), pesticides, hormones, surfactants, flame-retardants, plasticizers, industrial additives and agents, as well as their transformation products.
- Pharmaceuticals and Personal Care Products (PPCPs) are synthetic compounds widely used as consumer items such as cosmetics and therapeutic drugs across the globe. The inappropriate disposal of PPCPs in the environment has raised serious concerns regarding their potential adverse impacts on human and animal health.
- Pharmaceuticals and personal care products (PPCPs) consist of APIs, moisturizers, lipsticks, shampoos, hair colors, deodorants, toothpaste surfactants, and many household products that are mainly used for improving the quality of daily life and their degradation products have been inadvertently present in the aquatic environment since the 1970s.
- The soluble characteristics and bioactive nature of many pharmaceuticals and some personal care products may facilitate their biotransformation and depuration in biological features.
- The benefits of the intended use of pharmaceuticals and many personal care products are enormous. The consumption of drugs evokes specific biological responses in the host, after which 10–90% of excess unused doses and some conjugates metabolites are ultimately excreted and passed into the environment.
- Over 80% of PPCPs are released into the environment from the human body in the parent form un-metabolized, i.e., without transformation. These are fed through into the wastewater treatment plants for removal or decomposition during the water treatment process.
- Wastewater treatment plants (WWTPs) rely on biodegradation and adsorption methods using conventional adsorbents such as activated sludge for the removal of contaminating substances in aqueous matrices, but their use is limited.
- The chemical oxidation method offers a better technology in that the challenges associated with the use of conventional wastewater treatment plants, which include incomplete degradation of organic pollutants, sludge formation, the formation of secondary pollutants, and high cost, are eliminated.

## REVIEW QUESTIONS

1. `What are transformation products in the context of pharmaceuticals and personal care products?
2. How do chronic and acute exposures to PPCPs affect non-target organisms?

3. Discuss the challenges in evaluating the human health risks associated with long-term exposure to low concentrations of PPCPs.
4. What are the typical sources of PPCPs entering conventional wastewater treatment plants?
5. Explain how advanced oxidation processes (AOPs) contribute to the degradation of PPCPs.
6. Identify the key design parameters that influence the removal efficiency of PPCPs in horizontal subsurface flow constructed wetlands.
7. Describe the operational principles of vertical subsurface flow constructed wetlands for treating wastewater containing PPCPs.

## REFERENCES


1. Agencia Española de Medicamentos y Productos Sanitarios (AEMPS). (2006). Utilización de antiinflamatorios no esteroides (AINE) en España 1992–2006. Spanish Ministry of Health and Social Politics.
2. Agencia Española de Medicamentos y Productos Sanitarios (AEMPS). (2009). Uso de antibióticos en España (1992–2009). Spanish Ministry of Health and Social Politics.
3. Andreozzi, R., Marotta, R., & Paxéus, N. (2003). Pharmaceuticals in STP effluents and their solar photodegradation in aquatic environment. Chemosphere, 50(10), 1319–1330.
4. Ávila, C., Bayona, J. M., Aragón, C., Martín, I., Salas, J. J., & García, J. (2014a). Emerging organic contaminant removal in a full-scale hybrid constructed wetland system for wastewater treatment and reuse. Ecological Engineering. (submitted)
5. Ávila, C., Matamoros, V., Reyes-Contreras, C., Piña, B., Casado, M., Mita, L., Rivetti, C., Barata, C., García, J., & Bayona, J. M. (2014b). Attenuation of emerging contaminants in a hybrid constructed wetland system under different hydraulic loading rates and their associated toxicological effects in wastewater. Science of the Total Environment, 470-471, 1272–1280.
6. Ávila, C., Nivala, J., Olsson, L., Kassa, K., Headley, T., Bayona, J. M., & García, J. (2014c). Influence of media size, loading frequency and use of active aeration on the removal of emerging organic contaminants in vertical subsurface flow constructed wetlands. Science of the Total Environment. (submitted)
7. Ávila, C., Pedescoll, A., Matamoros, V., Bayona, J. M., & García, J. (2010). Capacity of a horizontal subsurface flow constructed wetland system for the removal of emerging pollutants: An injection experiment. Chemosphere, 81, 1137–1142.
8. Ávila, C., Reyes, C., Bayona, J. M., & García, J. (2013a). Emerging organic contaminant removal depending on primary treatment and operational strategy in horizontal subsurface flow constructed wetlands: Influence of redox. Water Research, 47, 315–325.
9. Ávila, C., Salas, J. J., Martín, I., Aragón, C., & García, J. (2013b). Integrated treatment of combined sewer wastewater and stormwater in a hybrid constructed wetland system in southern Spain and its further reuse. Ecological Engineering, 50, 13–20.

10. Balmford, A. (2013). Pollution, politics and vultures. Science, 339(6120), 653–654.

11. Barceló, D. (2003). Emerging pollutants in water analysis. Trends in Analytical Chemistry, 22(1), 14–16.

12. Barceló, D., & Petrović, M. (2007). Pharmaceuticals and personal care products (PPCPs) in the environment. Analytical and Bioanalytical Chemistry, 387, 1141–1142.

13. Behrends, L. L., Balley, E., Jansen, P., Houke, L., & Smith, S. (2007). Integrated constructed wetland systems: Design, operation, and performance of low-cost decentralized wastewater treatment systems. Water Science and Technology, 55, 155–161.

14. Benitez, F. J., Acero, J. L., Real, F. J., Roldan, G., & Rodriguez, E. (2013). Photolysis of model emerging contaminants in ultra-pure water: Kinetics, by-products formation and degradation pathways. Water Research, 47, 870–880.

15. Bergman, Å., Heindel, J. J., Jobling, S., Kidd, K. A., & Zoeller, R. T. (2013). State of the Science of Endocrine Disrupting Chemicals–2012. United Nations Environment Programme and the World Health Organization. Geneva, Switzerland.

16. Boulêtreau, S., Salvo, E., Lyautey, E., Mastrorillo, S., & Garabetian, F. (2012). Temperature dependence of denitrification in phototrophic river biofilms. Science of the Total Environment, 416, 323–328.

17. Buser, H. R., Poiger, T., & Müller, M. D. (1998). Occurrence and fate of the pharmaceutical drug diclofenac in surface waters: Rapid photodegradation in a lake. Environmental Science and Technology, 32, 3449–3456.

18. Cadby, P. H., Troy, W. R., & Vey, M. G. H. (2002). Consumer exposure to fragrance ingredients: Providing estimates for safety evaluation. Regulatory Toxicology and Pharmacology, 36, 246–252.

19. Carballa, M., Omil, F., & Lema, J. M. (2005). Removal of cosmetic ingredients and pharmaceuticals in sewage primary treatment. Water Research, 39, 4790–4796.

20. Carballa, M., Omil, F., Lema, J. M., Llompart, M., García-Jares, C., Rodríguez, I., Gómez, M., & Ternes, T. (2004). Behavior of pharmaceuticals, cosmetics and hormones in a sewage treatment plant. Water Research, 38, 2918–2926.

21. Chen, F., Ying, G. G., Kong, L. X., Wang, L., Zhao, J. L., Zhou, L. J., & Zhang, L. J. (2011). Distribution and accumulation of endocrine-disrupting chemicals and pharmaceuticals in wastewater irrigated soils in Hebei, China. Environmental Pollution, 159, 1490–1498.

22. Connell, D. W. (2005). *Basic concepts of Environmental Chemistry* (2nd ed.). Taylor & Francis: Boca Raton, FL, USA.

23. Cooper, P. F. (2003). Sizing vertical flow and hybrid constructed wetland systems. In *Proceedings of the 1st International Seminar on the Use of Aquatic Macrophytes for Wastewater Treatment in Constructed Wetlands* (pp. 1–11). Lisbon, Portugal. May 8–10.

24. Daughton, C. G. (2004a). Non-regulated water contaminants: Emerging research. *Environmental Impact Assessment Review, 24*(6), 711–732.

25. Daughton, C. G., & Ruhoy, I. S. (2013). Lower-dose prescribing: Minimizing "side

effects" of pharmaceuticals on society and the environment. *Science of the Total Environment, 443*, 324–337.

26. Díaz-Cruz, M. S., López de Alda, M. J., & Barceló, D. (2003). Environmental behavior and analysis of veterinary and human drugs in soils, sediments, and sludge. *Trends in Analytical Chemistry, 22*(6), 340–351.
27. Dordio, A. V., Estevao Candeias, A. J., Pinto, A. P., & Teixeria da Costa, C. (2009). Preliminary media screening for application in the removal of clofibric acid, carbamazepine, and ibuprofen by SSF-constructed wetlands. *Ecological Engineering, 35*(2), 290–302.
28. Dordio, A., Carvalho, A. J. P., Teixeira, D. M., Dias, C. B., & Pinto, A. P. (2010). Removal of pharmaceuticals in microcosm constructed wetlands using *Typha spp.* and LECA. *Bioresource Technology, 101*(3), 886–892.
29. Farré, M., Kantiani, L., Petrovi⊠, M., Pérez, S., & Barceló, D. (2012). Achievements and future trends in the analysis of emerging organic contaminants in environmental samples by mass spectrometry and bioanalytical techniques. *Journal of Chromatography A, 1259*, 86–99.
30. Faulwetter, J. L., Gagnon, V., Sundberg, C., Chazarenc, F., Burr, M. D., Brisson, J., Camper, A. K., & Stein, O. R. (2009). Microbial processes influencing performance of treatment wetlands: A review. *Ecological Engineering, 35*(7), 987–1004.
31. García, J., Aguirre, P., Barragán, J., Mujeriego, R., Matamoros, V., & Bayona, J. M. (2005). Effect of key design parameters on the efficiency of horizontal subsurface flow constructed wetlands: Long-term performance pilot study. *Ecological Engineering, 25*(4), 405–418.
32. García, J., Rousseau, D. P. L., Morató, J., Lesage, E., Matamoros, V., & Bayona, J. M. (2010). Contaminant removal processes in subsurface-flow constructed wetlands: A review. *Critical Reviews in Environmental Science and Technology, 40*(6), 561–661.
33. Halling-Sørensen, B., Nors Nielsen, S., Lanzky, P. F., Ingerslev, F., Holten Lützhøft, H. C., & Jørgensen, E. (1998). Occurrence, fate and effects of pharmaceutical substances in the environment – A review. *Chemosphere, 36*(2), 357–393.
34. Hijosa-Valsero, M., Matamoros, V., Martín-Villacorta, J., Bécares, E., & Bayona, J. M. (2010a). Assessment of full-scale natural systems for the removal of PPCPs from wastewater in small communities. *Water Research, 44*(5), 1429–1439.
35. Jeli⊠, A., Gros, M., Ginebreda, A., Cespedes-Sánchez, R., Ventura, F., Petrovi⊠, M., & Barceló, D. (2011). Occurrence, partition and removal of pharmaceuticals in sewage water and sludge during wastewater treatment. *Water Research, 45*(3), 1165–1176.
36. Kadlec, R. H., & Wallace, S. D. (2009). *Treatment Wetlands* (2nd ed.). CRC Press: Boca Raton, Florida.
37. Kayser, K., & Kunst, S. (2005). Processes in vertical-flow reed beds: Nitrification, oxygen transfer, and soil clogging. *Water Science and Technology, 51*(10), 177–184.
38. Kim, I., Yamashita, N., & Tanaka, H. (2009). Performance of UV and UV/$H_2O_2$ processes for the removal of pharmaceuticals detected in secondary effluent of a sewage treatment plant in Japan. *Journal of Hazardous Materials, 166*(2–3), 1134–1140.

39. Knowles, P., Dotro, G., Nivala, J., & García, J. (2011). Clogging in subsurface-flow treatment wetlands: Occurrence and contributing factors. *Ecological Engineering, 37*(1), 99–112.

40. Kosma, C. I., Lambropoulou, D. A., & Albanis, T. A. (2010). Occurrence and removal of PPCPs in municipal and hospital wastewaters in Greece. *Journal of Hazardous Materials, 179*(1–3), 804–817.

41. Llorens, E., Matamoros, V., Domingo, V., Bayona, J. M., & García, J. (2009). Water quality improvement in a full-scale tertiary constructed wetland: Effects on conventional and specific organic contaminants. *Science of The Total Environment, 407*(7), 2517–2524.

42. Mezcua, M., Gómez, M. J., Ferrer, I., Aguera, A., Hernando, M. D., & Fernández-Alba, A. R. (2004). Evidence of 2,7/2,8-dibenzodichloro-p-dioxin as a photodegradation product of triclosan in water and wastewater samples. *Analytica Chimica Acta, 524*(1), 241–247.

43. Pedescoll, A., Uggetti, E., Llorens, E., Granés, F., García, D., & García, J. (2009). Practical method based on saturated hydraulic conductivity used to assess clogging in subsurface flow constructed wetlands. *Ecological Engineering, 35*(8), 1216–1224.

44. Platzer, C., & Mauch, K. (1997). Soil clogging in vertical flow reed beds: Mechanisms, parameters, consequences and... solutions? *Water Science and Technology, 35*(5), 175–182.

45. Puigagut, J., Villaseñor, J., Salas, J. J., Bécares, E., & García, J. (2007). Subsurface-flow constructed wetlands in Spain for the sanitation of small communities: A comparative study. *Ecological Engineering, 30*(4), 312–319.

46. Rodríguez-Gil, J. L., Catalá, M., González, S., Romo, R., Valcárcel, Y., Segura, Y., Molina, R., Melero, J. A., & Martínez, F. (2010). Heterogeneous photo-Fenton treatment for the reduction of pharmaceutical contamination in Madrid rivers and ecotoxicological evaluation by a miniaturized fern spores bioassay. *Chemosphere, 80*(3), 381–388.

47. Rosal, R., Rodríguez, A., Perdigón-Melón, J. A., Petre, A., García-Calvo, E., Gómez, M. J., Agüera, A., & Fernández-Alba, A. R. (2010). Occurrence of emerging pollutants in urban wastewater and their removal through biological treatment followed by ozonation. *Water Research, 44*(2), 578–588.

48. Singer, H., Muller, S., Tixier, C., & Pillonel, L. (2002). Triclosan: Occurrence and fate of a widely used biocide in the aquatic environment: Field measurements in wastewater treatment plants, surface waters, and lake sediments. *Environmental Science and Technology, 36*(22), 4998–5004.

# Index

FSC
www.fsc.org
MIX
Papier | Fördert
gute Waldnutzung
FSC® C083411

Kolibri 360 GmbH
Ferdinand-Jühlke-Straße 7
99095 Erfurt, Deutschland
produktsicherheit@kolibri360.de